Life Cycle in the Natural Sciences and Traditional Cultures as a Complex System Self-Organization

Yury N. Kovalyov
National Academy of Culture and Arts Management, Ukraine

Nver M. Mkhitaryan
Institute of Renewable Energy, National Academy of Science of Ukraine, Ukraine & Emerald Palace Group, UAE

Andriy Y. Morozov
Kyiv National University of Trade and Economics, Ukraine

Yaroslava F. Zhukova
State Institution "Soils Protection Institute of Ukraine", Ukraine

IGI Global
Publishing Tomorrow's Research Today

Published in the United States of America by
IGI Global
701 E. Chocolate Avenue
Hershey PA, USA 17033
Tel: 717-533-8845
Fax: 717-533-8661
E-mail: cust@igi-global.com
Web site: https://www.igi-global.com

Library of Congress Cataloging-in-Publication Data

CIP PENDING

ISBN13: 9781668485095
EISBN13: 9781668485101

Vice President of Editorial: Melissa Wagner
Managing Editor of Acquisitions: Mikaela Felty
Managing Editor of Book Development: Jocelynn Hessler
Production Manager: Mike Brehm
Cover Design: Phillip Shickler

British Cataloguing in Publication Data
A Cataloguing in Publication record for this book is available from the British Library.

All work contributed to this book is new, previously-unpublished material.
The views expressed in this book are those of the authors, but not necessarily of the publisher.

Table of Contents

Preface

The events of the life cycle pass before our eyes: birth, childhood, maturity, senility, and death.

If pay attention, we will see that something similar happens with artificial objects, peoples and states, biological species, stars, etc. According to some ancient myths and modern scientific theories, even the Universe has a beginning and an end.

But let us return to humans.

Behind the ordinary evidence of events, questions are hidden - about the origin and meaning of life, why a human grows up, grows old, and dies, and whether there is something beyond the threshold of death.

Mankind has always sought to comprehend these secrets, starting from the times of existence as an independent species. Moreover, the feeling of death as a special phenomenon was inherent in its closest Homo relatives, the Neanderthals (Pomeroy et al., 2017).

Attitudes toward death have changed in different eras (Aries, 2013). In our time, the attitude towards death is characterized as very contradictory.

On the one hand, this is a rather dismaying topic, above all, those aspects, that are connected with the posthumous existence of consciousness – serious discussion in modern society and scientific circles is taboo.

On the other hand, in the products of mass culture, we observe an endless stream of apocalypses, catastrophes, massacres, wars, zombies, vampires, and demons, accustoming to the routine and, as it were, the vulgarity of death.

Our contemporaries see a way out in extending the body's life, creating "information copies" of a person, developing cyborgs (serious research is already underway here), cryonics, etc. - or in creating a great media name in the hope of long memory for descendants.

The life cycle research must begin long before the life events of a particular person. People are linked with the first living organisms through their near and distant ancestors by an unbroken evolutionary chain; the emergence of the first organisms is due to the presence of organic compounds, and they, in turn, are associated with simpler chemical substances.

Thus, the starting point is the appearance of life on Earth approximately 4 billion years ago, or even earlier, if we consider the stage of abiogenic synthesis. Probably, here one should look for the origins of not only life but also death.

There is another point of view on this problem. For many living organisms, mortality, at a certain stage of their life, stabilizes and ceases to depend on age; they die from the influence of external factors, not from senility. In this sense, they are biologically immortal. Then the origins of human mortality should be sought in the evolutionary process, as well as in the complex interactions with the environment that determine the structure of a human and the main events of his life cycle.

Both assumptions determine not only the chronological framework, but the research content: it is necessary to consider the data of cosmology, geology, biochemistry, biology, ecology, genetics, and the theory of evolution - and this list is far from complete!

A little less than a hundred years ago, one of the founders of modern mathematics foundations, A. Tarski, showed that the classes of true and provable statements do not coincide, that is, unprovability does not mean untruth (Tarski, 1936/1969). Consequently, the grains of truth can be sought not only in scientific data but also in the entire cultural heritage of mankind.

This legacy represents a huge amount of heterogeneous and paradigmatically incompatible knowledge and misconceptions. And, in addition to collecting them, which is a difficult task, they need to be critically assessed, selected, processed, and used to model key life cycle events. And, if the models allow extrapolation beyond the limits of human life, then it is possible to repeat the path of Gilgamesh in its modern sense. And if not, then the question of what lies beyond the threshold of death will remain unanswered until new methods of research are created.

So, the issue of the method is the key. What apparatus can be used for processing and modeling this "big data"?

The authors propose to use the theory of complex systems self-organization based on the axiomatic wave model of S-space (Kovalyov, 1996). Its advantage is the ability to model open heterogeneous complex systems in a dissipative environment.

This theory was used previously to model human consciousness and behavior in various historical eras (Kovalyov et al., 2020), as well as to solve applied problems of modeling and optimizing various complex open systems (Mkhitaryan et al., 2004; Shmelova, et al., 2018; Mkhitaryan et al., 2021). The potential of this apparatus for life cycle modeling has been theoretically considered (Kovalyov & Kalashnikova, 2023).

So, models and scenarios for the emergence and evolution of life, human structure, life cycle events, and a possible afterlife, are based on a formal mathematical model that describes the evolution and interactions of soliton-wave systems in a dissipative environment. This idea may seem too daring, but the authors were inspired by a late antique philosopher Proclus example, who presented the gods' relations as set-theoretic (Proclus, 1993).

One way or another, this assumption needs to be carefully verified. It is necessary to:

- Consider the conditions of abiogenic synthesis and the emergence of life - and check their compliance with the abstractions of the wave model.
- Consider the conditions of biogenic synthesis of complex organics, the appearance of the first organisms, the characteristics of metabolism, energy, and interactions with the environment - and check whether they correspond to the axioms of the wave model.
- Improve the theory of self-organization of complex systems to the tasks being solved.
- Constantly check the adequacy of the proposed mathematical apparatus when modeling evolution, human structure and its interaction with the environment, and life cycle events - do the models, scenarios, and forecasts correspond to known facts?

Now let us show how this program is implemented in each chapter.

In Chapter 1, *On the Edge of Life and the S-Space Abstractions Applicability Checking*, based on the available facts, 10 conditions for the emergence of life are formulated, which are compared with abstractions of the wave model.

We note that well-known facts and theories are presented in detail, and may be redundant to the theme discussed in the chapter. This is done to preserve the "big data" effect so the reader can operate specific information, seeing a wider context.

In Chapter 2, *The Cell as a Complex System and the S-Space Axiomatic Wave Model Adequacy Checking*, based on modern ideas about life processes and living organism's conditions, 13 regularities are formulated, which are checked for compliance with the axioms of the wave model. The correspondence of the axioms to known facts is shown. The proposed understanding of life as a border and open S-space does not remain abstract. Such an understanding allows us to define life and make assumptions about the places of its origin and intermediate forms between inanimate nature and living organisms.

Further, the means of description and verification methods specific to the S-space wave model are considered (Chapter 3, *Self-Organization of S-Space. Description Means and Construction Methods*). This is necessary to substantiate the theory of self-organization (Chapter 4) and create a method for working with heterogeneous artifacts and texts, belonging to different cultures and eras (Chapter 12).

A summary of the theory of self-organization of open complex systems is given in Chapter 4, *Self-Organization of S-Space Scenarios*. These scenarios are used to create various models.

So, the first part of the work is devoted to substantiating and presenting the modeling apparatus chosen by the authors.

In the next part, the authors proceed to such modeling. The sequence of a presentation corresponds to the first part of the work: first, modern ideas are presented, then they are generalized, and certain patterns are derived, then models are built and compared with these consistent patterns; if the result is satisfactory, further research is possible.

In Chapter 5, *Evolution Phenomena and Theories*, modern ideas about evolution at different levels are considered, and 15 of its regularities are formulated, including the mechanisms of reproduction and inheritance.

Chapter 6, *Evolution as a Self-Organization*, presents models of various evolutionary processes that correspond to the patterns discussed in Chapter 5. Thus, they can be described and explained within the framework of the theory of the self-organization of complex systems. It is concluded that the theory and apparatus are applicable for modeling evolutionary processes.

Chapter 7, *The Human Body System Structure and Self-Regulation*, considers modern data on the structure of a person and his subsystems, self-regulation, and interactions with the environment. These data are necessary for modeling the human structure.

In Chapter 8, *Human-Environment Model, Self-regulation and Rhythms Coordination*, the human structure general plan, subjective space, self-regulation mechanisms, and environment connections were modeled. These data are practically important to this work but also in testing human-machine systems operators, planning workloads under stress, creating information systems, etc. (Mkhitaryan, et al., 2004).

Chapter 9, *Life Cycle from Conception to Senility*, covers the major stages of the human life cycle, from conception to old age. Modern ideas are given. Models of the life cycle, various processes, and structures are substantiated. Known facts are used to verify them. It should be noted, that the proposed life cycle scenario in terms of its phases and their specific characteristics corresponds quite well to empirical data.

Chapter 10, *Near the Death*, deals in detail with near-death phenomena: this is the clinical picture of approaching death, near-death experiences, and similar trance states. The systematic analysis allows us to speak about their lifetime nature confidently. Another aspect is the problem of biological immortality. The modern understanding of the problem and possible ways to achieve biological immortality is briefly outlined. Based on all these data, the death phenomenon is modeled as a holistic soliton-wave system gradual destruction process, after which its components can exist as separate systems.

In all historical epochs, the problem of death becomes especially acute during wars, natural disasters, and epidemics. Today we are experiencing a coronavirus epidemic, natural disasters are becoming more frequent due to global climate change, and armed conflicts are constantly taking place in Asia, Africa, and now in Europe.

The authors hope, that the results of a life cycle analysis will interest the scientific community and a wide readership.

REFERENCES

Aries, P. (2013). *The Hour of Our Death*. Vintage.

George, A. (2016). *The Epic of Gilgamesh*. Penguin.

Kovalyov, Y., Mkhitaryan, N., & Nitsyn, A. (2020). *Self-organization of the Human Mind and the Transition from Paleolithic to Behavioral Modernity*. IGI Global. DOI: 10.4018/978-1-7998-1706-2

Kovalyov, Y. M., & Kalashnikova, V.V. (2023). Human life cycle modeling. *Suchasni problemy modeljuvannja, 25,* 110-122.

Pomeroy, E., Mirazón Lahr, M., Crivellaro, F., Farr, L., Reynolds, T., Hunt, C. O., & Barker, G. (2017). Newly discovered Neanderthal remains from Shanidar Cave, Iraqi Kurdistan, and their attribution to Shanidar 5. *Journal of Human Evolution*, 111, 102–118. DOI: 10.1016/j.jhevol.2017.07.001 PMID: 28874265

Shmelova, T., Sikirda, Y., Rizun, N., Salem, A.-B. M., & Kovalyov, Y. (2018). *Socio-Technical Decision Support in Air Navigation Systems: Emerging Research and Opportunities*. IGI Global. DOI: 10.4018/978-1-5225-3108-1

Tarski A. (1969). Truth and Proof. *Scientific American,220*, 63–77. (Original work published 1936)

Introduction

Interest in the events of the life cycle has been inherent to people at all times. In one way or another, the culture of all the peoples of the Earth includes fairy tales, myths, epic and literary works, sometimes developed religious ideas and magical rituals, as well as artifacts and monuments reflecting certain events.

However, modern understanding of such issues as the chemical basis, origin, and evolution of life and life cycle, methods of improving the quality and prolongation of life, etc. are associated with the progress of natural sciences and modeling methods.

Among all the stories associated with the life cycle, the theme of post-mortem existence has always stood out: the human consciousness cannot come to terms with the death of loved ones and own death.

At the same time, modern ideas about the life cycle and individual finitude differ significantly from traditional ones.

For traditional ideas, despite their numerous variations and features, two ideas were characteristic:

1. Even if a person's body is mortal, then at least his soul can exist forever;
2. If the soul exists forever, earthly life should provide the best conditions for its posthumous existence.

The modern scientific worldview denies the existence of an eternal soul, which leads to the "everyday" conclusion that – it is necessary to get maximum pleasure from life, trying to prolong the body's (preferably young and healthy) life as much as possible.

However, such a perspective does not satisfy a developed consciousness: a person will never be able to come to terms with the death of loved ones and his own death.

And a person again turns to religion, mythology, fairy tales - or to those humanities sciences, whose subject area is the analysis of such ideas.

The availability of materials is another important feature of our era: everyone is free to choose the "optics" for their research, be it within the framework of religious teaching, personal mystical experience, science, or, for example, works of art. In the same way, within the framework of one paradigm, a person can study, for example, physiological, medical, social, and other aspects of childhood or aging.

The experience of mankind accumulated over thousands of years can rightfully be characterized by the fashionable term - "big data"!

A person, who wants to understand the problems of life and death impartially, falls into an epistemological trap: what to accept as facts and postulates, how to process contradictory and incompatible data, what conclusions to draw, what worldview, and what way of life to consider appropriate?

For such readers, our research will be interesting and useful. Considering, as follows from the above, not only the difference in the subject area but also the large volume and heterogeneity of the materials used, it is advisable to divide it into two parts:

1. Life Cycle in the Natural Sciences as a Complex System Self-Organization.
2. Post-Mortem Existence Within the Complex Systems Self-Organization Theory Framework and in Traditional Cultures.

Here we will talk about the first part only.

Progress in understanding the life cycle could be achieved through new research. What exactly do the authors propose?

First, this is a new research method. A human being is an open complex system. If we compare the mathematics abstractions with real open complex systems characteristics, we can see that they are not only dissonant to each other but also opposite (Table 7, Chapter 1). Hence, it is need to develop a new, more suitable apparatus for modeling complex systems, including abstractions, axiomatic definitions, description tools, and verification methods. This is the wave model of S-space. It can work with heterogeneous and irreducible data, necessary for the life cycle studying.

Further, the S-space does not remain unchanged under external influences, but evolves, stratifying into sets and elements - waves and solitons - and in the absence of such influences, the reverse process begins, called convolution. These scenarios of stratification and convolution are a kind of Ariadne's thread, allowing one to go through the life cycle events, understanding them as the general laws of the evolution of complex systems evolution.

The event of death can be interpreted as the destruction of an integral system, and the posthumous existence of the body (after all, it does not disappear at the moment of death), and the "soul" existence can also be considered as the convolution scenarios. This is described in detail, with individual phases of the process, predictions, and conclusions, in this book.

But what about the reliability of theoretical constructs?

The authors pay attention to this from the very beginning.

The origin of the life cycle is logically linked to the origin of life as such - the authors analyze the available scientific data on the conditions on the Ancient Earth and the first living organisms and compare them with the abstractions and axioms of the wave model of S-space, concluding that the proposed apparatus can be used. Also, within the framework of the proposed paradigm, considerations are made about the definition of life, the driving forces of the process of complication at a time when living organisms and natural selection did not yet exist, the places of origin of life, and the forms of the first living organisms that may be of interest to relevant specialists.

Further, the life cycle evolved together with living organisms, and the next set of modeling and testing is associated with comparing self-organization scenarios with both the phenomena of the evolutionary process and modern theories of evolution. These tests confirm the possibility of considering evolution as a specific implementation as a realization of general scenarios of self-organization of complex systems, including known forms of reproduction and inheritance.

The description and understanding of evolution are not complete without the evolution of human consciousness and behavior analysis. Therefore, it is worth paying attention to (Kovalyov et al., 2020), dedicated to such an analysis based on the theory of self-organization of complex systems. This book can be considered as a third part of our research.

The same format (scientific facts, scenarios, models, their comparison, and conclusions) is preserved in the following chapters. Human interaction with the environment - structure, levels, and channels of interaction with the environment, self-regulation, and biorhythms - are considered within the framework of the theory of self-organization and soliton-wave models.

The features and models of all life cycle periods and the life cycle as a whole are considered. Note that the estimated duration of each of them, obtained from theoretical considerations, correlates well with the data given by different authors. The aging process based on known facts is modeled as the S-space convolution. This scenario helps us to explore strategies for life extension, as well as the stages of dying. The event of death is defined as the destruction of an entire system.

All this has important practical significance: the task of improving the quality of life is set in pedagogy, design, sociology, and medicine, etc.; the processes of aging and life extension are the subject of gerontology; processes associated with the death of an organism are studied by thanatology, etc.

The outlined range of topics, data, methods, and models, as well as their verifications, constitutes the content of *Life Cycle in the Natural Sciences as a Complex System Self-Organization.*

The authors invite the reader on a fascinating journey lasting – if the dating of the origin of life on Earth is correct – more than 4 billion years!

The authors hope that their research, methods, and conclusions will be interesting not only to specialists in the fields of mathematics, biochemistry, biology, evolutionary theory, physiology, medicine, gerontology, and anthropology (data from these sciences are used and discussed in the book) but also to the inquiring reader who makes an effort to gain systematic and comprehensive knowledge about the life cycle.

REFERENCES

Kovalyov, Y., Mkhitaryan, N., & Nitsyn, A. (2020). *Self-organization of the Human Mind and the Transition from Paleolithic to Behavioral Modernity*. IGI Global. DOI: 10.4018/978-1-7998-1706-2

Chapter 1
On the Edge of Life and the S-Space Abstractions Applicability Checking

ABSTRACT

The emergence of the life cycle is logically connected with the emergence of life conditions. Therefore, the first task is to present the conditions of the Ancient Earth: the age of the Solar System and the Earth, the composition and parameters of the Earth's crust, atmosphere, hydrosphere, energy sources, and characteristics of vital substances. These data allow us to evaluate modern ideas of life emergence, and evolution. On this basis, 10 conditions for the emergence of life are formulated. The second task is to justify the adequacy of the apparatus used in the life cycle study, for which a comparison of 10 conditions with the abstractions of the wave model of S-Space is carried out. The comparison showed their compliance, which allows us to proceed to further tests and research.

BACKGROUND

To estimate the S-space abstractions adequacy, it is necessary to have realistic ideas about the conditions for the emergence of life on Ancient Earth:

- Consider the biochemical foundations of life (for the work, present the structure of proteins, lipids, carbohydrates, and amino acids).
- Determine the boundaries: the time from the formation of the Earth to the appearance of the first living organisms, as well as the dating of the stages of evolution of living organisms, which will be used in the following chapters.

DOI: 10.4018/978-1-6684-8509-5.ch001

- Consider the conditions on Ancient Earth, understanding that they are derived from the evolution of the Solar system and the universe.
- Check the conformity of the reactions underlying the modern theories of abiogenic synthesis with the conditions of the Ancient Earth and modern times.
- Formalize the conditions for the emergence of life and compare them with abstractions of the S-spatial wave model.

Let's make two preliminary remarks:

1. The authors do not conduct independent research in this area, and base their reasoning on facts and reconstructions, accepted by the scientific community.
2. Therefore, the authors use monographs, handbooks, and textbooks as sources; where actual data are needed, or alternative points of view need to be mentioned, modern scientific publications are used.

There are various phenomenological definitions of life, based on lists of attributes (there is no reason to consider these lists complete); these attributes taken in isolation can also apply to non-living objects (McKay & Chris, 2004). At the same time, the biochemical understanding of life is at least partially confirmed experimentally.

In biochemistry, the basic idea is "carbon life", according to which carbon, oxygen, nitrogen, and hydrogen are necessary for the emergence of life. Hypotheses about the silicon basis of life (Kan & Jennifer, 2016), as well as nitrogen, boron, etc., are considered less probable. It is thought, they were not realized on Ancient Earth (Nikitin, 2013a; 2018).

The chemical composition of cells, the structure of proteins, lipids and carbohydrates, sugars and amino acids, and other vital substances are well known (Bilich & Kryzhanovsky, 2009; Pimenova & Pimenov, 2003). Reactions leading to their synthesis are also known (Gusev & Mineeva, 2003), but under laboratory conditions. Therefore, it is necessary to check how these reactions and laboratory conditions correspond to the situation on Ancient Earth.

The age of the Universe is 13.799 ± 0.021 billion years (Planck Collaboration, 2016). The Solar system formed about 4.57 billion years ago (Bouvier, & Meenakshi, 2010); the age of the Earth is 4.54 billion years (Dalrymple & Brent, 2001) or, according to the International Stratigraphic Commission, 4.6 billion years. The first living beings appeared on Earth 3.7 billion years ago (Pearce et al., 2018); according to other sources, 4.1 billion years ago (Bell et al, 2015).

There is a vast literature devoted to the Ancient Earth conditions reconstruction (Britt & Roy, 2002; Bogdanov et al., 1978; Wilde et al., 2001; Hyde et al., 2000) and so on. Sources use data from comparative planetology, research of objects in the Solar system by spacecraft; radioisotope dating; structure of geological rocks; paleontology, and others.

Assessing the Ancient Earth conditions and events dating, it should be noted that they are reconstructions based on modern research methods. Even 150 years ago, intervals of billions of years would have seemed incredibly long.

There are several hypotheses of life origin based on biogenetic or abiogenetic concepts. The first group includes the stationary state hypothesis the panspermia hypothesis (Wickramasinghe, 2011), and the Martian origin of Earth life (McKay et al., 1996; Gibson et al., 1999).

The stationary state hypothesis, extended either to the specific conditions of the Earth or to the entire Universe, is based on the postulate that living matter is qualitatively different from non-living matter and can only come from living matter. At the same time, the differences between living matter and non-living matter are not formulated; the conditions in the Universe and on Earth cannot be called stationary.

The hypothesis of panspermia in the form of an assumption about the possibility of transferring life through outer space was put forward by J. Berzelius; S. Arrhenius substantiated the fundamental possibility of the transfer of bacterial spores under light pressure (Arrhenius, 1908). Carriers can also be natural (asteroids, comets, dust accumulations, etc.) or created by representatives of extraterrestrial civilizations. The panspermia hypothesis, focusing on how life spreads, does not answer the question of its origin.

Abiogenetic theories include the hypothesis of spontaneous generation of life (which appeared in Antiquity and is currently not relevant), the hypothesis of biochemical evolution by Oparin–Haldein (Oparin, 1924), the theory of the iron-sulfur world by K. Wächtershäuser (Wächtershäuser, 2006), the theory of the zinc world by A. Mulkidjanian (Mulkidjanian, 2009) and others (Peretó, 2005). In (Lurquin, 2003; Luisi, 2006; Rauchfuss, 2008; Koonin, 2011; Markov, 2010; Nikitin, 2013a-f; 2018) the sequence of the emergence of life is considered. The decisive success of these theories would be the experimental creation of a living organism.

Based on all the above data, the authors formulated 10 conditions for the emergence of life. It is convenient to compare these conditions with the S-space wave model abstractions (Kovalyov, 1996).

VITAL SUBSTANCES

The chemical composition of organic matter is a clue that allows you to select the necessary reactions, checking them for compliance with the conditions of the Ancient Earth.

According to modern theories, the first complete living organisms are those that can reproduce on their own. The smallest structure that provides such an opportunity today is the cell. Therefore, it is necessary to start with cells to determine the vital substances.

The chemical composition of cells includes 86 chemical elements; 25 are necessary for a normal life; 16-18 are absolutely necessary (Table 1).

Table 1. Chemical composition of the cell

Inorganic substances		Organic substances		
Water 70–80%	Mineral salts 1–1,5%	Proteins 10–20%	Lipids 1–5%	Carbohydrates 0.2–2%
		Nucleic acids 1–2%	Adenosine triphosphoric acid 0.1–0.5%	Biologically active substances

(Bilich & Kryzhanovsky, 2009)

Listed on the right side of Table 1, proteins, lipids, carbohydrates, nucleic acids, adenosine triphosphoric acid, enzymes, vitamins, and other substances that provide biogenic synthesis, inheritance, energetics, and metabolism, could be the result of abiogenic synthesis. Many of them, believed to be of biological origin, were later synthesized in laboratories.

Proteins (over 170 of them were found in cells) are high-molecular organic non-periodic polymers, the monomers of which are 20 types of interchangeable and irreplaceable α-amino acids (Figure 1). Proteins consist of carbon, hydrogen, nitrogen, oxygen, and sulfur. Some proteins form complexes with other molecules containing phosphorus, iron, zinc, and copper. Peptides are organic substances consisting of amino acid residues connected by a peptide bond.

Figure 1. Formation of protein macromolecules from monomeric amino acids (Glossary. Biology. Structure of proteins, n / a)

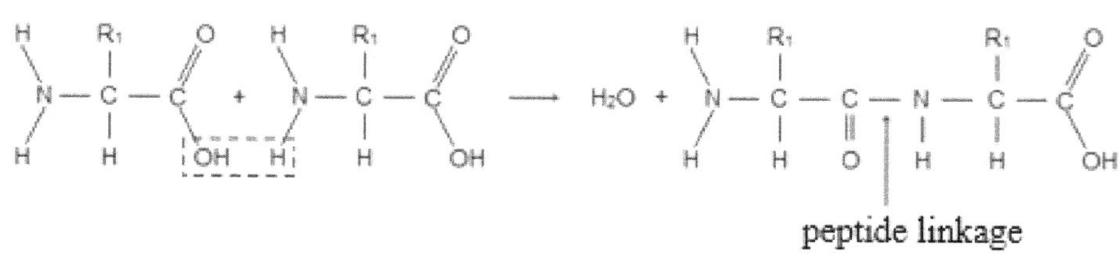

There are four levels of the spatial organization of proteins:

- Primary - the sequence of amino acid residues in the polypeptide chain that makes up the protein molecule.
- Secondary-ordered folding of the polypeptide chain into a spiral.
- Tertiary - packing of polypeptide chains into globules.
- Quaternary - characteristic of complex proteins, the molecules of which are formed by two or more globules.

Functions of proteins: building, transport, regulatory, protective, motor, signal, storage, energy, catalytic. About ten thousand different proteins have been found in the human body, which differs both from each other and from the proteins of other organisms. Proteins have a large molecular weight, for example, hemoglobin - 152,000, myosin - 500,000. For comparison: the molecular weight of alcohol is 46, acetic acid - 60, and benzene - 78. (Pimenova & Pimenov, 2003).

Lipids (Figure 2) are a collection of water-insoluble organic compounds, that can be extracted from cells with organic solvents. Simple lipids are mainly esters of higher fatty acids and triglycerides. Fatty acids have a carboxyl group (-COOH) and a radical - a chain of a different number (from 14 to 22) of $-CH_2-$ groups, by which they differ from each other. Complex lipids are phospholipids, glycolipids, lipoproteins, and others. Lipids have functions of energetic, structural, storage, protective, regulatory, and catalytic, and a source of metabolic water.

Figure 2. The structure of lipids (Pimenova & Pimenov, 2003)

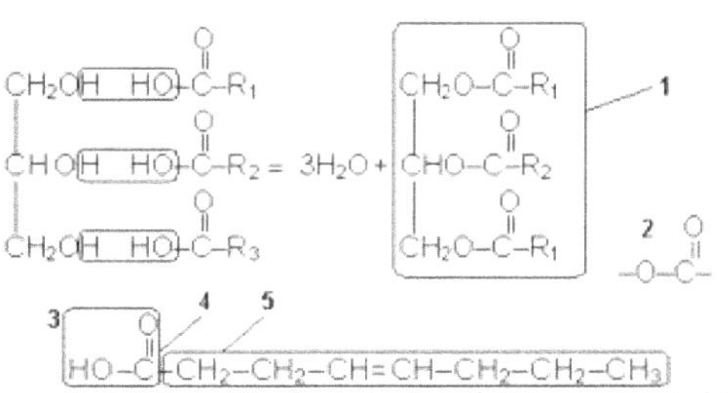

1 - triglyceride; 2 - ester bond; 3 - unsaturated fatty acid; 4 - hydrophilic head; 5 - hydrophobic tail

Carbohydrates (Figure 3) are organic compounds whose composition in most cases is expressed by the general formula $Cn(H_2O)m$ (n and m $\geq$ 4). Carbohydrates are divided into monosaccharides, oligosaccharides, and polysaccharides. Monosaccharides, depending on the number of carbon atoms, are divided into trioses (3), tetroses (4), pentoses (5), hexoses (6), and heptoses (7 atoms). So, ribose and deoxyribose belong to the group of pentoses and are part of the nucleotides of RNA and DNA, ribonucleoside triphosphates and deoxyribonucleoside triphosphates, glucose belongs to the group of hexoses, can exist in the form of α-glucose or β-glucose. Oligosaccharides are carbohydrates formed because of a condensation reaction between several (from two to ten) monosaccharide molecules. Depending on the number of monosaccharide residues, disaccharides, trisaccharides, and so on are distinguished, for example, sucrose (cane or beet sugar) is a disaccharide consisting of glucose and fructose residues. Polysaccharides - carbohydrates formed because of the polycondensation reaction of many (several tens or more) molecules of monosaccharides, for example, starch and glycogen are polymers, the monomers of which are α-glucose, cellulose is a polymer, the monomer of which is β-glucose, inulin is a polymer, a monomer which is fructose. Their functions are energetic, structural, accumulative, and protective.

Figure 3. The structure of carbohydrates (Pimenova & Pimenov, 2003)

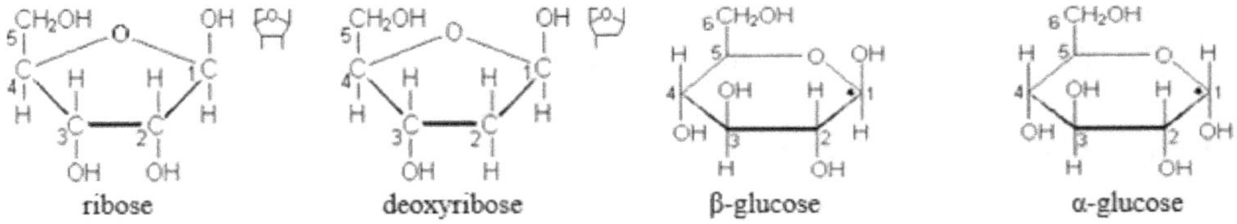

Some organic substances have a relatively simple structure, while others, like DNA and RNA, are separated from the former by a complexity barrier. The process of their synthesis by living organisms can also be a barrier (Chapter 2).

Consider the simple low molecular weight compounds synthesis (amino acids, sugars, adenosine triphosphoric acid (ATP), and nitrogenous bases that are part of RNA and DNA).

Amino acids (Figure 4) are monomeric compounds containing both amine ($-NH_2$) and carboxyl (–COOH) functional groups and a radical or R-group (the rest of the molecule), which is different for different types of amino acids.

Figure 4. The structure of amino acids (Amino acids, n / a)

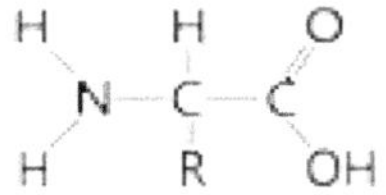

Amino acids in solution can act as both acids and bases. In aqueous solutions, amino acids exist in different ionic forms. Depending on the position of the amino group concerning the carboxyl group, α, β, γ, and so on amino acids are distinguished. Natural amino acids contain an amino group only in the α-position. Amino acids are classified depending on the nature of the side chains, according to the number of amine and carboxyl groups, according to the nature of the charge of the side radicals and their polarity of the amino acid, according to the biological significance (non-essential - glycine, alanine, serine, cysteine, tyrosine, aspartic, acid, glutamic acid, asparagine, glutamine, essential - valine, leucine, isoleucine, threonine, lysine, methionine, phenylalanine, tryptophan, semi-essential - arginine and histidine Let's give a table of amino acids that make up proteins (Figure 5).

Figure 5. The 20 amino acids that serve as the building blocks for proteins (Amino Acids Reference Guide, n / a)

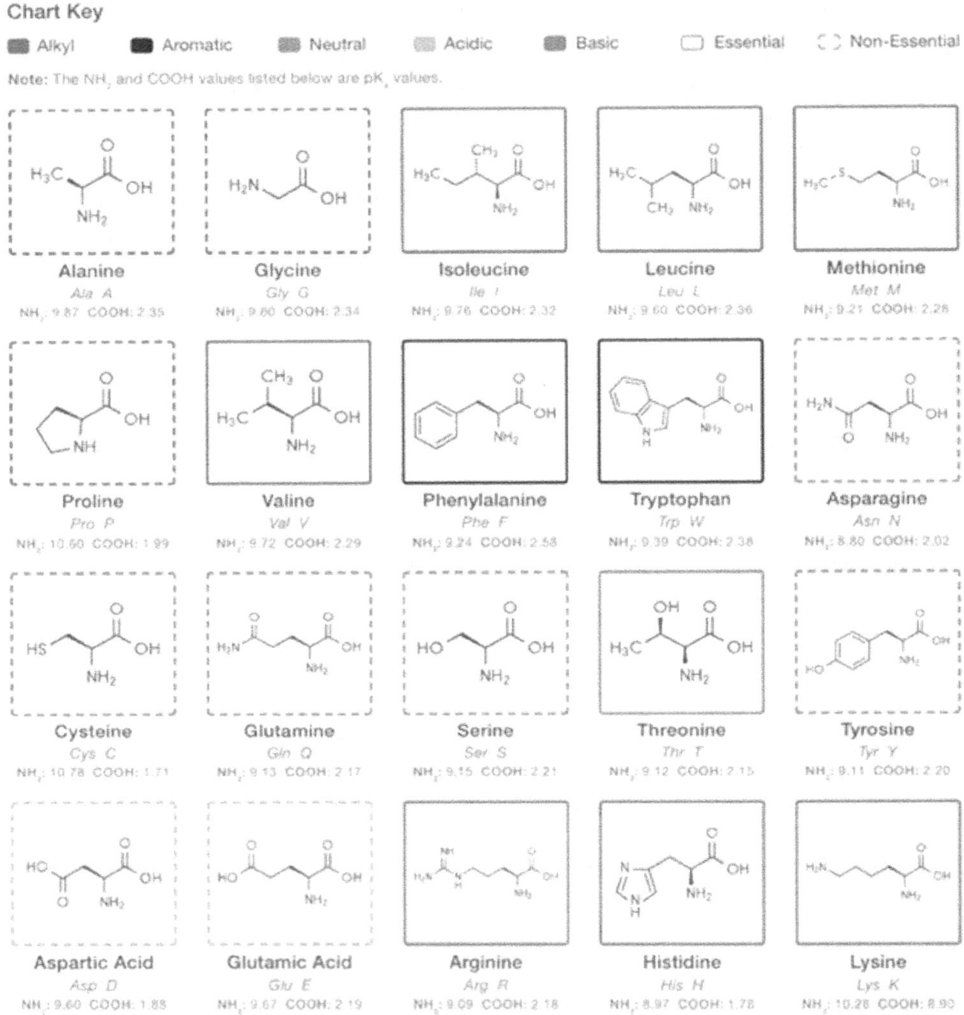

Nucleotides (Figure 6): guanine; cytosine; thymine (5-methyl-uracil) does not occur in RNA (except tRNA) and takes the place of uracil in DNA; uracil, found in bacteriophages in DNA, takes the place of thymine in RNA. These compounds also perform other important functions.

Figure 6. The chemical structure of adenosine triphosphate. Автор: derivative work: Batterytime ATP_structure.svg: (User: Mysid, н/д)

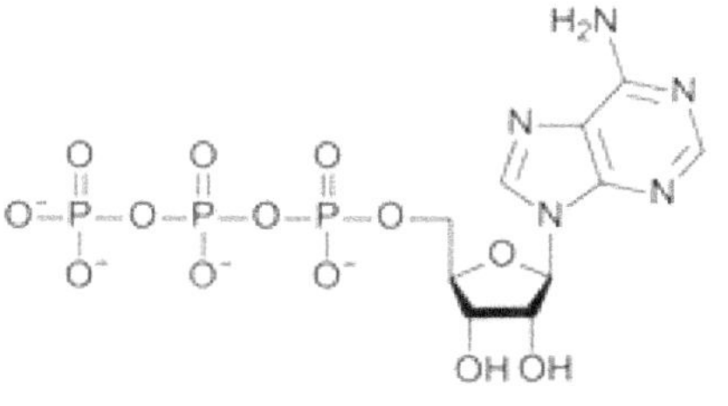

Thus, reactions of abiogenic synthesis should at least lead to the formation of 20 amino acids, 2 sugars (ribose and deoxyribose), adenosine triphosphoric acid, 5 nitrogenous bases (adenine, guanine, cytosine, thymine, and uracil).

The ***adenosine triphosphoric acid (ATP) structure*** is shown in Figure 7.

Many organic molecules, including amino acids and sugars necessary for life, have chiral symmetry, that is, they can exist in two mirror-symmetric forms that differ from each other as the right hand differs from the left (D- and L-molecules). In abiogenic synthesis, an equal amount of D- and L-forms of organic molecules is formed (racemic mixture).

Figure 7. Nucleotides (Pino, 2015)

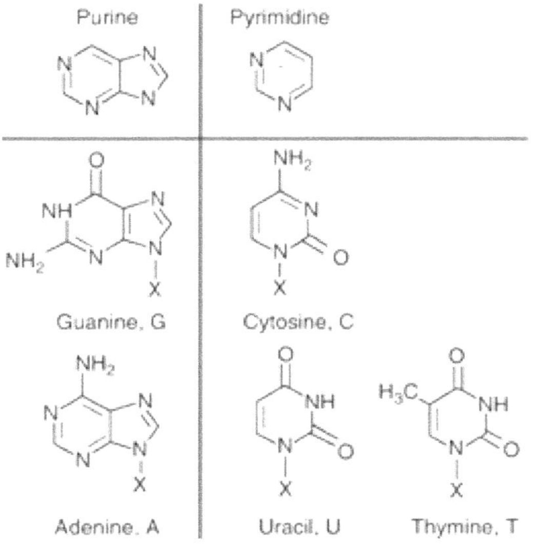

However, in living organisms, proteins consist exclusively of L-amino acids, and deoxyribose and ribose sugar residues included in DNA and RNA molecules have D-forms (Nikitin, 2013c; Vorontsov, 2019). All cellular mechanisms are tuned to the perception of just such isomers. For example, during protein synthesis, tRNA binds only to the L-amino acid, while the enzyme D-amino acid oxidase cleaves the D-forms and leaves their mirror counterparts intact. A certain group of enzymes carries out the enantiomeric control, searching for and destroying randomly formed "unnatural" isomers. In the biosynthesis of DNA, RNA, and enzymes, the enantiomeric configuration of future units is controlled very precisely: less than one error in a billion. And this is not accidental, since the inclusion of D-amino acids leads to a violation of the secondary and tertiary structure of the protein, and L-sugars - to the formation of nucleic acids that cannot form complementary pairs and a double helix. After the death of organisms, amino acids, and sugars lose their chiral purity, creating a racemic mixture containing 50% D- and 50% L-isomers (Nikitin, 2013c). Hence it follows that the chiral purity of the products is one more criterion for selecting the necessary reactions of abiogenesis.

ANCIENT EARTH: CHRONOLOGY

The age of the Universe - the time elapsed since the beginning of its expansion - is determined based on theoretical interpretations of the observed redshift, as well as observations of globular clusters and non-evolved objects. In some cases (in the presence of cometary and meteorite matter, as well as the soil of meteorites, comets, and planets of the Solar system), it is possible to use radioisotope methods.

According to the current ΛCDM (Lambda-Cold Dark Matter) model, the age of the Universe is 13.799 ± 0.021 billion years (Planck Collaboration, 2016).

Accounting for "dark matter" and "dark energy", introduced to explain the discrepancy between theory and observational data, in the case of their real existence, can lead to a revision of the theoretical positions and a change in the estimate of the age of the Universe. However, for the time being, the existing estimates should be accepted, especially since they are consistent with the age of the bodies of the Solar system.

It is believed, that the Solar system was formed by the gravitational contraction of a gas and dust cloud about 4.57 billion years ago (Bouvier & Meenakshi, 2010). By the time the Solar system was formed, the Universe was already like the modern one - the dramatic events of the Big Bang and the "first three minutes" were left behind (Weinberg, 1993). Space and time, matter and field, basic physical interactions, and so on already existed and were separated.

To determine the absolute age of the ancient Earth, radioisotope methods are used - uranium-lead, lead-lead, potassium-argon, argon-argon, rubidium-strontium, samarium-neodymium, rhenium-osmium, lutetium-hafnium, and radiocarbon. The problem is the exchange of matter between the object and the environment after the formation of the object, as well as the uncertainty of its initial isotopic and elemental composition. If a certain amount of the daughter isotope were present in the object when it was formed, the calculated age would be overestimated. If the child isotope left the object, it would be underestimated. For the solution, the isochron method is used, as well as samples of materials that are slightly affected by the environment, particularly zircon crystals (Rink & Thompson, 2015).

The oldest zircon crystals in the Jack Hills in Western Australia are at least 4.404 billion years old (Wilde et al., 2001).

The age of the Earth is 4.54 billion years according to modern estimates (Dalrymple & Brent, 2001) and about 4.6 billion years according to the International Stratigraphic Commission (Figure 8).

By the time of the emergence of life in the Solar system and on Earth, the conditions were far from stationary: 4.1 - 3.8 billion years ago there was an intense fall of meteorites - the so-called late heavy bombardment (Britt & Roy, 2002). Perhaps this fact testifies in favor of the panspermia hypothesis. Also, during this period, the strongest volcanic eruptions occurred.

After the living organisms' appearance, it became possible to combine radioisotope and other dating methods, such as paleontological finds, sedimentation rate, and estimation of the sediments, carried by river, volume.

For the radiocarbon method, often used to estimate the age of organic remains, there must be no violations of the carbon isotope ratio at the initial moment: the decay product does not differ from ordinary nitrogen, and the age is determined only based on measurements of the undecayed fraction of the parent isotope. Thus, knowledge of the features of the initial isotopic composition of the object and the history of its exchange of matter with the environment is critically important. For obvious reasons, this is not always possible.

According to paleontological findings, the first living beings appeared on Earth 3.7 billion years ago (Pearce et al., 2018); according to other sources, 4.1 billion years ago (Bell et al., 2015). The evolutionary scale is shown in Figure 9.

Note that the dating of the remains of living beings, and the rate of accumulation and removal of sedimentary rocks largely depend on the validity of preliminary assumptions and give different definitions of age.

There are alternative scientific ideas about the structure and evolution of the Earth, as well as its age. For example, the hypothesis of V. Larin about the original hydride Earth allows for the abiogenic origin of coal, and hence the exclusion of this classical means of paleontological dating with a corresponding revision of the chronostratigraphic scale (Larin, 1975, 1993; Digonsky & Ten, 2006). But until this hypothesis is generally accepted, it is advisable to traditional dating.

The presence of a time scale makes it possible to roughly estimate the duration of biochemical evolution from 450 to 900 million years. Detailed calculations are presented by (Pearce et al., 2018), which give an interval from 200 to 800 million years.

Figure 8. Age of the Earth (Official chronostratigraphic scale of the International Stratigraphic Commission, n / a)

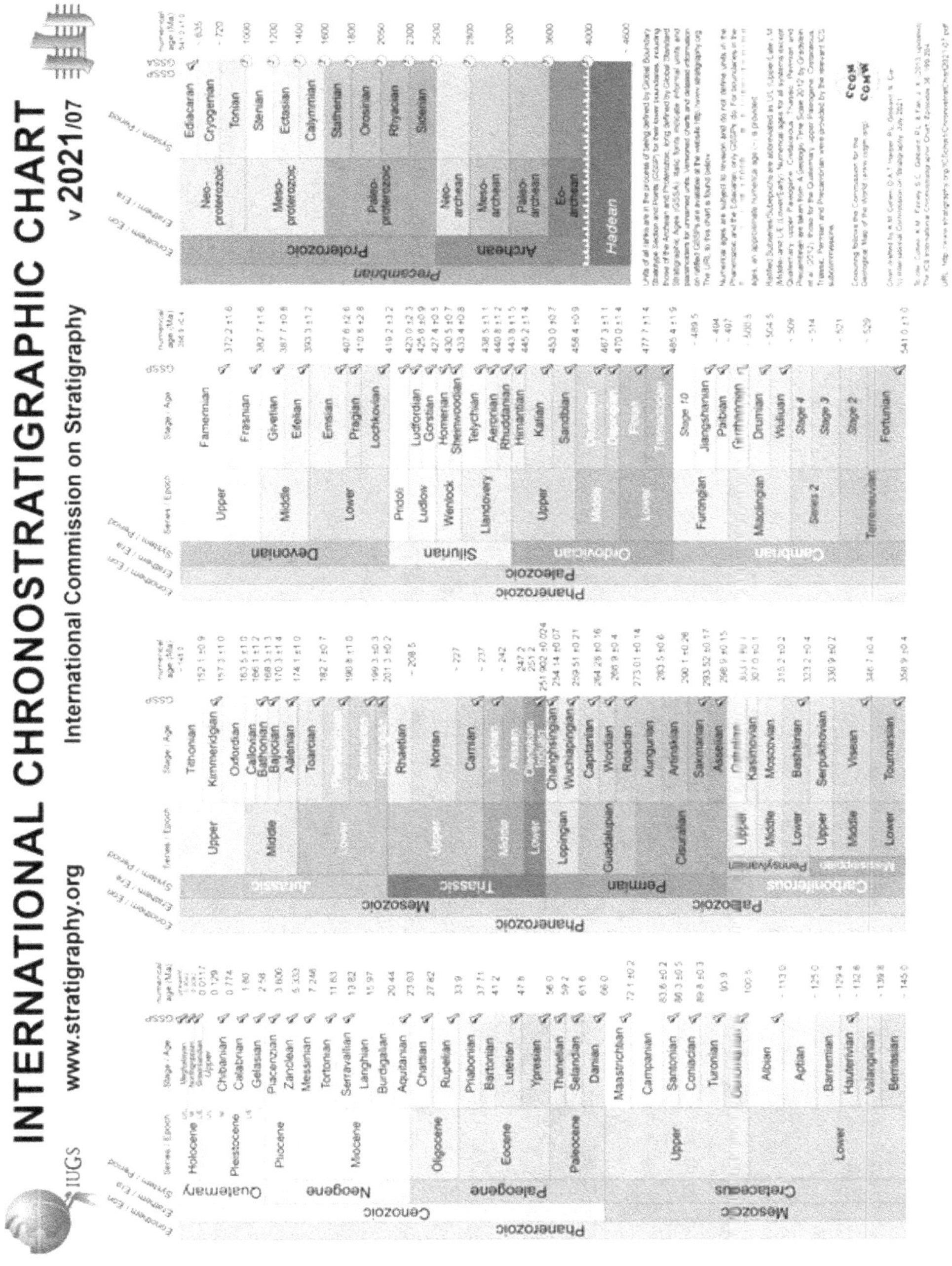

Figure 9. Timeline of the evolutionary history of life (Wikipedia, n / a)

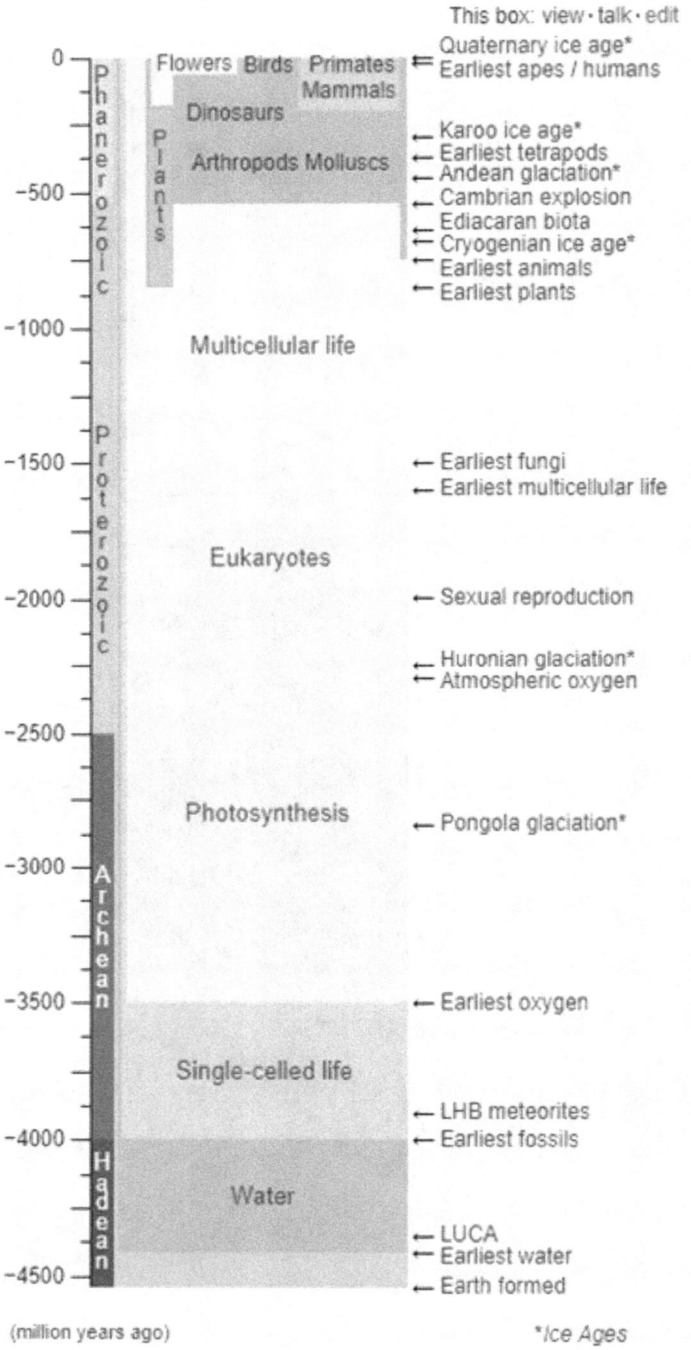

The fossilized remains of the first living beings are shown in Figure 10.

Figure 10. Photomicrograph of pyritized stromatolites from the 3.5-billion-year-old Dresser Formation (Scientists find the 'smoking gun,' signs of life from 3.5B years ago, n / d)

ANCIENT EARTH: CRUST, ATMOSPHERE, HYDROSPHERE, ENERGY SOURCES

Composition of Ancient Rocks

Table 2 presents data on the prevalence of chemical elements in the modern earth's crust. In ancient times, their composition could differ from the modern one due to radioactive decay, coming from the bowels of the Earth, and the fall of meteorites and comets.

Table 2. The prevalence of chemical elements in the earth's crust:

Element	% mass	Element	% mass	Element	% mass
Oxygen	49.5	Fluorine	0.065	Tin	0.004
Silicon	25.3	Sulfur	0.05	Cobalt	0.003
Aluminum	7.5	Barium	0.05	Lead	0.0016
Iron	5.08	Chlorine	0.045	Arsenic	0.0005
Calcium	3.39	Strontium	0.04	Boron	0.0003
Sodium	2.63	Rubidium	0.031	Uranus	0.0003

continued on following page

Table 2. Continued

Element	% mass	Element	% mass	Element	% mass
Potassium	2.4	Zirconium	0.02	Bromine	0.00016
Magnesium	1.93	Chromium	0.02	Iodine	0.00003
Hydrogen	0.97	Vanadium	0.015	Silver	0.00001
Titanium	0.62	Nitrogen	0.01	Mercury	0.000007
Carbon	0.1	Copper	0.01	Gold	0.0000005
Manganese	0.09	Nickel	0.008	Platinum	0.0000005
Phosphorus	0.08	Zinc	0.005	Radium	0.0000000001

(Prevalence, n / a)

Samples of rocks that are synchronous or preceding the appearance of life on Earth are extremely few, which makes it impossible to restore the full picture.

The Isua formation in southwest Greenland belongs to the oldest rocks. The possible presence of life in them is determined in the following way. Atmospheric carbon dioxide can include the stable isotope ^{12}C and the stable isotope ^{13}C, but the enzyme that fixes carbon dioxide for photosynthesis preferentially captures CO_2 molecules with the lighter ^{12}C isotope. Therefore, photosynthesizing organisms have a more stable isotope ^{12}C in their composition compared to atmospheric CO_2, which makes it possible to determine the biogenic origin of carbon. This is the case for Isua rocks, but a physical or chemical process of isotope separation during metamorphism cannot be excluded (Fedo & Whitehouse, 2002).

The formation consists of metamorphosed basalt and sedimentary rocks, the age between 3.7 and 3.8 billion years. The structure of the rock shows that it was in the water at the time of its formation. A disproportion of light and heavy isotopes of carbon and zinc have been established in the rocks that make up the formation. Their presence in iron ores may indicate that these rocks were formed in the presence of photosynthetic organisms, and the oxygen released by them oxidized dissolved iron to form ores.

If we imagine igneous rock as the latest analog of the most ancient rocks, then the common oxides of igneous formations are SiO_2, Al_2O_3, Fe_2O_3, FeO, MgO, CaO, Na_2O, and K_2O, H_2O. (Saranchina, 1998). Igneous rocks are classified according to the content of silica SiO2 (ultrabasic - SiO_2 in the rock is less than 45%, basic - from 45% to 54%, medium - from 54 to 65%, acidic - more than 65%) and alkalis $Na_2O + K_2O$ (rocks of normal, subalkaline and alkaline series).

Meteorites can be another analog. The most common are stone meteorites (92.8% of falls). They consist mainly of silicates: olivines $(Fe, Mg)_2[SiO_4]$ (from fayalite $Fe_2[SiO_4]$ to forsterite $Mg_2[SiO_4]$) and pyroxenes $(Fe, Mg)_2Si_2O_6$ (from ferrosilite $Fe_2Si_2O_6$ to enstatite $Mg_2Si_2O_6$). Iron meteorites account for 5.7% of the falls. They mainly consist of two iron-nickel minerals - kamazite with a nickel content of up to

7.5% and taenite with a nickel content of 27% to 65%. (Wikipedia, Meteorite, n/a). It is believed that meteorites and comets reflect the composition of protoplanetary primary matter, but traces of complex organics found in meteorites cast doubt on this idea.

Composition of the Earth's Atmosphere

It is believed that the primary atmosphere consisted of hydrogen and helium captured from a protoplanetary cloud. Active volcanic activity has led to the saturation of the atmosphere with carbon dioxide, ammonia, and water vapor. Thus, a secondary atmosphere was formed, devoid, due to the lack of oxygen, of the ozone layer, and therefore permeable to ultraviolet radiation. Further, light hydrogen and helium leaked into interplanetary space, and chemical reactions under the influence of ultraviolet radiation, lightning discharges, and other factors. The study of the most ancient rocks gave the following composition of the second atmosphere of the Earth - 98% CO_2, 1.5% N_2, and small fractions of other gases, mainly argon and SO_2 (Nikitin, 2013b). It turned out to be very similar to the modern atmospheres of Venus (~ 96.5% carbon dioxide (CO_2), ~3.5% nitrogen (N_2), 0.018% sulfur dioxide (SO_2), 0.007% argon (Ar), 0.003% steam (H_2O), 0.0017% carbon monoxide (CO), 0.0012% helium (He), 0.0007% neon (Ne) and Mars (95.32% carbon dioxide, 2.7% nitrogen, 1.6% argon, 0.145% oxygen, 0.08% carbon monoxide) (Wikipedia. Venus. Mars).

For comparison, the modern composition of the Earth's atmosphere is presented in Table 3.

Table 3. Composition of the near Earth's atmosphere

Element	% by volume	Element	% by vol	Element	% by vol
Nitrogen	78.08	Inert gases	0.94	Water vapor, dust, microorganisms, ammonia, sulfur dioxide, etc.	less than 0.01
Oxygen	20.95	Carbon dioxide	0.03		

(Composition of the Earth's atmosphere, n/a)

Water and Oceans

The process of the water origin, the oceans' formation, and the reasons for the existing water distribution in the Solar System, have not yet received a generally accepted justification. It is believed that the sources of water can be both chemical processes in the bowels of the Earth, and comets and meteorites that bombarded it. A study of Australian zircon crystals 4,4 billion years old showed, that they had come into contact with liquid water (Wilde et al., 2001). Therefore, the assumption that there were already bodies of water on Earth at that time seems justified. Later, 2.5-2.2 billion years ago (the Siderian period of the Paleoproterozoic era), the existence of the Panthalassa-0 super ocean, which opposes the Pangea-0 supercontinent, is assumed quite confidently. It is believed that the primordial ocean was hot. Due to the high partial pressure of carbon dioxide in the atmosphere, which reached 5 bar, its waters were saturated with carbonic acid, many various metals, especially iron in the form of $FeCl_2$ chloride, and characterized by an acidic reaction, pH $\approx$ 3-5 (Bogdanov et al., 1978).

For comparison, the modern composition of the seawater is presented in Table 4.

Table 4. The main chemical elements in seawater content according to V.I. Vernadsky and A.P. Vinogradov

Element	% mass	Element	% mass	Element	% mass
Oxygen	85.80	Calcium	$4.5\ 10^{-2}$	Strontium	$1\ 10^{-3}$
Hydrogen	10.67	Sulfur	$9\ 10^{-2}$	Boron	$4.5\ 10^{-4}$
Chlorine	2.00	Potassium	$3.8\ 10^{-2}$	Fluorine	$1.0\ 10^{-4}$
Sodium	1.07	Bromine	$6.5\ 10^{-3}$	Silicon	$2.0\ 10^{-5}$
Magnesium	$1.4\ 10^{-1}$	Carbon	$3.5\ 10^{-3}$		

(Content of elements in water, n/d)

Elements and simple compounds in earth, water, and air are potential participants in abiogenic synthesis. Only some of them are involved for reasons. Given the uneven distribution in the surface layers of the planet and the different ability to enter further reactions, water, which is a universal solvent and, after electrolysis (for example, during thunderstorms), a source of hydrogen and oxygen, is needed.

Energy Sources

Energy sources on Ancient Earth were like modern ones: solar radiation, the heat of the Earth's interior, in particular, thermal ground and underwater sources, geysers, volcanic eruptions, energy from the fall of meteorites and comets, wind energy, and thunderstorms. There was also radiation.

The ratio of power, periodicity, and other characteristics of these sources with modern ones were variable, which can be illustrated by the climate history. The Earth has experienced both periods of global glaciation (Hyde et al., 2000) and global warming (the Carboniferous period 360-290 million years ago, characterized by a warm and humid climate). The reasons for these changes are not entirely clear: they can be the passage of the Solar system through dust accumulations, processes on the Sun, a change in the Earth's orbit, processes in the Earth's interior, and a change in the composition of the atmosphere.

THE ORIGIN OF LIFE THEORIES

Several theories of life origin and their connection with abiogenic synthesis reactions have been mentioned above. Their validity is determined by:

- The possibility of the necessary chemical reactions (the initial substances must be present in one of the environments of the Ancient Earth, and catalysts or the reactions must be autocatalytic).
- The efficiency of reactions - the resulting compounds should be accumulated and used for further synthesis faster than decomposing.
- Observance of the conditions for the beginning and maintenance of reactions.
- Ensuring chiral purity.
- Transport of substances.
- Availability of energy sources with certain properties.
- Sufficient time to obtain vital substances.

The Oparin-Haldane hypothesis (Oparin, 1941/1924) was proposed when the conditions on Ancient Earth seemed very approximate. It is based on the premise of "carbon life", the thesis of the emergence of life at a certain stage of matter evolution, and the studies of colloidal solutions of proteins and polysaccharides conducted by A. Oparin. Under certain conditions, dissolved proteins gather into droplets (coacervates), which can grow and divide by absorbing dissolved substances from the external environment, like cells. Oparin suggested that the atmosphere of Ancient Earth was anoxic (otherwise there would have been rapid oxidation of organic matter), and organic substances abiogenic synthesis was possible in it. J.B.S. Haldane developed the idea of a "primordial soup". Various chemical reactions occur in the ocean under the lightning discharges, solar ultraviolet radiation, and volcanic eruptions influence. The primordial soup must contain all 25 amino acids, nitrogenous bases, ATP, and sugars for the proteins and nucleic acids. The sequence of the life-origin events is shown in Figure 11.

Figure 11. The sequence of the origin of life, according to the Oparin-Haldane hypothesis

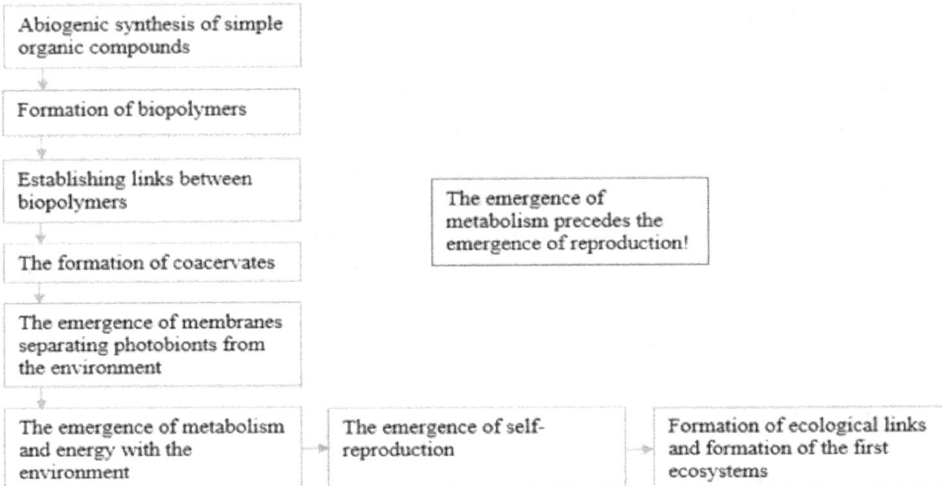

Miller and Urey conducted an experiment simulating the formation of a "primordial soup" (Figure 121). The following was placed in a sealed glass installation: a reaction flask filled through a tap with mixtures of gases (CH_4, NH_3, H_2S, CO_2, H_2O, H_2), corresponding, as it was then believed, to the atmosphere of the ancient Earth and containing tungsten electrodes; heated flask with water; fridge; trap. Discharges simulating lightning were generated in a fusion reaction flask (Miller & Urey, 1959).

Figure 12. Scheme of the S. Miller apparatus (4 February 2008. Own work from Image: MUexperiment.png by Carny)

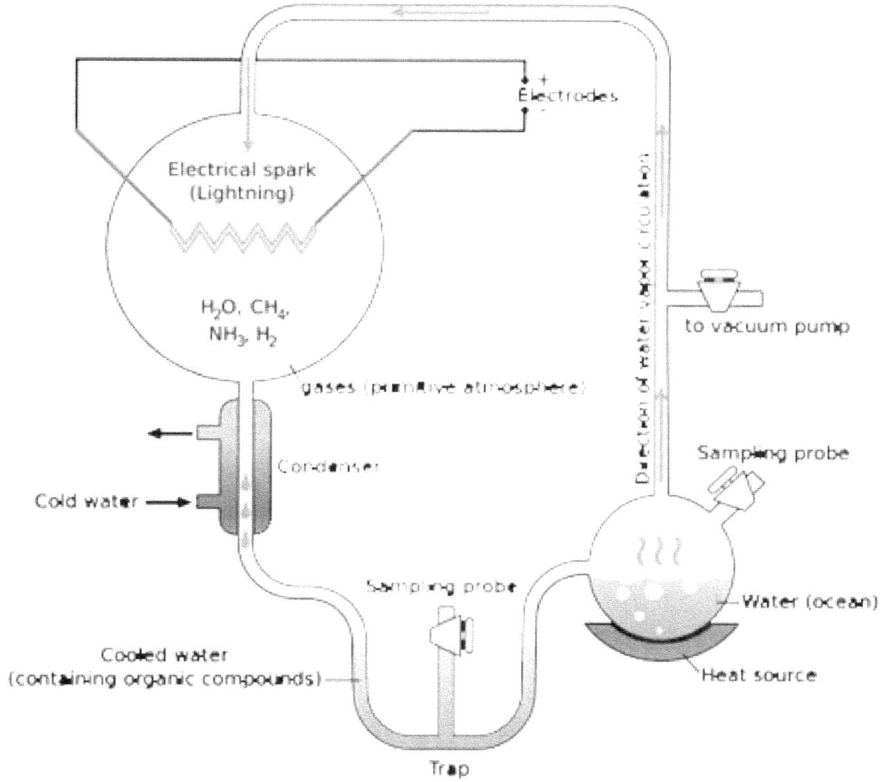

As a result, 4 amino acids, 1 nitrogenous base (adenine, in low yield), and cytosine precursors (cyanoacetylene and cyanoacetaldehyde), as well as formaldehyde, from which sugars could be obtained, were obtained in water. So, was not obtained the required composition of the "primary soup". The composition of gases used in the experiment did not correspond to the atmosphere of the ancient Earth; therefore, the synthesis of amino acids, carbohydrates, and lipids without a carbon dioxide-reducing agent would be impossible. The parameters of the electric discharges did not correspond much to the real ones. The required degree of chiral purity was also not ensured. Another problem was the above-mentioned instability of organic compounds. For example, the half-life of ribose at 0°C and pH=7 is 44 years, adenine - 80 years at 37°C, and cytosine - 340 years at 25°C, which makes a long-term accumulation of bases and sugars in sufficient quantities and concentrations in the primary ocean unlikely. The situation with amino acids is similar (Vorontsov, 2019).

The answer to these inconsistencies was the search for reactions corresponding to the conditions of the Ancient Earth, providing a chiral frequency and, if necessary, the reduction of carbon dioxide. Consider the abiogenic synthesis of simple organic compounds

Amino acids synthesis is like the Miller and Urey experiment but uses other mixtures of gases and other energy sources. The most amino acids have been obtained (Gusev & Mineeva, 2003; Jiang et al., 2014) (Table 5). The problems were also similar: inconsistency in the composition of gases, lack of chiral purity, low efficiency, etc.

Methods for obtaining L-isomers of amino acids have also been developed (Nikitin, 2013c). Upon evaporation of a solution containing a slightly higher proportion of L-isomers, racemic crystals precipitate first, and a solution with a significantly predominant L-isomer (up to 90%) is obtained. The main optically active intermediate (and autocatalyst) of the Butlerov reaction, glyceraldehyde, behaves similarly.

Table 5. Summary results of abiogenic synthesis of amino acids

Reactants	Phase	Energy source	Amino acids detected
CH_4, NH_3, H_2O, H_2	gas	electric discharges	alanine, arginine, asparagine, aspartic acid, cysteine, cysteamine, diaminosuccinic acid, glutamine, glutamic acid, glycine, histidine, leucine, lysine, ornithine, proline, serine, taurine, valine
NH_3, CO_2, H_2O, H_2			
CH_4, NH_3, CO_2, H_2O, H_2, N_2		X-rays	
CH_4, NH_3, H_2O		ultraviolet	
NH_3, HCN, H_2O	water	warm (70° C)	
CH_4, NH_3, H_2O	gas	β-rays	
CH_2O, N_2, H_2O	water	sunlight	
NH_3, H_2S, H_2O		fast electrons	

(Gusev, Mineeva, 2003)

It is possible under certain conditions to obtain chirally pure amino acids from a mixture of equal amounts of both isomers. If an aspartate solution is heated in presence of salicylaldehyde and acetic acid to 100–130° C, pure crystals of one optical isomer are formed. Aspartate is one of two amino acids whose optical isomers crystallize separately. Salicylaldehyde in an acidic medium catalyzes the transition of isomers in solution to each other, therefore, small random deviations at the beginning of crystallization lead to the complete conversion of the mixture into a pure L- or D-isomer. Another way for optical isomers separation is adsorption on a certain mineral's surface. Thus, calcite crystals retain L-amino acids more strongly on some faces, and D-isomers on others. Influences to ensure chiral purity

Synthesis of sugars. The Butlerov reaction. In modern experiments, formaldehyde, which can be obtained in the experiments of Miller and Urey, is often used as the starting material (Nikitin, 2013d). During the Butlerov reaction, an aqueous solution of formaldehyde (CH_2O) with the addition of $Ca(OH)_2$ or $Mg(OH)_2$ turns into a complex mixture of sugars with slight heating (Figure 13).

Figure 13. Synthesis of sugars on the base of Butlerov reaction

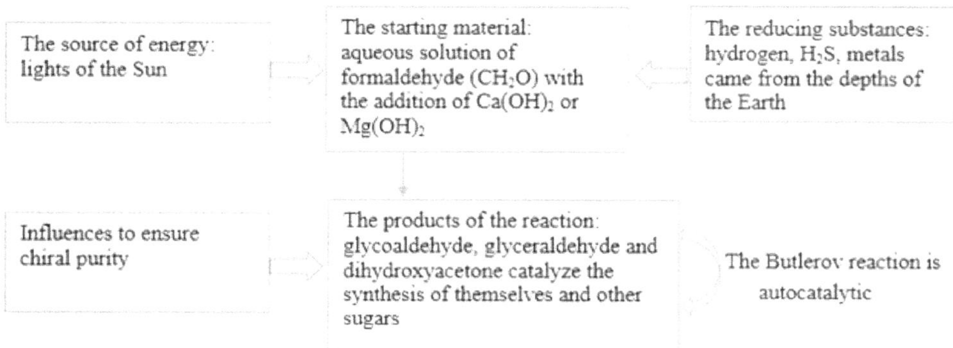

In the resulting mixture, the sugars, characteristic of living cells, are mixed with a variety of seven-, eight-, nine-carbon, and more complex sugars, which prevents the subsequent stages of synthesis. Ways to selectively accumulate the necessary sugars have been found:

- When soluble silicates, for example, Na_2SiO_3, are added, complexes with four- and six-carbon sugars are formed, which precipitate and do not further participate in the reaction. This is how erythrose, threose, glucose, and mannose accumulate.
- When hydroxyapatite $Ca_3(PO_4)_2 \bullet Ca(OH)_2$ is added, ribose is practically selectively deposited on its surface.
- Boric acid salts selectively precipitate ribose from the reaction mixture.
- The complex of proline amino acid with zinc ion stops the reaction at the stage of five- and six-carbon sugars, besides, the complex of L-proline with zinc selectively synthesizes D-sugars.
- Several other amino acids, such as glutamine and leucine, also provide a stereospecific synthesis of D-sugars in the presence of L-amino acids but do not stop it at the stage of ribose and six-carbon molecules.

Thus, the necessary sugars synthesis is theoretically possible; their chiral purity is associated with the presence of chirally pure amino acids, especially L-proline. Considering the half-life of amino acids, synthesis must be very fast.

Synthesis of nitrogenous bases and ATP can occur in different ways (Nikitin, 2018; Gusev & Mineeva, 2003). Adenine and guanine are formed from hydrocyanic acid by freezing its aqueous solution, ultraviolet irradiation, or heating. All four nitrogenous bases are synthesized in high yield from formamide on the surface of TiO_2 particles under ultraviolet irradiation (Senanayake, 2006; Saladino et al., 2007), adenine, cytosine, and uracil are synthesized on the surface of montmorillonite (a kind of clay) or iron oxides when heated (Constanzo et al., 2007).

K. Ponnamperuma (Ponnamperuma et al., 1963) showed that when a mixture of aqueous solutions of adenine and ribose is irradiated with UV at a temperature of 40° C in the presence of phosphoric acid, a condensation reaction occurs, leading to the formation of adenosine. If the reaction is carried out by adding ethyl metaphosphate to the reaction mixture, AMP, ADP, and ATP are also formed. Phosphorus compounds in these chemical syntheses play the role of catalysts and can be directly included in the reaction products. The abiogenic synthesis of ATP, which is the result of several relatively simple chemical reactions, indicates the possible early appearance of this compound.

The genetic information preservation is ensured by various nucleic acid bases. The use of adenine, guanine, cytosine, and uracil/thymine may be due to their resistance to UV radiation (Mulkidjanian et al., 2003; Mulkidjanian & Galperin, 2007).

Synthesis of nucleosides and nucleotides. So, theoretically, there is a sequence of reactions that ensures the formation of substances necessary for the nucleoside synthesis - associations of purine and pyrimidine nitrogenous bases with sugar, and then nucleotides - monomeric units of RNA.

For adenosine, this possibility was shown in 1963 (Ponnamperuma et al., 1963). Later, the possibility of synthesizing guanosine was shown, as for adenosine with low efficiency (Vorontsov, 2019).

Next, adenosine must attach a phosphate molecule to itself, which is also shown in (Ponnamperuma, Mariner, and Sagan, 1963). The excited triplet state of adenine locates the unpaired electron on the amino group, and this form easily forms a high-energy phosphoramide bond with phosphate. Next, the phosphate is transferred to the 5'-hydroxyl group of the ribose. The diphosphate chain of ADP is of suitable length for efficient transfer of the third phosphate group. However, this method for the activated nucleotide synthesis is not suitable for other nitrogenous bases. Guanine attaches ribose upon irradiation but practically does not react with phosphate. Pyrimidines do not react with ribose either (Nikitin, 2018).

A synthesis for cytosine and uracil was proposed in 2009 (Powner et al., 2009). By mixing cyanamide, glyceraldehyde, and glycolaldehyde in one system, activated pyrimidine nucleotides were obtained. Phosphate selectively catalyzes the synthesis of 2-amino-oxazole, (arabinose-amino-oxazoline), suppressing possible side reactions. The product is then reacted with cyanoacetylene to give an arabinose anhydronucleoside. In an ordinary aqueous solution, acidity decreases, which leads to hydrolysis of intermediate products and side reactions, but phosphate maintains an acidic medium, directing the reaction towards the formation of arabino-anhydronucleoside. To convert into cyclic cytidine monophosphate, the reaction mixture must heat. The phosphorylation catalyst is urea, formed from cyanamide during one of the side reactions. To rid the reaction side products and convert some of the cytosines into uracil, the ultraviolet illumination of a solution is sufficient (Figure 14).

Figure 14. Pyrimidine ribonucleotide assembly options. Previously assumed synthesis of b-ribocytidine-29,39-cyclic phosphate 1 (blue; note the failure of the step in which cytosine 3 and ribose 4are proposed to condense together) and the successful new synthesis described here (green). p, pyranose; f, furanose (Powner et al., 2009)

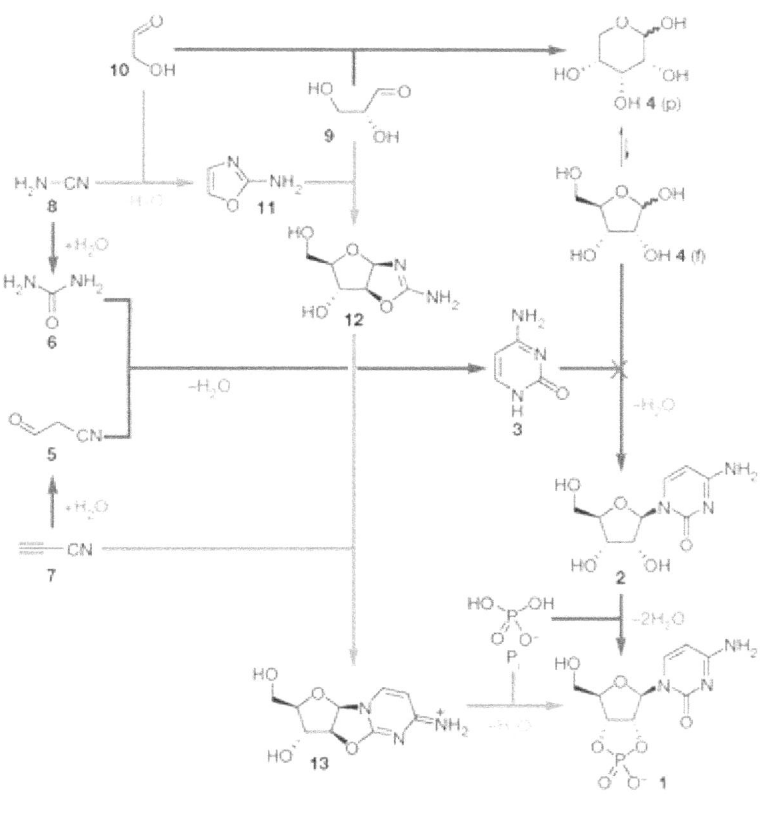

In 2011, stereospecific nucleotides were synthesized by adding a chiral-enriched solution of various amino acids to the Sutherland system. This was achieved by partial crystallization and photochemical processes involving polarized ultraviolet light (Hein et al., 2011; Hein & Blackmond, 2012). Other processes leading to enrichment in one type of optical isomer have also been studied. Meteorites contain amino acids with a chiral enrichment of up to 18% and with an excess of precisely L-isomers. Amino acids interfere with the Sutherland synthesis at the stage of the reaction of 2-amino oxazole with glyceraldehyde, and a ternary product is formed. This reaction is stereospecific: a pair of glyceraldehyde with an amino acid of the same chirality reacts four times faster than a pair of different chirality. In Hayne's experiments, chirally pure ribonucleotides were thus obtained, starting with as little as 1% chiral- enriched proline.

To carry out such a synthesis, an exact ratio of the starting substances is required, as well as a change in the temperature regime and ultraviolet irradiation synchronized with certain stages, which at more advanced stages becomes fatal to life, as critics drew attention to (Vorontsov, 2019).

Moreover, the feasibility of the entire sequence, which depends on many conditions and requires the consistency of many parameters for a limited time in nature rather than laboratory conditions, is highly questionable. But the long duration of biochemical evolution (from 450 to 900 million years) allows many attempts, and the fact of the existence of life inspires biochemists, showing that success has nevertheless been achieved.

Another response to the results of the Miller and Urey experiment was the search for more suitable places for the origin of life than the open ocean.

It is believed that the appearance of phototrophs with the ability to use light energy was preceded by chemotrophs - organisms that receive energy because of redox reactions. Chemosynthesis-based ecosystems exist on the ocean floor and in other dark places, most of which use oxygen as an oxidant. The chemical elements and compounds from which energy is extracted as an oxidation result include inorganic (molecular hydrogen, sulfur, and others) and organic (carbohydrates, fats, proteins, paraffin, and other simpler compounds).

There are two theories of abiogenic synthesis of organics, linking the reduction of CO_2, energy metabolism, and the content of metal ions in living matter.

The "iron-sulfur world" by K. Washterhauser suggests the origin of life in an iron-sulfur environment, on underwater geothermal sources, which are also called "black smokers" (Figure 15).

Figure 15. Black smoker hydrothermal vent at 2,980m depth, Mid-Atlantic Ridge (Kate Larkin, n / d)

Modern underwater geothermal sources emit water with a temperature of up to 400 °C, which does not boil due to high pressure. Precipitates of metal sulfides fall out of it. Upon contact of geothermal water with ocean water, the first one cools, and black sulfides first fall out of it: FeS, CuS, NiS. With further cooling, in the temperature range of 200-300 °C, ZnS, and MnS precipitate out of the water. In addition to transition metal sulfides, geothermal water is enriched with hydrogen sulfide, potassium, and magnesium ions, and has an alkaline reaction. In terms of their concentrations, the cytoplasm of cells is more similar to geothermal water than to ocean water.

Wächtershäuser experimentally tested many of the chemical reactions that take place in this environment (Wächtershäuser, 2006). The main source of reducing equivalents is the reaction $H_2S + FeS > FeS_2 + 2[H]$, in which iron sulfide is converted to pyrite, and hydrogen atoms remain adsorbed on its surface. With the help of these hydrogen atoms on the surface of iron sulfide, nitrogen fixation, CO_2 reduction to methyl mercaptan, and reductive amination of alpha-keto acids can occur. Geothermal waters contain carbon monoxide, which enters into various chemical reactions on the surface of iron and nickel sulfides with thiomethyl acetate and pyruvic acid formation. Negatively charged organic acids are adsorbed on the surface of sulfides and can accumulate in high concentrations. More complex chemical reactions, leading to the synthesis of various organic products also occur

(Nikitin, 2013d). In the presence of nickel sulfidein in geothermal sources, the diversity of products increases - CO_2 and CO are converted into acetic, pyruvic, and other acids, as well as into thiomethyl acetate CH-CO-S-CH. The latter, reacting with phosphoric acid, gives acetyl phosphate, a high-energy substance that can phosphorylate organic molecules, including converting ADP to ATP.

Negatively charged organic acids remain adsorbed on the surface of sulfides and can accumulate in high concentrations. More complex chemical reactions occur there, for example, the glycine peptide cycle, in which urea derivatives, hydantoins, and purine nitrogenous bases are formed (Figure 16).

Figure 16. Dipeptide cycle of glycine (Wächtershäuser, 2006)

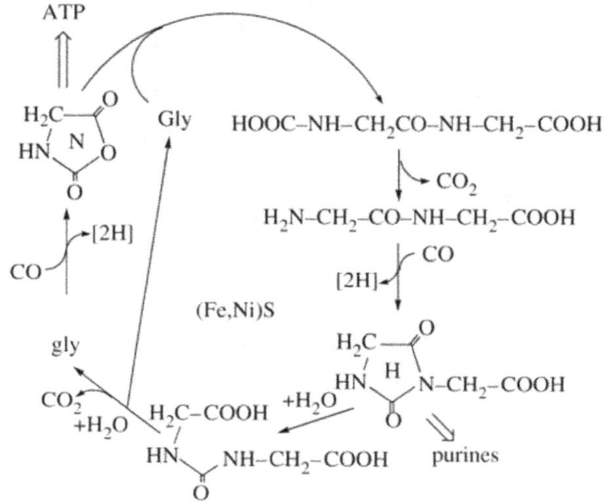

Thus, under the conditions of a "black smoker", the various organics synthesis is possible.

A. Mulkidjanian's "Zinc World" (Mulkidjanian, 2009; Mulkidjanian & Galperin, 2009) suggests the origin of life in the "Zinc World" at geothermal springs. It follows from the ability of zinc and manganese sulfides to the photochemical reduction of various substances. ZnS, MnS, TiO_2, and WO_3 absorb near ultraviolet. Their suspension in water reduces CO2 to formic acid and dicarboxylic acids, zinc sulfide is the most effective reducing agent with a quantum yield of up to 80%. In this case, a positive charge accumulates on the crystal. If there are no suitable reducing agents, then photo corrosion of zinc sulfide occurs with the release of zinc ions into solution and the formation of molecular sulfur; like iron sulfides, zinc sulfide in the light can reduce nitrogen to ammonia, which then combines with CO_2

to give formamide, which can polymerize to nitrogenous bases under UV radiation. Reductive amination of keto acids occurs under the same conditions (Nikitin, 2013d).

The reliability of both theories must be confirmed by the possibility of realizing the fixation and reduction of CO_2 under the conditions of the Ancient Earth, as is characteristic of living organisms. These are the Calvin cycle (plants and most photosynthetic bacteria), the Krebs cycle (in some anaerobic photosynthetic bacteria), and the reductive acetyl-CoA pathway (in some chemosynthetic bacteria and archaea). Another check is compliance with the content of metals in the cells. Both checks testify in favor of Mulkidzhanyan's "zinc world" (Nikitin, 2013d; 2018).

However, further checks are needed. In 2019, Maruyama (Maruyama et al., 2019) analyzed various hypotheses in terms of site selection, energy source, and other conditions (Figure 17).

Figure 17. Proposed sites for the birthplace of life on Earth. (1) Classic concepts, including Darwin's warm little pond with a primordial soup, or a shallow marine tidal flat, (2) Panspermia or Neo-Panspermia, (3) Mars, (4) a mid-oceanic ridge hydrothermal system or in an alkaline hydrothermal vent, (5) a hydrothermal environment with island arc volcanoes, and (6) a nuclear geyser system on the Hadean primordial continent (Maruyama et al., 2019)

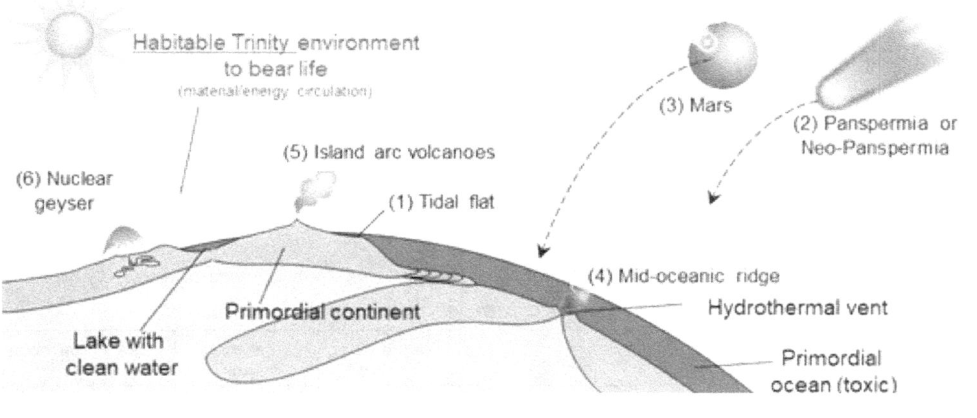

As a result, Maruyama substantiated the hypothesis of the origin of life in systems of nuclear geysers (Table 6).

Other theories have been proposed with their advantages and disadvantages, for example:

- **The primordial pizza model** (Maynard & Szathmáry, 1995). It is assumed that life originated as a thin layer of organic molecules on the surface of the clay, the particles of which are composed of thin aluminosilicate layers. Clay can swell because water molecules, like other cations and anions, are embedded between these layers, increasing the distance. Experiments have shown that clay quite strongly binds amino acids, nucleotides, proteins, RNA, and other biological molecules. On the surface of clay crystals and between their layers, they accumulate in high concentration even from a very dilute solution. On clay, nucleotides spontaneously line up exactly as needed for their "crosslinking" into an RNA chain. Long RNA molecules very rarely completely separate from the clay. However, they can slowly move along its surface, detaching one end or the other. Mathematical modeling of the interaction of different RNAs on a flat mineral surface showed that under such conditions, groups of different molecules are easily formed, and linked by mutual assistance, and the reproduction of parasites is limited and does not lead to the extinction of cooperating RNAs.

- **The "primary mayonnaise" model** (Vellas, 1988). It is assumed that primitive analogs of cell membranes have existed since ancient times, even before the advent of self-copying RNA. In other words, the entire world of RNA existed inside protocells - small fatty vesicles. The "primary mayonnaise" theory has fewer supporters than the "primary pizza" theory because there is a nutritional problem for protocells: nucleotides pass through membranes very poorly. In modern cells, there are special transport proteins for this, but an adequate solution for the uptake of nucleotides by primitive protocells has not yet been found. Because of the effective separation of RNA molecules into cooperating groups in the "primary mayonnaise" model, scientists are in no hurry to reject it.

Table 6. Nine requirements for the emergence of life. For the emergence of life on the planet, there are several essential requirements to be fulfilled

	Environmental factors	Nuclear geyser system	Hydrothermal system	Mars	Universe
1	Energy source (ionizing radiation + thermal energy)	Yes	No	Yes	Yes
2	Supply of nutrients (P, K, KREEP, etc.)	Yes	?	Yes	No
3	Supply of life constituent elements (C, H, O, N)	Yes	?	Yes	Yes
4	The concentration of reducing gas	Yes	No	?	No
5	Dry/wet cycle	Yes	No	Yes	No
6	Na-poor water	Yes	No	Yes	No
7	Non-toxic water environment	Yes	No	?	No

continued on following page

Table 6. Continued

	Environmental factors	Nuclear geyser system	Hydrothermal system	Mars	Universe
8	Diversified environments (Ocean: pH, salinity, heavy metals, Atmosphere: Temperature & Pressure, Continent: varied geology)	Yes	No	?	No
9	Cyclic change	Yes	No	No	No

(Maruyama et al., 2019)

It is possible to combine these theories: clay particles, as it turned out, help the formation of membrane bubbles, while the resulting bubble surrounds the clay particle from all sides (Nikitin, 2018).

The considered abiogenic synthesis of simple compounds is the first of eight stages in the emergence of life according to Oparin. However, and this applies to all hypotheses considered, a lot of problems were found related to the search for reactions corresponding to the conditions of the ancient Earth, the place where life originated, the characteristics of energy sources and reducing substances, the chirality and stability of organic compounds, the yield of products, and the overall efficiency. Another problem is the synchronization of heating and cooling, "inclusions" and exclusions of ultraviolet and radioactive radiation with various stages of complex reactions, provided in the laboratory, but difficult to achieve in natural conditions. The attempt to move from the amino acids, sugars, and nitrogenous bases synthesis to nucleosides and nucleotides shows, how much these problems are aggravated. The abundance of competing hypotheses shows that a satisfactory solution has not yet been found.

Therefore, the panspermia hypothesis is of increasing interest, especially since a few data testify in its favor:

1. Presence of organic matter in cosmic dust (Wickramasinghe & Allen, 1980; 2016; Allen & Wickramasinghe, 1981).
2. The presence of water ice, volatile fractions, carbonates, polynuclear aromatic hydrocarbons, sulfides, and other components in the cometary substance, detected by the analysis of the collision of the probe dropped by the Deep Impact probe with Comet Tempel (Deep Impact made it possible to study the composition of comet Tempel-1., 2006) were discovered during the Rosetta mission during the study of comet Churyumov-Gerasimenko; (Fray et al., 2016).
3. Detection of molecules of various sugars in meteorites, including ribose (NASA, First Detection of Sugars in Meteorites Gives Clues to Origin of Life, 2019; Furukawa et al., 2019).
4. Detection of the hemolithin protein in the Acfer 086 meteorite (McGeoch et al. 2002).
5. DNA resistance to space travel, tested during suborbital flights (DNA showed high resistance to extreme space conditions, 2014).

6. As a result of the experiment on the Foton-M4 satellite, the possibility of the survival of microorganisms on materials simulating the bases of meteorites and asteroids under space flight conditions has been shown. After the apparatus landed, out of 11 thermophilic and 4 spore-forming bacteria, one line of bacteria survived and continued multiplying under terrestrial conditions (Russian scientists have proved the possibility of bringing life to Earth by meteorites, n / d).
7. A class of bacteria that can survive in space conditions has been discovered, and a way of such survival has been determined (Ott et al, 2020).
8. "Suspicious" in terms of the presence of microorganisms (or traces of their vital activity) are several Martian meteorites (NASA\JPL Mars Meteorite Home Page (JPL), n / d; McKay et al., 1996; Gibson et al., 1999), as well as the Orgueil meteorite found in France as early as the 19th century. Here are the photographs, which, according to (Hoover et al., 2019), are the remains of living creatures of extraterrestrial origin, found during the study of the Orgueil meteorite (Figure 18). However, their arguments were not accepted by most of the scientific community.

Figure 18. Orgueil meteorite: 1 - Trazinofit (green algae); 2 - Unknown nuclear organism on the right - unknown nuclear organism (Hoover et al.,2019)

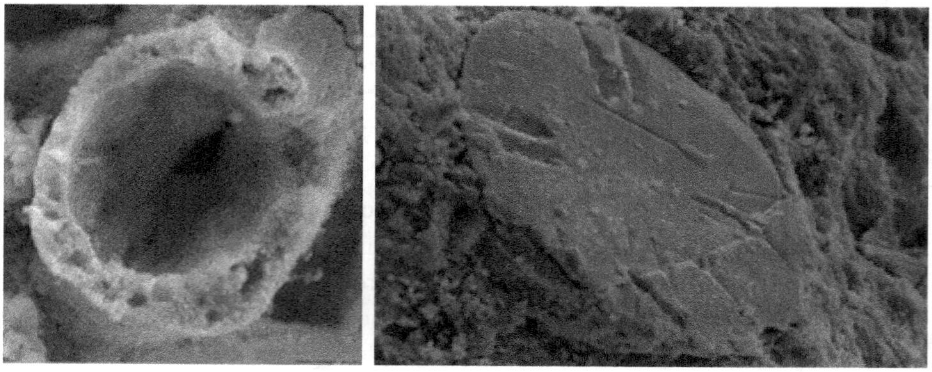

Convincing proof of the panspermia hypothesis would be a successful experiment on the biochemical synthesis of living organisms, free from the limitations associated with attempts to reproduce the conditions on Ancient Earth (but within the foreseeable time - which may already contradict real conditions). However, such a "homunculus" has not yet been created.

Nevertheless, the hypothesis that assumes the transfer of complex organics, catalytic substances, and other materials, the presence of which can create favorable conditions for the start of biochemical evolution not from zero, but from more advanced stages, already looks quite reasonable. But it is necessary to consider the time factor: if complex organics are present in meteorites and comets now, then was it present at the time of the emergence of life on Earth, when the Earth was subjected to intense meteorite bombardment (about 4 billion years ago)?

"ZONE OF LIFE" AND ABSTRACTIONS OF S-SPACE

The above, in terms of systems theory, looks like this. Since the beginning of the expansion of the Universe, the formation of the Solar system and the Earth, numerous dissipative environments and boundary spaces have arisen in which the exchange of various substances and energies takes place, and the synthesis and decay processes and the unification also occur at the supramolecular level, an example of which are crystals. A few processes have a cyclic and wave character. Some processes release energy, but in most cases, an external source is needed. The transition from synthesis to decay can be due to:

- Immanent causes, such as radioactive decay.
- External causes, for example, the destruction of rocks due to the fall of a meteorite.
- Processes wave character, for example, from compression to expansion change.
- The influence of the supersystem, for example, the metamorphization of rocks during the evolution of the Earth.

Thus, the prerequisites for life and death are already observed for inorganic substances. On the other hand, there are very stable, practically "immortal" structures, like zircon crystals, which may exist for billions of years.

Let us group the above conditions for the emergence of life so that it is convenient to compare them with abstractions of the wave model.

The Condition for the Emergence of Life and the Abstractions of the Wave Model

1. *Homogeneity of the environment.* Life arose in that part of the Universe where there is practically no antimatter left, there is a unified physics, globally homogeneous and isotropic space, and so on. This homogeneity is also transferred

to living beings - they consist of the same atoms and obey the same laws as the Universe.

2. *Differentiation of the environment.* At the same time, there is a separate substance and field, time, physical interactions, chemical compounds, and so on, interacting with each other. The diversity of such factors and their qualitative heterogeneity are the prerequisites for the complexity of life.
3. *Symmetry of the environment.* According to Noether's theorem (Wikipedia. Noether's theorem, n / a, n / d), certain symmetries correspond to different conservation laws. Symmetry stabilizes the environment, and a certain constancy of conditions for a long time is necessary for the efficient flow of abiogenic synthesis.
4. *Asymmetry of the environment.* The property of chiral purity is important. In the symmetrical to the right and left connections environment, the useful functions of connections are not realized. Probably, our part of the Universe (or our entire Universe, if we accept the hypothesis of multi-worlds) would not have evolved without the separation of matter from antimatter. Further, if symmetry stabilizes the medium, then asymmetry causes the transfer of various substances and the spread of fields from one area to another.
5. *The dissipativity of the environment* follows from the previous condition and is an important independent factor. Even systems of abiogenic synthesis are complex open systems, not to mention living beings. The transfer of energy, matter, and fields creates "free" resources, necessary for the complex systems' self-organization, and the living organisms' existence, which are "parasitic" on these resources.
6. *The periodicity of the environment states* is manifested in the change of day and night, seasons, and astronomical phenomena - unfavorable living conditions are replaced by favorable, and vice versa. Both create a non-equilibrium, dynamic environment, are selection factors, and initiate the creation of adaptation mechanisms.
7. *The limited values of vital parameters.* There are certain limits to temperature changes, radiation levels, concentrations of substances, and similar factors that make abiogenic synthesis possible. For life, which is based on silicon or boron, a similar picture should be expected, due to the conditions for the feasibility of key biochemical reactions. The distances from the star to the planets where life is possible are limited - naturally, considering the characteristics of the star's radiation ("life zone"). Similarly, a limited area is allocated around underwater thermal springs, and so on. The time of existence of the "life zone" is limited - the time remaining before the explosion of a star, the fall of a large meteorite, etc. Such examples testify to the important role of boundary spaces for life and death manifestations.

8. *The external initiating factors*. And yet, the periodic presence of favorable factors for life emergence is not enough - an external initiator is needed - an electric discharge, energy, substances, or radiation of thermal sources. The role of external factors reaches its maximum in the panspermia hypothesis - whether it be the "delivery" of catalysts, genetic material, or living organisms.
9. *The possibility of self-organization as a factor of evolution*. Self-organization is a common property of complex open systems, regardless of their nature. This fully applies to processes of abiogenic synthesis; the mechanism of self-organization can be, for example, autocatalytic reactions, or chirality, that separates compounds involved in abiogenic synthesis from a racemic mixture of chemically identical compounds. Self-organization, as a general law of the existence of complex systems in a dissipative environment, is an evolutionary factor: changes have a "path" in the form of a certain scenario of self-organization and an "aim" in the form of its next stage. "Purposefulness" of self-organization in comparison with the "randomness" of evolution significantly reduces the time of occurrence of progressive changes. Note that self-organization occurring in a limited region of the Universe does not violate the second law of thermodynamics, since "chaos" continues to increase outside this region.
10. *The possibility of selection and inheritance*. Usually, these factors are already associated with living beings, however, even in abiogenic synthesis, compounds are selected according to certain "chemistry" (for example, iron-sulfur or zinc worlds), resistance to ultraviolet radiation, and so on. The same applies to inheritance - the environment that has changed by a reaction cycle serves as the basis for the reactions of the "next generation". The selection and inheritance of living organisms are to some extent based on the mechanisms worked out in the process of abiogenic synthesis.

S-space wave model abstractions. It is necessary to introduce and illustrate notions and definitions used in the S-space description (Figure 19).

Figure 19. Universal, Object, Subject, and limit S-space (Kovalyov et al., 2020)

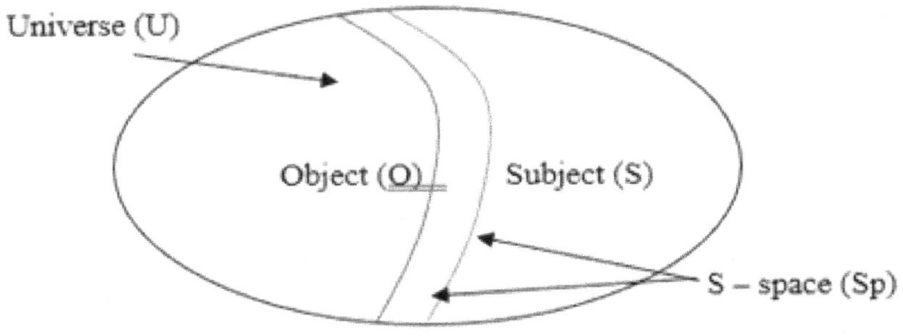

Let us represent Universal (U) which contains everything in the form of an ellipse and divide it into two parts which we call Object (O) and Subject (S). These names have no fixed meaning and may be interpreted differently depending on the situation (as variables x and y in mathematics). As Universal includes everything, hence border between Object and Subject which reflects their differences cannot be introduced from outside, and whereas nothing can be added to Universal, this border doesn't bring in anything new. If we attribute the mark «+» to differences between O and S, differences between S and O must have the opposite mark «-», then the border which contains both differences will be equal to «0», i.e., both conditions will be fulfilled. Let us call *S-space* such a symmetrical border exposed to the influence of S and O and vice versa, which exists since they both exist. The appropriate abstraction of the S-space is a wave (Figure 20).

Figure 20. Wave nature of S-space, potential, modalities of existence

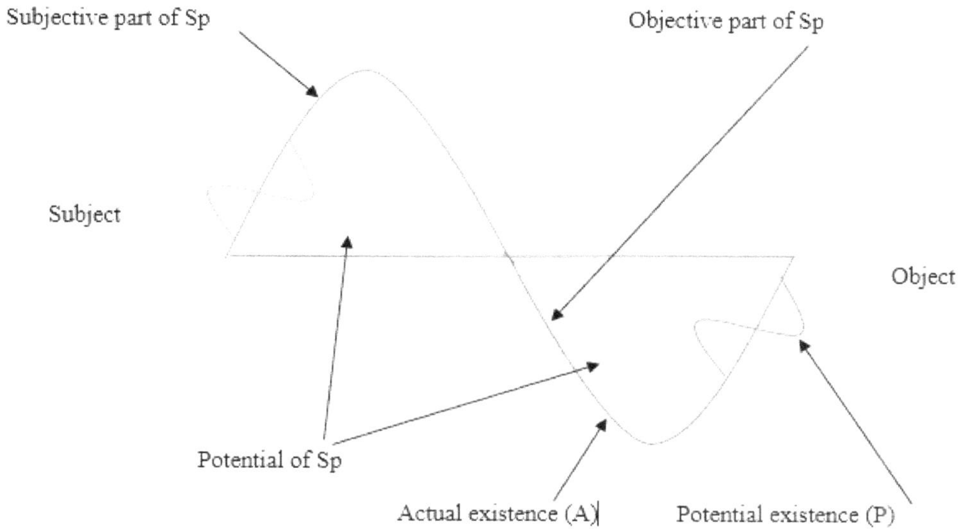

A wave of S-space "grasps" a particular part of Universal; we will name this part *potential* (π) of S-space and give it a different interpretation depending on the situation. Subjective and objective parts of Sp are symmetrical and it is one of the integrity properties of Sp.

Next, the modalities of actual existence (***A***) is introduced, which means the existing Sp, or S-sets, or S-elements and potential existence (***P***), the possibility of their occurrence (as opposed to non-existence). Figure 18 shows that a big wave of Sp exists actually and small waves - potentially – they emerge due to the existence first, of Sp, and second, an influence of Subject and Object.

Interaction of S and O – *observation* (***H***) – changes the existence modality of Sp and causes its *stratification* (St) into actually existing S- sets and S-elements. An absence of ***H*** triggers a backward process to potential existing – the *convolution* (Co) of Sp.

S-space (°) is a carrier of sets and elements; *S-sets* (□) – is a potentially or actually existing structure, that coordinates potentially or actually existing and relatively homogenous S-elements.

S-elements include *wave* (∪) – potentially or actually existing element, which is marked by immanent alterability, extension, relative instability, ability to interact, condition modalities, capability to transfer in soliton and *soliton*[1] (•) – a single wave which can transfer in ∪. Entry ∪ may also describe multiple waves; • – multiple solitons; ∪• – waves and solitons. Elements may have only one component (prototype – a sound wave of one frequency), or several perpendicular components (as an electromagnetic wave that contains perpendicular electric and magnetic elements). Accordingly, single-component, and multi-component waves and soliton waves are distinguished.

S-space is described by four groups of axiomatic statements (Chapter 2).

Thus, wave model abstractions implement all the following conditions:

1) *The homogeneity of the environment* corresponds to the S-space wave nature; when modeling within the subject or object parts of the S-space, signature homogeneity is added.
2) *The differentiation of the environment* corresponds to the Universe division and the subjective and objective parts of the S-space.
3) *Symmetry of the environment.* The subjective and objective parts are "skew-symmetric".
4) *Asymmetry of the environment.* The subject and object parts have different signatures.
5) *The dissipativity of the environment* also follows from the previous paragraph: different signatures create the basis for the emergence of interaction.
6) *The periodicity of the states of the environment* follows directly from the wave character of the S-space abstractions.
7) *The limited values of vital parameters* correspond to the boundary nature of the S-space. If we talk about the time of existence of waves on the surface of the subjective and objective parts of Sp, then it is obvious that it cannot exceed the time of existence of these parts themselves. This corresponds to "death" not only from external influences but also from internal factors.
8) *The external initiating factors* are expressed in the influence of S and O on Sp.
9) *The possibility of self-organization as a factor of evolution* is expressed in the formation of differentiation of the subjective and objective parts of the S-space under the influence of the influences of S and O.
10) *The possibility of selection and inheritance.* The actual effects of S and O select a specific scenario of self-organization from their set; each stage of self-organization depends on the previous one, using both its potential and elements.

FUTURE RESEARCH DIRECTIONS

First, it is necessary to carry out a similar check of the axiomatic definitions that specify the stated concept. The S-space axioms will be compared with the key characteristics of the organisms' vital activity (Chapter 2).

CONCLUSION

The vital substances, dating, and conditions on Ancient Earth, including the composition of the crust, atmosphere, and hydrosphere, as well as possible energy sources, are considered. It is noted that one of the causes of living organisms' mortality appears at the stage of abiogenic synthesis, and it is the instability of organic compounds and their exposure to external influences. It follows, that natural or artificial strategies for immortality (Chapters 2, 5, 9) achieving should provide for the restoration of destroyed vital substances, or their replacement with the same undamaged substances, or their replacement with persistent substances, modified to perform life support functions.

Another possible factor in mortality is, paradoxically, the life cycle itself, the "idea" of which is adaptability to various periodic phenomena.

Note, the chirality of organic compounds, which can be called relative homogeneity with the external environment: being chemically identical to the substances of the racemic mixture, they stand out from them by one geometric property.

Hypotheses of the origin of life, including the hypothesis of panspermia, are also considered from the point of view of their compliance with the conditions on Ancient Earth.

As a result of the analysis of modern ideas about abiogenic synthesis, 10 conditions for the emergence of life were determined. Their comparison with the abstractions of the S-space shows a complete correspondence. Thus, the main task of this chapter is solved, and now it is possible to check the correspondence of the wave model of the S-space axioms (Chapter 2).

Let us discuss the paradigmatic meaning of the S-space concept.

Death is always associated with destruction, decay, and simplification; life is associated with growth, synthesis, and complication. But if decay and death seem natural, explanations are easily found for them (for example, in the form of an increase in entropy). Complications and life seem to be a miracle, although they are observed just as often. Of course, for living things, complexity can be explained by natural selection - even if evolution sometimes leads to simplification of structure - but what makes simple molecules assemble into complex compounds through sophisticated reactions that require the many factors synchronization? Self-organization, but how and why does it work?

Perhaps both disintegration and synthesis are not what they seem at first glance. The expansion of the Universe is called the Big Bang, but the structure that appears to be simplified during the explosion demonstrates an increasing complication. Of course, the Big Bang is just a metaphor, and many details of the Universe's evolution are still unclear. However, for example, a plane crash is an event that leads to disintegration and simplification. However, an investigation into the causes sometimes reveals a chain of events and errors so unlikely in themselves, and even more so in their sequence, that this can be called the self-organization of a catastrophe. And isn't this like the self-organization of the abiogenic syntheses?

Such paradoxes show some connection between decay and synthesis, death and life, but how to describe it, and what methods can be used to study it?

After the victory of the nominalists in the centuries-old disputes of the scholastics, their point of view found the embodiment in the theory of sets: a set is completely determined by its elements. There is no place for self-organization here: during synthesis, no new quality can be created, and during disintegration, no quality can be lost. The disintegrated set does not affect its former elements in any way, they are primary and can be freely combined into any other sets, according to other features. This means that someone must organize this association - an idea that inevitably leads to creationism.

Creationism is not needed in S-space: sets are not determined by elements but define themselves, and the connection of elements with sets is not lost. The set and its elements are of different quality since the characteristics of the waves that make them up are different - and they will interact differently with the same external factor. But, in addition to inherent properties, S-sets, and S-elements are subject to the influence of external factors. All this together determines the scenarios for stratification and convolution of the S-space. Let's compare these characteristics with the properties of geometric objects fixed by the axiomatic definitions of classical mathematics (Table 7).

Table 7. Comparison of complex systems and space, sets, elements, and operations properties

Living organisms	Space, sets, elements, operations
Living organisms as complex systems appear in the dissipative region on the border of environments. They are open, heterogeneous, emergent ones with variable parameters at different evolution stages. They can be self-organizing and self-regulative. Their behaviours' may be non-linear and teleological. Time of existence is limited.	**Space** is absolute, homogeneous, isotropic, with a fixed dimension, no self-organization, and self-regulation. The existence time is unlimited
components are dependent on a system, and from each other, immanently variable, not additive	**Sets are** independent of space and from each other, immanently non-variable, additive
elements are immanently long, dependent, variables, may be complex systems	**elements** (points) are not extended, independent, non-variable, simple
interoperability of systems, components, and elements with each other and with the environment is nonlinear, the nonlinearity threshold exists, there is a synergy	**functions** allow decomposition. **operations** are linear
measurement (a special case of interaction) changes value, measurable, and coordinate system	**the measure** does not change the measurable value and coordinate system

As one can see, the classical axioms differ from the observed living systems properties; they are opposite to them in all respects. Therefore, the study of the manifestations of life will be continued with the help of the wave model of S-space.

REFERENCES

Adelberg, E. A., & Magee, P. T. (2015, July 8). *Amino acids*. AccessScience. https://doi.org/DOI: 10.1036/1097-8542.028100

Allen, D. A., & Wickramasinghe, D. T. (1981). Diffuse interstellar absorption bands between 2.9 and 4.0 μm. *Nature*, 294(5838), 239–240. DOI: 10.1038/294239a0

Amino Acids Reference Guide. (2021). https://beta-static.fishersci.com/content/dam/fishersci/en_US/images/amino-acids/AminoAcidsReference.pdfwww.stratigraphy.org/ICSchart/ChronostratChart2021-07.pdf

Arrhenius, S. (1908). *Worlds in the Making: The Evolution of the Universe*. Harper & Row.

Bell, E. A., Boehnk, P. T., Harrison, M., & Mao, W. L. (2015). Potentially biogenic carbon preserved in a 4.1-billion-year-old zircon. *Proceedings of the National Academy of Sciences of the United States of America*, 112(47), 14518–14521. DOI: 10.1073/pnas.1517557112 PMID: 26483481

Bouvier, A., & Meenakshi, W. (2010). The age of the Solar system is redefined by the oldest Pb–Pb age of a meteoritic inclusion. *Nature Geoscience*, 3(9), 637–641. DOI: 10.1038/ngeo941

Dalrymple, G. B. (2001). The age of the Earth in the twentieth century: A problem (mostly) solved. *Special Publication - Geological Society of London*, 190(1), 205–221. DOI: 10.1144/GSL.SP.2001.190.01.14

Gibson, E. K., Westall, F., Jr., McKay, D. S., Thomas-Keprta, K., Wentworth, S., & Romanek, C. S. (1999). Evidence for ancient Martian life. *NASA Johnson Space Center.*

Fedo, C. M., & Whitehouse, M. J. (2002). Metasomatic Origin of Quartz-Pyroxene Rock, Akilia, Greenland, and Implications for Earth's Earliest Life. *Science*, 296(5572), 1448–1451. DOI: 10.1126/science.1070336 PMID: 12029129

Fray, N., Bardyn, A., Cottin, H., Altwegg, K., Baklouti, D., Briois, C., Colangeli, L., Engrand, C., Fischer, H., Glasmachers, A., Grün, E., Haerendel, G., Henkel, H., Höfner, H., Hornung, K., Jessberger, E. K., Koch, A., Krüger, H., Langevin, Y., & Hilchenbach, M. (2016). High-molecular-weight organic matter in the particles of comet 67P/Churyumov–Gerasimenko. *Nature*, 538(7623), 72–74. DOI: 10.1038/nature19320 PMID: 27602514

Furukawa, Y., Chikaraishi, Y., Ohkouchi, N., Ogawa, N. O., Glavin, D. P., Dworkin, J. P., Abe, C., & Nakamura, T. (2019). Extraterrestrial ribose and other sugars in primitive meteorites. *Proceedings of the National Academy of Sciences of the United States of America*, 116(49), 24440–24445. DOI: 10.1073/pnas.1907169116 PMID: 31740594

Gibson, E. K.Jr, Westall, F., McKay, D. S., Thomas-Keprta, K., Wentworth, S., & Romanek, C. S. (1999). *Evidence for ancient Martian life. Mail Code SN2*. NASA Johnson Space Center.

Hoover, R. B., Rozanov, A. Yu., Krasavin, E. A., Ryumin, A. K., & Kapralov, M. I. (2018). Diatoms in the Orgueil Meteorite. *Paleontological Journal*, 52(13), 1647–1650. DOI: 10.1134/S0031030118130051

Hein, J. E., Tse, E., & Blackmond, D. G. (2011). A route to enantiopure RNA precursors from nearly racemic starting materials. *Nature Chemistry*, 3(9), 704–706. DOI: 10.1038/nchem.1108 PMID: 21860459

Hein, J. E., & Blackmond, D. G. (2012). On the Origin of Single Chirality of Amino Acids and Sugars in Biogenesis. *Accounts of Chemical Research*, 45(12), 2045–2054. DOI: 10.1021/ar200316n PMID: 22353168

Hyde, W. T., Crowley, T. J., Baum, S. K., & Peltier, W. R. (2000). Neoproterozoic 'snowball Earth' simulations with a coupled climate/ice-sheet model. *Nature*, 405(6785), 425–429. DOI: 10.1038/35013005 PMID: 10839531

Jiang, L., Dziedzic, P., Spacil, Z., Zhao, G.-L., Nilsson, L., Ilag, L. L., & Córdova, A. (2014). Abiotic synthesis of amino acids and self-crystallization under prebiotic conditions. *Scientific Reports*, 4(1), 6769. Advance online publication. DOI: 10.1038/srep06769 PMID: 25346284

Kan, S. B., Lewis, R. D., Chen, K., & Arnold, F. H. (2016). Directed evolution of cytochrome c for carbon–silicon bond formation: Bringing silicon to life. *Science*, 354(6315), 1048–1051. DOI: 10.1126/science.aah6219 PMID: 27885032

Koonin, E. V. (2011). *The Logic of Chance: The Nature and Origin of Biological Evolution*. FT Press.

Kovalyov, Y., Mkhitaryan, N., & Nitsyn, A. (2020). *Self-organization of the human mind and the transition from paleolithic to behavioral modernity*. IGI Global. DOI: 10.4018/978-1-7998-1706-2

McGeoch, M. W., Dikler, S., & McGeoch, J. E. M. (2002) Hemolithin: a Meteoritic Protein containing Iron and Lithium. https://arxiv.org/ftp/arxiv/papers/2002/2002.11688.pdf

NASA. (2019). *First Detection of Sugars in Meteorites Gives Clues to Origin of Life*. (n / a).

NASA (n.d.). Mars Meteorite Home Page (JPL). *NASA*.

Maruyama, S., Kurokawa, K., Ebisuzaki, T., Sawaki, Y., Suda, K., & Santosh, M. (2019). Nine requirements for the origin of Earth's life: Not at the hydrothermal vent, but in a nuclear geyser system. *Geoscience Frontiers*, 10(4), 1337–1357. DOI: 10.1016/j.gsf.2018.09.011

Maynard, S. J., & Szathmáry, E. (1995). *The Major Transitions in Evolution*. Oxford University Press.

McKay, D. S., Gibson, E. K.Jr, Thomas-Keprta, K. L., Vali, H., Romanek, C. S., Clemett, S. J., Chillier, X. D., Maechling, C. R., & Zare, R. N. (1996). Search for past life on Mars: Possible relic biogenic activity in Martian meteorite ALH84001. *Science*, 273(5277), 924–930. DOI: 10.1126/science.273.5277.924 PMID: 8688069

McKay, C. P. (2004). What Is Life–and How Do We Search for It in Other Worlds? *PLoS Biology*, 2(9), E302. DOI: 10.1371/journal.pbio.0020302 PMID: 15367939

Miller, S. L., & Urey, H. C. (1959). Organic Compound Synthesis on the Primitive Earth. *Science*, 130(3370), 245–251. DOI: 10.1126/science.130.3370.245 PMID: 13668555

Miller-Urey experiment. (2008). *File: Miller-Urey experiment-en.svg - Wikipedia*. https://en.m.wikipedia.org/wiki/File:Miller-Urey_experiment-en.svg#/media/File%3AMiller-Urey_experiment-it.svg

Mulkidjanian, A. Y., Cherepanov, D. A., & Galperin, M. Y. (2003). Survival of the fittest before the beginning of life: Selection of the first oligonucleotide-like polymers by UV light. *BMC Evolutionary Biology*, 3(1), 12. DOI: 10.1186/1471-2148-3-12 PMID: 12773209

Mulkidjanian, A. Y., & Galperin, M. Y. (2007). Physico-chemical and evolutionary constraints for the formation and selection of first biopolymers: Towards the consensus paradigm of the abiogenic origin of life. *Chemistry & Biodiversity*, 4(9), 2003–2015. DOI: 10.1002/cbdv.200790167 PMID: 17886857

Mulkidjanian, A. Y. (2009). On the origin of life in the zinc world: 1. Photosynthesizing, porous edifices built of hydrothermally precipitated zinc sulfide as cradles of life on Earth. *Biology Direct*, 4(1), 26. DOI: 10.1186/1745-6150-4-26 PMID: 19703272

Mulkidjanian, A. Y., & Galperin, M. Y. (2009). On the origin of life in the zinc world. 2. Validation of the hypothesis on the photosynthesizing zinc sulfide edifices as cradles of life on Earth. *Biology Direct*, 4(1), 27. DOI: 10.1186/1745-6150-4-27 PMID: 19703275

Larin, N., & Warren, H. C. (1993). *Hydridic Earth: the New Geology of Our Primordially Hydrogen-rich Planet*. Polar publishing.

Lurquin, P. (2003). *The Origins of Life and the Universe*. Columbia University Press. DOI: 10.7312/lurq12654

Luisi, P. L. (2006). *The Emergence of Life: From Chemical Origins to Synthetic Biology*. Cambridge University Press. DOI: 10.1017/CBO9780511817540

NASA. (2019). *First Detection of Sugars in Meteorites Gives Clues to Origin of Life*. NASA.

Ott, E., Kawaguchi, Y., Kölbl, D., Rabbow, E., Rettberg, P., Mora, M., Moissl-Eichinger, C., Weckwerth, W., Yamagishi, A., & Milojevic, T. (2020). Molecular repertoire of Deinococcus radiodurans after 1 year of exposure outside the International Space Station within the Tanpopo mission. *Microbiome*, 8(150), 150. Advance online publication. DOI: 10.1186/s40168-020-00927-5 PMID: 33121542

Pearce, B. K. D., Tupper, A. S., Pudritz, R., & Higgs, P. G. (2018). Constraining the Time Interval for the Origin of Life on Earth. *Astrobiology*, 18(3), 343–364. DOI: 10.1089/ast.2017.1674 PMID: 29570409

Peretó, J. (2005). Controversies on the origin of life. *International Microbiology*, 8(1), 23–31. PMID: 15906258

Pino, S. (2015). Nucleotide. In *Encyclopedia of Astrobiology* (pp. 1736–1738). Springer Berlin Heidelberg. DOI: 10.1007/978-3-662-44185-5_1085

Planck Collaboration. (2016). Planck 2015 results: XIII. Cosmological parameters. *Astronomy and Astrophysics.*, 594, 3.

Ponnamperuma, C., Mariner, R., & Sagan, C. (1963). Formation of adenosine by ultra-violet irradiation of a solution of adenine and ribose. *Nature*, 198(4886), 1199–1200. DOI: 10.1038/1981199a0 PMID: 13972132

Powner, M. W., Gerland, B., & Sutherland, J. D. (2009). Synthesis of activated pyrimidine ribonucleotides in prebiotically plausible conditions. *Nature*, 459(7244), 239–242. DOI: 10.1038/nature08013 PMID: 19444213

Rauchfuss, H. (2008). *Chemical Evolution and the Origin of Life* (Mitchell, T. N., Trans.). Springer.

Rink, W. J., & Thompson, J. W. (Eds.). (2015). *Encyclopedia of Scientific Dating Methods*. Springer. DOI: 10.1007/978-94-007-6304-3

Scientists find the 'smoking gun,' signs of life from 3.5B years ago. (n.d.). https://www.foxnews.com/science/scientists-find-smoking-gun-signs-of-life?fbclid=IwAR0CLOHoDiKNLFOaL3Qc0uP98KtB2vo-0iid1y8cUoMOIMRMVx8WQo5vy5k

Saladino, R., Costanzo, G., Crestini, C., Ciciriello, F., & Di Mauro, E. (2007). Formamide as the main building block in the origin of nucleic acids. *BMC Evolutionary Biology*, 7(4), 694–720. DOI: 10.1186/1471-2148-7-S2-S1 PMID: 17767725

The chemical structure of adenosine triphosphate. (n.d.). *File:ATP structure revised.png*. https://commons.wikimedia.org/w/index.php?curid=5041461

Vellas, F. (1988). Mayonnaise and the Origin of Life: Thoughts of Minds and Molecules. *Biochemical Education*, 16(1), 50. DOI: 10.1016/0307-4412(88)90031-3

Weinberg, S. (1993). *The First Three Minutes: A Modern View of the Origin of the Universe*. Basic Books.

Wickramasinghe, D. T., & Allen, D. A. (1983). Three components of 3–4 μm absorption bands. *Astrophysics and Space Science*, 97(2), 369–378. DOI: 10.1007/BF00653492

Wächtershäuser, G. (2006). From volcanic origins of chemoautotrophic life to Bacteria, Archaea and Eukarya. *Philosophical Transactions of the Royal Society of London. Series B, Biological Sciences*, 361(1474), 1787–1808. DOI: 10.1098/rstb.2006.1904 PMID: 17008219

Wickramasinghe, C. (2011). Bacterial morphologies supporting cometary panspermia: A reappraisal. *International Journal of Astrobiology*, 10(1), 25–30. DOI: 10.1017/S1473550410000157

Wikipedia.org. (2008). Retrieved August 12, 2024. Scheme of the S. Miller apparatus. https://en.wikipedia.org/wiki/File:Miller-Urey_experiment-en.svg

Wilde, S. A., Valley, J. W., Peck, W. H., & Graham, C. M. (2001). Evidence from detrital zircons for the existence of continental crust and oceans on the Earth 4.4 Gyr ago. *Nature*, 409(6817), 175–178. DOI: 10.1038/35051550 PMID: 11196637

ADDITIONAL READING

Bogdanov, Yu. A., Kaplin, P. A., & Nikolaev, S. D. (1978). *Proyskhozhdenye y razvytye okeana* [Origin and development of the ocean]. Mysl.

Digonsky, S. V., & Ten, V. V. (2006). *Neyzvestnyj vodorod: rolj vodoroda v polymorfyzme tverdykh tel, processy tverdofaznogho vosstanovlenyja oksydov y spekanyja poroshkov* [Unknown hydrogen: the role of hydrogen in the polymorphism of solids, processes of solid-phase reduction of oxides and sintering of powders]. Nauka.

Larin, V. N. (1975). *Ghypoteza proyskhozhdenyja ghydrydnoj Zemly* [Hypothesis of an original hydride Earth]. Nedra.

Markov, A. (2010). *Rozhdenye slozhnosty: эvoljucyonnaja byologhyja seghodnja. Neozhydannыe otkrytyja y novye voprosy* [The Birth of Complexity: Evolutionary Biology Today. Unexpected discoveries and new questions]. Publishing house Astrel.

Nikitin, M. A. (2013a). Elementy zhyzny: pochemu ne kremnyj y ne ftor? [Elements of life: why not silicon and not fluorine?] *Chemistry and Life 1.*https://www.hij.ru/read/1947/

Nikitin, M. A. (2013b). *Ystoryja voprosa.* [Question history]. *Chemistry and life 2*_https://elementy.ru/nauchno-populyarnaya_biblioteka/431888/Khimiya_i_zhizn_2_2013

Nikitin, M. A. (2013c). Problema khyraljnoj chystoty. [The problem of chiral purity]. *Chemistry and Life 3*https://www.hij.ru/read/issues/2013/march/2167/

Nikitin, M. A. (2013d). Fotokhymyja y «chernye kuryljshhyky» [Photochemistry and "black smokers"]. *Chemistry and Life 4*https://hij.ru/read/2527/

Nikitin, M. A. (2013e). Sposoby vosstanovlenyja ughlekyslogho ghaza [Ways of carbon dioxide recovery]. *Chemistry and life 6*https://hij.ru/read/2646/

Nikitin, M. A. (2013f). Proyskhozhdenye synteza belka y ghenetycheskyj kod. [The origin of protein synthesis and the genetic code]. *Chemistry and Life 7.* https://hij.ru/read/2721/

Nikitin, M.A. (2018). *Proyskhozhdenye zhyzny. Ot tumannosty k kletke.* [Origin of life. From nebula to cell]. Alpina dokumentaljnaja lyteratura.

Oparin, A. I. (1941). *Voznyknovenye zhyzny na Zemle* [The emergence of life on Earth]. Publishing House of the Academy of Sciences of the USSR. (Original work published 1924)

Polishchuk, V. R. (1979). Kak uvydetj molekulu. [How to see a molecule]. *Chemistry*.

Saranchina G. M. (1998). *Porodoobrazujushhye myneraly: Metod opredelenyja krystallooptycheskykh konstant, kharakterystyka myneralov* [Rock-forming minerals: Method of determining crystal-optical constants, characteristics of minerals]. Publishing House of St. Petersburg. un-ta.

Vorontsov, V. (2019). *Proyskhozhdenye zhyzny* [The Origin of Life]. Publishing Solutions.

KEY TERMS AND DEFINITIONS

Modalities of Existence: The potential and actual existence of S-space, S-sets, and S-elements, and the rules of their mutual transition, depend on external factors. The modality of potential existence (***P***) is the possibility of the existing Sp, S-sets, or S-elements (as opposed to non-existence). The modality of actual existence (***A***) means the real existing Sp, S-sets, or S-elements.

Observation (*H*): Interaction of S and O, which changes the existence modality of Sp. ***H*** causes Sp *stratification* (St) into actually existing S-sets and S-elements. An absence of ***H*** triggers a backward process to potential existing – the *convolution* (Co) of Sp.

Potential of S-Space: (π): A particular part of Universal, "grasps" by S-space wave. This part is named and has a different interpretation depending on the situation. Subjective and objective parts of Sp are symmetrical and it is one of the integrity properties of Sp.

S-Elements: Waves and solitons.

Soliton: (•): A single wave which can transfer in ∪. For the first time, a psychical prototype of soliton was described by G.C. Russel, as a single wave, that behaves as a particle in many aspects (localization, motion without deformation, energy, and impulse transfer, it retains structure within contact).

S-Sets ( ): Potentially or actually existing structure, that coordinates potentially or actually existing and relatively homogenous S-elements.

S-Space: (Sp, °): Axiomatic wave model of a complex system. Sp is a carrier of sets and elements. The axiomatic wave model includes 17 axioms that define the structure of space, state, and elements interactions and the procedure for measuring.

Wave (∪)**:** Potentially or actually existing element, marked by immanent alterability, extension, relative instability, ability to interact, condition modalities, capability to transfer in soliton. The notation ∪ may also describe multiple waves; • – multiple solitons; ∪• – waves and solitons. Elements may have only one component (prototype – a sound wave of one frequency), or they can contain several perpendicular components (as an electromagnetic wave that contains perpendicular electric and magnetic elements). Accordingly, single-component and multi-component waves and solitons are distinguished.

ENDNOTE

1 For the first time psychical prototype of soliton wave was described by G.C. Russel, as a single wave, which behaves as a particle in many aspects (localization, motion without deformation, energy and impulse transfer, it retains structure within contact).

Chapter 2
The Cell as a Complex System and the S-Space Axiomatic Wave Model Adequacy Checking

ABSTRACT

The verification of the proposed scientific apparatus adequacy is continued and an example of its use when formulating the definition of life is given. The processes of nucleic acids and proteins synthesis, metabolism, and energy of cells, associated with the life cycle and death, are described. The existing ideas about the transition from inanimate matter to living organisms and the initial stages of evolution, including difficulties and problems, are considered. Based on these facts and theories, 13 characteristics of life are identified. Their comparison with the wave model of S-space formulations showed the adequacy of the wave model axioms. A preliminary definition of life is given within the framework of its systemic understanding, assumptions are made about intermediate forms and places of formation of the first living organisms.

BACKGROUND

The main task is to check the axiom adequacy. To do this, modern ideas are presented and analyzed, then characteristics of life are formulated, and then axioms are compared with these characteristics.

Many researchers give their life definitions, describe life functions, and offer theories of its origin. Sources can be divided into several groups.

DOI: 10.4018/978-1-6684-8509-5.ch002

First, publications, devoted to the substantiation of signs of life. There are "classic" lists, which are included in educational literature and encyclopedias (The Characteristics of Life, 2021, n/d; The 7 Unmistakable Characteristics of Life, n/a), and definitions that have not become generally accepted. All of them are quite similar to each other. The similarity is explained by the implicit "geocentricity" of ideas about living organisms - on a carbon basis, consisting of cells, etc. Interest in life and its signs has increased in recent decades in connection with programs for the planets of the solar system instrumental study. One of the main objectives of such missions is the search for extraterrestrial life (Habitability and Biology, 2013, n/a).

In publications (Benner, 2010; Forterre, 2010) devoted to the definition of life in general or specifically the life of cells (Bose, 2019), the existing definitions of life and its signs are criticized and clarified. In (Brunk & Marshall, 2021) the systemic character of life is emphasized.

General definitions and features are also offered, which, according to the authors, are not tied to the features of earthly life (Koshland, 2002). It should be said that the authors of such publications do not come to a consensus; debatable, for example, the question of whether viruses are living beings.

Secondly, these are publications describing various life processes at the cell level (Barton et al., 2007; Semak et al., 2011; Bilich & Kryzhanovsky, 2008). The prerequisites for death (Chapter 1) are internal (instability of organic compounds) and external (unfavorable environmental conditions). Life processes can be seen as responses to the threat of death, especially self-repair and reproduction.

Cells, separated by a "complexity barrier" from inanimate nature, are interesting in that they are the simplest of known living organisms, capable of self-reproduction and largely self-sufficient. Therefore, it is convenient here to track the properties of open complex systems by comparing them with the axioms of S-space. The biosynthesis of nucleic acids and proteins (Timin, 2018; Hames & Hooper, 2005; Aryal et al., 2018), cell energetics (Bilich & Kryzhanovsky, 2008; Urry, Cain, Wasserman, Minorsky, Reece, 2016; Kostakis & Tibbetts, 2020), exchange processes with the environment (Gennis, 1989; Thomas & Ran, 2007; Woese et al., 1990; Koonin, 2003; Forterre et al 2005; Glansdorff et al., 2008; Weiss et al., 2020).

Thirdly, these are publications devoted to attempts to find an intermediate link between inanimate matter and living beings. If we take the presence of reproduction as a necessary feature of living organisms, then viruses are not alive - cells are required for reproduction. On the other hand, there is a huge difference in the complexity of the viruses, viroids, and prions, not to mention abiogenic compounds and cells. To solve this problem, the RNA world hypothesis was proposed in (Rich & Woese, 1967), and later developed by L. Orgel (Orgel, 1968).

According to this hypothesis, the RNA world is an intermediate stage in the emergence of life on Earth, when ribonucleic acid molecules perform the function of genetic information storing and catalyzing chemical reactions. Subsequently, modern DNA-RNA-protein life arose from them, isolated from the external environment by a membrane. After the discovery of the catalytic functions of RNA by T. R. Check, as well as the ability of RNA to create a double strand and self-replicate (Johnston, et al, 2001), the hypothesis received a certain experimental basis.

However, the RNA world hypothesis has not yet been fully confirmed, leaving room for criticism. The objections mainly boil down to two points:

- Discrepancy between the conditions on the Ancient Earth and the conditions for the formation of RNA; simultaneously, their occurrence is assumed in other places of the Universe (Spirin, n/d).
- There is a need for other "intermediate links" (Shapiro, 2007). Returning to Oparin's ideas Figure 10, Chapter 1), Shapiro proposes the emergence of metabolism as an "intermediate link".

Fourthly, these are publications devoted to studying the transition from the RNA world to DNA-RNA heredity. There is a hypothesis (Forterre, 2005; Forterre, 2006; Villarreal, 2005; Koonin, 2009) about viruses as an intermediate stage of such a transition (viral-protein world). Therefore, possible scenarios for their formation are presented (Mulkidjanian et al., 2008; Mulkidjanian et al., 2009). This problem is related to the characteristics of the supposed LUCA (Last Universal Common Ancestor), the primordial ancestor of bacteria and archaea (Forterre et al., 2005). The evolution of ancient prokaryotes and eukaryotes led to the modern forms the emergence of unicellular and multicellular organisms. There is, however, another point of view on the origin of viruses – they are the product of regressive cell evolution (Dimmock, Easton, Leppard, 2007) and cannot be an intermediate link.

Fifth, the interactions of living organisms with the environment and especially the evolutionary mechanisms associated with the Great Oxygen Event (Holland, 2006; Blaustein, 2016; Hodgskiss et al., 2019; Chen et al., 2020; Olejarz et al., 2021).

After summarizing the characteristics of the systems and processes mentioned above, we will check whether the axioms of the S-space wave model correspond to them (Kovalyov, 1996).

After checking the axioms, it is possible to return to the definitions of life and the problem of an intermediate link between non-living matter and living organisms. The authors give a preliminary definition and suggest liquid crystals as an intermediate form.

SIGNS OF LIFE

Consider the generally accepted characteristics of living organisms.

1. According to (The Characteristics of Life, 2021, n/d) there are:
   a. Order (organisms are highly organized, coordinated structures with one or more cells. Even single-celled organisms are remarkably complex).
   b. Sensitivity or response to stimuli (organisms respond to diverse stimuli.).
   c. Reproduction (single-celled organisms reproduce by first duplicating their DNA, and then dividing it equally as the cell prepares to divide to form two new cells. Multicellular organisms often produce specialized reproductive germline cells that will form new individuals).
   d. Growth and development (organisms grow and develop following instructions coded in genes).
   e. Regulation (even the smallest organisms are complex and require multiple regulatory mechanisms to coordinate internal functions, respond to stimuli, and cope with environmental stresses).
   f. Homeostasis (the ability of an organism to maintain constant internal conditions).
   g. Energy processing (all organisms use a source of energy for their metabolic activities).
2. According to (Characteristics of Life, 2021, n/d) the list looks like this: response to the environment; growth and development; reproduction; keeping things constant; complex chemistry; cells.
3. Another list (The 7 Unmistakable Characteristics of Life, n/a) gives the following signs: cells; homeostasis; heredity; use of energy; reproduction; response to the environment; evolution, and adaptation.
4. Attempts are being made to generalize these signs, to give them a universal character and free them from the restrictions caused by the characteristics of earthly life. For example, (Koshland, 2002) gives the following list:
   a. Program (an organized plan that describes the constituent parts of living systems and the interactions between them in time), which is understood as DNA.
   b. Improvement (adaptation to environmental changes is realized through mutations and selection).
   c. Separation (isolation from the environment is achieved through a membrane or skin; organs and systems are inside the body, and so on).
   d. Energy (the processes of metabolism, regeneration, growth, etc.) requires energy from the external environment; despite various transformations, most of it is ultimately provided by the sun.

e. Regeneration (during chemical reactions inside the body, the reserves of certain substances are depleted; organs and cells lose their efficiency over time. To compensate for such changes, regeneration is necessary, taking the form of relatively simple diffusion and complex protein resynthesis).

When evaluating such lists, the following should be noted.

The line between living and non-living matter is indistinguishable. For example, the properties of self-reproduction and inheritance are implemented in non-living computer viruses. At the same time, living mules cannot reproduce (but interspecific hybrids of wolves and dogs demonstrate this ability).

Most features directly correspond to the properties of complex systems. Perhaps only a "complexity barrier" separates living matter from non-living matter. According to (Nikitin, 2013g), bacteria have a genome of over a million nucleotides encoding over a thousand proteins. Its operation requires molecular mechanisms of protein synthesis (ribosome), DNA synthesis (replication fork), energy supply (at least 12 glycolysis enzymes, as well as an electron transport chain on the membrane), regulation, and control (transcription factors and signal proteins). No simpler self-replicating systems have been found - viruses need cells to reproduce. Further, the protein synthesis system in cells consists of ribosomes (each of 3 RNA molecules with a total length of up to 5000 nucleotides), a set of about 40 transfer RNAs, aminoacyl-tRNA synthase enzymes, elongation and termination factors, and other auxiliary proteins. It is also a complex system and simpler protein synthesis variants are unknown.

The lists ignore the duality of matter, which manifested itself before the emergence of life: the presence of matter and field, corpuscular-wave dualism, etc., which should have some correspondence in living organisms.

The simultaneously existing homogeneity and isolation of life with the environment are not considered: homogeneity is the use of chemical elements and compounds of the environment and isolation is the chiral purity of biological compounds.

Ignore such key life cycle events as birth and death. And here are pertinent questions that will be constantly present in this chapter and throughout the work.

If reproduction and inheritance are necessary for natural selection, which requires living cells, how do cells become beings? What is considered the beginning of the life of an organism? Even before fertilization, there is the father's sperm and the mother's oocytes. And even if we consider the completion of fertilization as the formation of a new system, then how to consider the development of consciousness?

Why is the well-established system of a living organism destroyed at death? And when a person dies, do individual cells remain alive for some time? On the contrary, the death of individual old and diseased cells does not lead to the death of a human - regeneration processes replace such cells with new ones. But the time comes, regeneration starts slowly and incorrectly - and the human dies. What is the reason for this time?

CELL: SYNTHESIS OF NUCLEIC ACIDS AND PROTEINS

With these questions in mind, let's continue the analysis of Chapter 1 and consider the formation of complex organic compounds already at the cellular level.

The structure of nucleic acids (Figure 1). Recall that nucleic acids are polymeric molecules consisting of nitrogenous bases (purine - adenine, guanine - and pyrimidine - cytosine, uracil, thymine), 2 types of pentose (ribose and deoxyribose), as well as phosphoric acid residues. Depending on the presence of deoxyribose or ribose, deoxyribonucleic (DNA) and ribonucleic (RNA) acids are distinguished, respectively; thymine is found only in DNA, while uracil is found in RNA. The structure of DNA, which is a double-stranded helix, is shown in Figure 1[1]. DNA helices can be right-handed or left-handed. But the right-handed helices are the most stable. The diameter of the double helix is 2 nm, and the double helical structure repeats at an interval of 3.4 nm, corresponding to ten base pairs (Hames & Hooper, 2005).

Figure 1. The structure of DNA and RNA (Mackenzie, 2022)

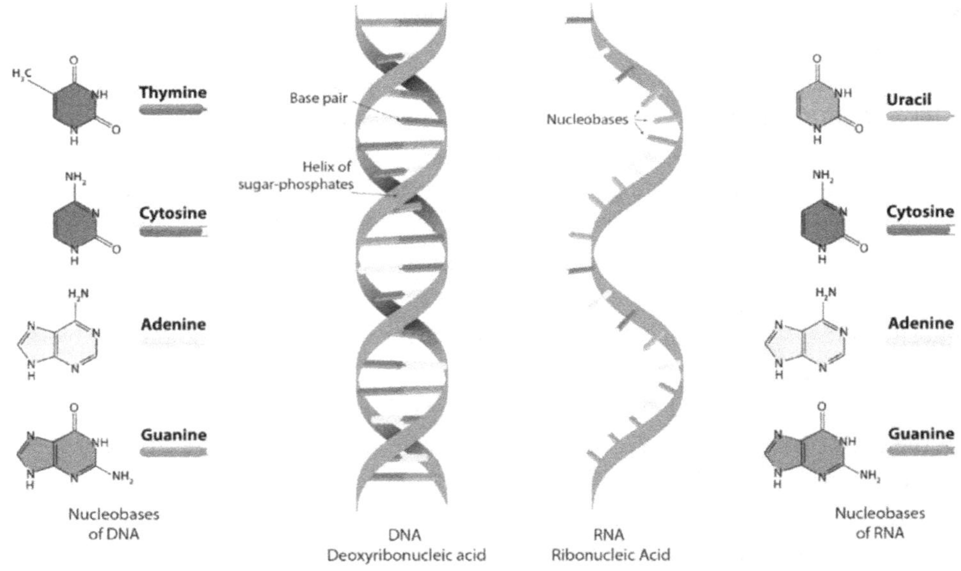

Nuclear DNA. Two double-stranded DNA molecules form one chromosome. The chromosomes are most clearly visible before and during mitosis; in resting cells, the chromosomal material looks indistinct and is distributed throughout the nucleus. In this state, he received the name "chromatin" (60% protein (histones and acidic proteins), 35% DNA, and about 5% RNA). Chromatin is arranged in the spherical particle form - nucleosomes, connected by a DNA strand. The nucleosome is a complex of a segment of a DNA molecule and eight histone molecules. In contact with histones, the DNA strand wound around the histone nucleus.

In addition to nucleosomes, there are 2 more levels of DNA folding in the nucleus: fibrils consisting of a chain of nucleosomes, and fibers formed when fibrils are twisted into a superhelix. Due to the presence of nucleosomes, a decrease in the chromosome size is achieved, and then the fibril is laid in a superhelix and is now a "super superhelix" (Timin, 2018).

Mitochondrial DNA is located in the mitochondria. Contains 100-1,000 copies per cell. According to (Sharma, 2021), mitochondrial DNA chromosomes usually have closed, circular structures, and contain 16,569 nucleotides. Mitochondrial DNA is haploid, coming only from the mother. The mutation rate for mitochondrial DNA is generally higher than nuclear DNA, but it is a stable molecule and holds more complex information for longer periods.

DNA is crucial as genetic material in most living organisms because it carries genetic information from generation to generation. Its major functions include storing genetic information, directing protein synthesis, determining genetic coding; directly responsible for metabolic activities, evolution, heredity, and differentiation.

Ribonucleic acid (RNA) is a sequence of ribonucleoside monophosphates - AMP, GMP, CMP, and UMP, linked by 5'-3'-phosphodiester bonds. According to the structure of RNA, it differs from DNA in a single-stranded structure, a significantly lower molecular weight, and the presence of uracil instead of thymine and ribose instead of deoxyribose/ RNA is a single-stranded helix.

Most RNA molecules are single stranded but an RNA molecule may contain regions that can form complementary base pairing where the RNA strand loops back on itself (RNA secondary structure). Ribosomal RNAs (rRNAs) and transfer RNAs (tRNAs) exhibit substantial secondary structure, as do some messenger RNAs (mRNAs) (Aryal, 2018).

The following types of RNA are present in the cell:

Messenger RNA (mRNA) is a linear sequence of nucleotides. It accounts for about 5% of the total RNA in the cell. As part of post-transcriptional processing in eukaryotes, the 5' end of mRNA is capped with a guanosine triphosphate nucleotide, which helps in mRNA recognition during translation or protein synthesis. Similarly, the 3' end of an mRNA has a poly-A tail or multiple adenylate residues added to it, which prevent mRNA enzymatic degradation. Both 5' and 3' mRNA ends impart stability to the mRNA. mRNA transcribes the genetic code from DNA into a form that can be read and used to make proteins. mRNA carries genetic information from the nucleus to the cytoplasm of a cell (Aryal, 2018);

Ribosomal RNA (rRNA) is involved in building ribosomes (found in the ribosomes and accounts for 80% of the total RNA in the cell). Ribosomes consist of the small ribosomal subunits that read the RNA and the large subunits that join amino acids to form a polypeptide chain. Each subunit comprises one or more ribosomal RNA (rRNA) molecules and ribosomal proteins. Different rRNAs include small rRNAs and large rRNAs, which denote their presence in the small and large subunits of the ribosome. rRNAs combine with proteins in the cytoplasm to form ribosomes, as the sites of protein synthesis, and have the enzymes needed. These structures travel along the mRNA molecule during translation and facilitate the assembly of amino acids to form a polypeptide chain. They bind to tRNAs and other molecules that are key for protein synthesis. rRNA directs the translation of mRNA into proteins (Aryal, 2018).

Transfer RNA (tRNA) includes 73-93 nucleotides (bacteria and eukaryotes). They carry amino acids from the cytosol to the ribosomes. The guanine nucleotide is located at the 5' end of the tRNA, and the C-C-A triplet is at the 3' end. tRNAs have a clover leaf structure which is stabilized by strong hydrogen bonds between the nucleotides (Figure 2). Four loops are distinguished in the clover leaf, each of which has its function (Aryal, 2018):

- Anticodon - connects to the codon of matrix RNA in the ribosome.
- T-loop - responsible for binding to the ribosome.
- D-loop - responsible for binding to aminoacyl-tRNA synthase.
- Acceptor - binds the transferred amino acid.

Transfer RNA brings or transfers amino acids to the ribosome that correspond to each three-nucleotide codon of rRNA. The amino acids then can be joined together and processed to make proteins.

Other functions include RNA editing, gene regulation, and RNA interference. These processes are carried out by a group of small regulatory RNAs, which include small nuclear RNA, microRNA, and small interfering RNA (Wang; Farhana, 2021).

Small nuclear RNAs (snRNA) are non-coding RNAs that are responsible for splicing introns (Wang; Farhana, 2021). The snRNAs join with proteins to form small nuclear ribonucleoproteins (snRNP), most commonly containing U1, U2, U4, U5, and U6 snRNA molecules. Spliceosome assembly and activity begin once U1 of the snRNP binds a complementary sequence on a pre-mRNA transcript's 5' splice site. Introns are then removed from the pre-mRNA transcript by the spliceosome complex and mature mRNA forms.

Figure 2. Cloverleaf structures of tRNA (Uploaded by Victor Serebrov, 1998)

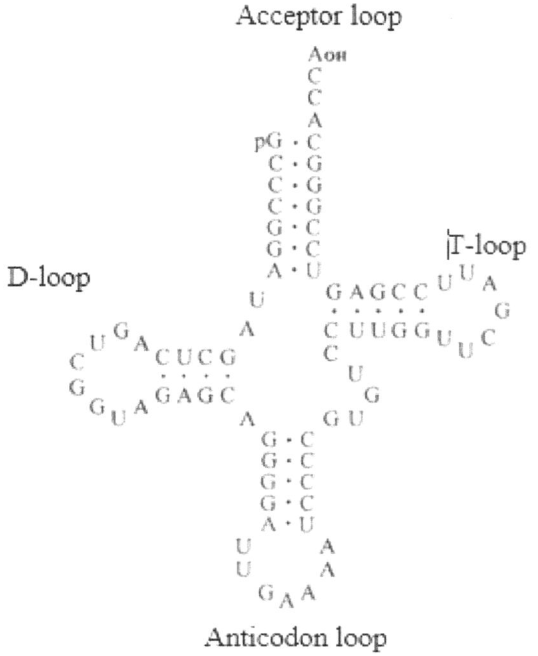

MicroRNAs (miRNA) are non-coding RNAs mainly involved in gene regulation (Wang; Farhana, 2021). They are mostly processed from introns and are transcribed into primary miRNA from the host gene by RNA polymerase II. miRNAs that bind to an untranslated region (3'UTR) on mRNAs suppress translation, while miRNA binding to promoter regions can upregulate transcription. miRNAs can also function similarly to hormones. They are released into the extracellular fluid and taken up by target cells for cellular activity regulation. mRNAs as ideal biomarkers for various diseases.

Small Interfering RNAs (siRNA) are double-stranded, non-coding RNAs that inhibit gene expression through RNA interference (Wang & Farhana, 2021). They interfere with gene expression by degrading mRNA and preventing the translation of proteins. With the assistance of Dicer, siRNAs form from long double-stranded RNAs. Once fully formed, siRNA binds to an RNA-induced silencing complex (RISC) and cleaves mRNA through a catalytic RISC protein.

The transfer of hereditary information from the mother to the daughter cell uses DNA (replication). For genetic information to be used by the cell, RNA is needed, which is formed on the DNA matrix (transcription). Further, RNA is directly involved in all stages of the synthesis of protein molecules (translation), which provide the structure and activity of the cell. Based on the previous, the central dogma of molecular biology has a place: the transfer of genetic information is carried out only from nucleic acids (DNA and RNA). The recipient of the information can be another (DNA or RNA) nucleic acid.

The structure of proteins was discussed in Chapter 1.

DNA synthesis-replication (Figure 3) is the most complex and exact process, which for humans is characterized by the following parameters. For each cell in the body, it is necessary to copy about 3 billion base pairs of DNA in a strictly defined period of the cell's life. In total, 4 phases of the cell cycle are distinguished: mitosis, synthetic, pre-synthetic, and post-synthetic (Timin, 2018; Semak et al., 2011). The change in the cell cycle phases is regulated by cyclins - proteins, the level of which changes during the cell cycle, ensuring progress through its stages. DNA synthesis occurs during the synthetic phase of the cell cycle when the cell is preparing to divide. The replication mechanism is semi-conservative, i.e., a child copy is synthesized on each strand of maternal DNA.

Figure 3. DNA replication in Prokaryotes (DNA replication in Prokaryotes, 2014)

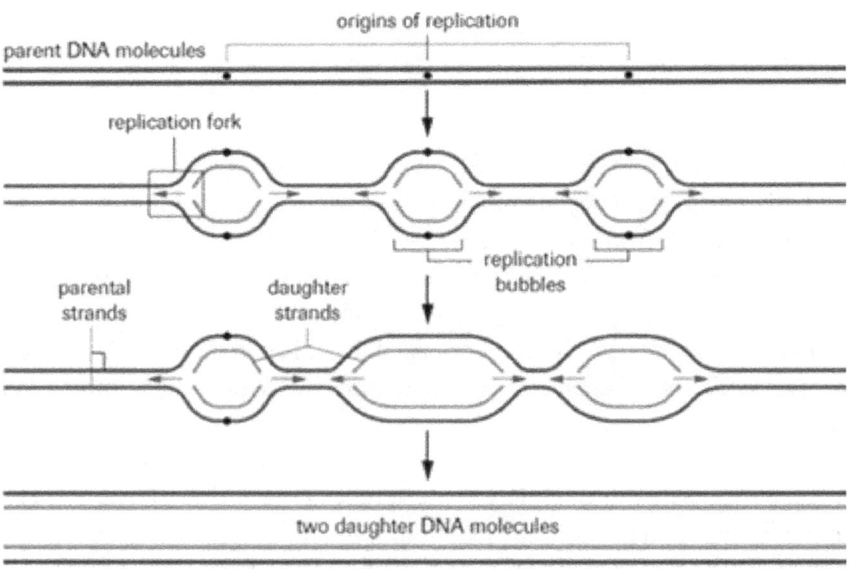

Like any template biosynthesis, replication requires several components:

- Matrix - in its role is the maternal DNA strand.
- Growing chain - a daughter strand of DNA.
- Substrates for synthesis – dATP, dGTP, dCTP, TTP.
- Energy source – dATP, dGTP, dCTP, TTP.
- Enzymes - DNA polymerase, DNA primase, DNA helicase, DNA ligase, and topoisomerase.

DNA synthesis begins in certain places (ori points), which in mammals on each DNA are about 100. Replication spreads from these sites in both directions along the DNA, strands with the formation of replication “bubbles”. Each has two replication “forks” in which unwinding and direct DNA synthesis occurs. In this case, replication forks move away from each other (Figure 3).

The synthesis of a new DNA chain proceeds from the 5' end to the 3' end, i.e. The 5'-end of the new DNA remains free, and the following nucleotides are attached to the 3'-hydroxyl group of the previous nucleotide.

In the replication fork in the direction 5' → 3', only one strand is continuously synthesized (i.e., by the usual addition of subsequent nucleotides to the previous ones through C3 and C5), namely, the one for which the direction of synthesis coincides with the direction of movement of the replication fork and corresponds to the direction of the mother thread 3' → 5'. As the DNA unwinds and the replication fork moves, sites are opened on this parent strand where the leading daughter strand can continue to elongate.

The 5' → 3' direction for the other daughter DNA strand is opposite to the movement of the replication fork. Therefore, the synthesis of this lagging strand (in the direction 5' → 3') is possible only after the unwinding of a part of the DNA and the release of the site for synthesis.

Thus, the synthesis of daughter DNA proceeds fragmentarily. The DNA segments synthesized on the lagging strand were called Okazaki fragments after the Japanese researcher of this process.

The sequence of replication steps is shown in Figure 4.

Figure 4. The main steps of DNA replication (Timin, 2018)

DNA topoisomerases, being in front of the replication fork, cut the DNA molecule to facilitate its unwinding and unwinding

↓

DNA helicases, following topoisomerases, unwind and unwind the DNA molecule

↓

DNA-binding proteins bind unraveled DNA strands and stabilize them, preventing them from rejoining each other

↓

DNA polymerase δ, coordinated with the speed of the replication fork, elongates the leading strand of the daughter DNA in the 5'→3' direction on the template of one of the strands of the parent DNA (speed 100 base pairs per second)

↓

DNA polymerase α attaches to another DNA strand immediately after unwinding and synthesizes a primer in the 5' → 3' direction - an RNA sequence on a DNA template with a length of 10 to 200 nucleotides. The enzyme is then removed from the DNA strand

↓

DNA polymerase ε carries out the synthesis of a fragment (length 150-200 nucleotides) of the lagging strand of the daughter DNA in the continuation of the primer. It works until it encounters the primer of the previous Okazaki fragment synthesized earlier. This enzyme is then removed from the chain

↓

DNA polymerase δ takes the place of DNA polymerase ε, moves in the same direction 5' → 3' and removes primer ribonucleotides, simultaneously inserting deoxyribonucleotides in their place. The enzyme works until the primer is completely removed, after which it leaves the chain

↓

DNA ligase ligates the segment synthesized by DNA polymerase ε and the segment of the lagging strand of the daughter DNA inserted by DNA polymerase δ (Okazaki fragments)

Error correction. During DNA copying, most DNA polymerases "check" to see if they are adding the correct nucleotide. If it is found that the wrong nucleotide has been added, DNA polymerase will remove and replace it and then continue DNA synthesis.

DNA repair. After the creation of a new DNA, the mechanism of repair is launched - the removal and replacement of mismatched nucleotides remaining as a result of replication, as well as the correction of small insertions and deletions resulting from the "slipping" of the polymerase from the original chain. The protein complex recognizes the mismatched nucleotide and binds to it. Another complex cuts the DNA in the mismatch area, and another group of enzymes cleaves off the incorrect nucleotide and a small region around it. DNA polymerase then fills the gap with the correct nucleotides, and the DNA ligase enzyme stitches the gaps together.

If DNA is damaged under the influence of external factors (ultraviolet radiation, radiation, chemicals) or spontaneous processes, other types of repairs are used:

- Direct reparation. DNA damage, caused by chemical reactions, is "fixed" by enzymes in the cell.
- Excision repair. One or more DNA nucleotide damage is often corrected by removing and replacing the damaged area. Only the damaged base is removed in the case of basis excision repair. In the case of nucleotide excision repair, as in the case of mismatch repair, which we discussed above, entire nucleotides are removed.
- Repair of double-strand breaks. There are two main ways: non-homologous end joining and homologous recombination. They are used to repair DNA double-strand breaks (when the entire chromosome splits in two).

RNA synthesis - transcription (Timin, 2018). Transcription requires 5 elements:

- Matrix - one of the DNA chains.
- Growing chain – RNA.
- Substrate for synthesis - ribonucleotides (UTP, GTP, CTP, ATP).
- Source of energy - UTP, GTP, CTP, ATP.
- RNA polymerase enzymes and protein transcription factors.

There are three stages of transcription (Figure 5).

Figure 5. The steps of transcription (Ankush01, 2021)

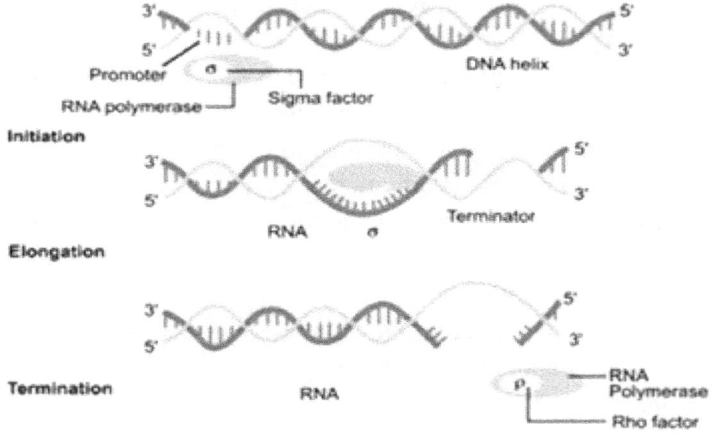

Initiation. The promoter contains the transcription start signal, the TATA-box, a specific sequence of DNA nucleotides that binds the first initiation factor, the TATA factor. The TATA factor provides for the attachment of RNA polymerase to the DNA strand that will be used as a template for transcription (matrix DNA strand). Since the promoter is asymmetric (TATA), it binds RNA polymerase in only one orientation, which determines the direction of transcription from the 5' end to the 3' end (5' → 3'). For binding of RNA polymerase to the promoter, one more initiation factor, the σ-factor, is required, but immediately after the seed RNA fragment (8-10 ribonucleotides long) synthesis, the σ-factor is detached from the enzyme. Other initiation factors unwind the DNA helix before RNA polymerase.

Elongation. Protein elongation factors ensure the movement of RNA polymerase along DNA and unwind the DNA molecule over about 17 nucleotide pairs. RNA polymerase moves at a rate of 40-50 nucleotides per second in the 5' → 3' direction. The enzyme uses ATP, GTP, CTP, and UTP simultaneously as a substrate and energy source.

Termination. RNA polymerase will stop when it reaches the stop codons. With the help of a protein termination factor, the so-called ρ-factor, the enzyme and the synthesized RNA molecule, which is the primary transcript, the precursor of mRNA or tRNA or rRNA, are separated from the DNA matrix.

RNA processing. After synthesis, primary RNA transcripts are passive and undergo a series of changes during activation, which is called processing. In eukaryotes, all types of pre-RNA are processed; in prokaryotes, only rRNA and tRNA precursors are processed.

Protein synthesis – translation (Timin, 2018; Semak et al., 2011; Litwack, 2021; Alberts et al., 2002). After the information from DNA to messenger RNA transferred, protein synthesis begins. Each mature mRNA carries information about only one polypeptide chain. If the cell needs other proteins, then it is necessary to transcribe mRNA from other DNA regions. Protein biosynthesis or translation occurs on ribosomes - intracellular protein-synthesizing organelles. Synthesis requires:

- Matrix - messenger RNA.
- Growing chain - polypeptide.
- Substrate for synthesis - 20 proteinogenic amino acids.
- Energy source - GTP.
- Ribosomal proteins, rRNA, and protein factors.

There are three stages of translation, called the same as the stages of transcription.

Initiation. mRNA, GTP, small and large subunits of the ribosome, three protein initiation factors (IF-1, IF-2, IF-3), methionine, and tRNA for methionine are required.

In the beginning, there is the first complex - mRNA + small subunit + IF-3 - and the second complex - methionyl-tRNA + IF-2 + GTP. They then combine with the large subunit of the ribosome. Protein initiation factors are actively involved in the process, and GTP is an energy source. After the assembly of the complex, the initiating methionyl-tRNA binds to the first AUG codon of a messenger RNA and is in the peptidyl center of the large subunit. The A-center (aminoacyl center) remains free, it will be used at the elongation stage for the aminoacyl-tRNA binding.

Elongation. 20 vital amino acids are needed, tRNA for all amino acids, protein elongation factors, and GTP. Elongation is a cyclic process. The second and subsequent elongation cycles include such steps (Figure 6).

Chain elongation occurs at a rate of approximately 20 amino acids per second.

Figure 6. The steps of elongation (of translation)

Attachment of aminoacyl-tRNA to the mRNA codon - the amino acid is inserted into the A-center of the ribosome. The energy source is GTP

↓

The enzyme peptidyltransferase transfers methionine from methionyl-tRNA from the P-center to the second aminoacyl-tRNA in the A-center with the formation of a peptide bond between methionine and the second amino acid. The activated COOH group of methionine binds to the free NH_2 group of the second amino acid. The energy source is the macroergic bond between the amino acid and tRNA.

↓

The translocase enzyme moves the mRNA relative to the ribosome: the first codon AUG is outside the ribosome, the second codon becomes opposite the P-center, and the third codon is opposite the A-center. The energy of GTP is needed. Since the tRNA attached to it moves along with the mRNA, the initiating first tRNA leaves the ribosome, and the second tRNA with the dipeptide is placed in the P-center.

The elongation cycle (reactions 1,2,3) is repeated as many times as the amino acids must be included in the polypeptide chain.

The second repetition of the cycle - begins with the attachment of the third aminoacyl-tRNA to the third mRNA codon, amino acid-3 moves to the A-center. Then the transferase reaction is repeated and a tripeptide is formed that occupies the A-center, after which it shifts to the P-center

↓

The fourth aminoacyl-tRNA enters the empty A-center and everything repeats →

Termination. Protein synthesis ends when the ribosome reaches special termination codons on mRNA - stop codons UAA, UAG, and UGA. When these codons enter the ribosome, protein factors are activated, which sequentially catalyze hydrolytic cleavage of the polypeptide from the final tRNA; separation from the P-center of the last, already empty, tRNA; and dissociation of the ribosome. The source of energy is GTP. Since the lifespan of messenger RNA is short, the cell must have time to get the maximum number of copies, for which each mRNA can have several ribosomes that stand one after another and synthesize peptide chains. Such formations are called polyribosomes.

Since, most often, polypeptide chains are formed in an inactive form, *processing* is necessary (Figure 7).

Next, protein folding occurs - laying an elongated polypeptide chain into a regular three-dimensional spatial structure (Chapter 1), for which a group of auxiliary proteins - chaperones - is used. They prevent the interaction of newly synthesized proteins with each other, isolate the hydrophobic regions of proteins from the cytoplasm, remove them inside the molecule, and properly position the protein domains.

The description of protein translation is illustrated in Figure 8

Figure 7. The steps of processing

1. Removal from the N-terminus of methionine or even a few amino acids by specific aminopeptidases
2. Formation of disulfide bridges between cysteine residues
3. Partial proteolysis - removal of part of the peptide chain, as is the case with insulin or proteolytic enzymes of the gastrointestinal tract
4. Attachment of a chemical group to the amino acid residues of the protein chain: phosphoric acid; carboxyl group; methyl group; hydroxyl group; iodine
5. Inclusion of a prosthetic group: carbohydrate residues; hema
6. Association of protomers into a single oligomeric protein, for example, hemoglobin, lactate dehydrogenase, creatine kinase

Figure 8. Localization of protein biosynthesis in the cell. (Kep17, 2020:)

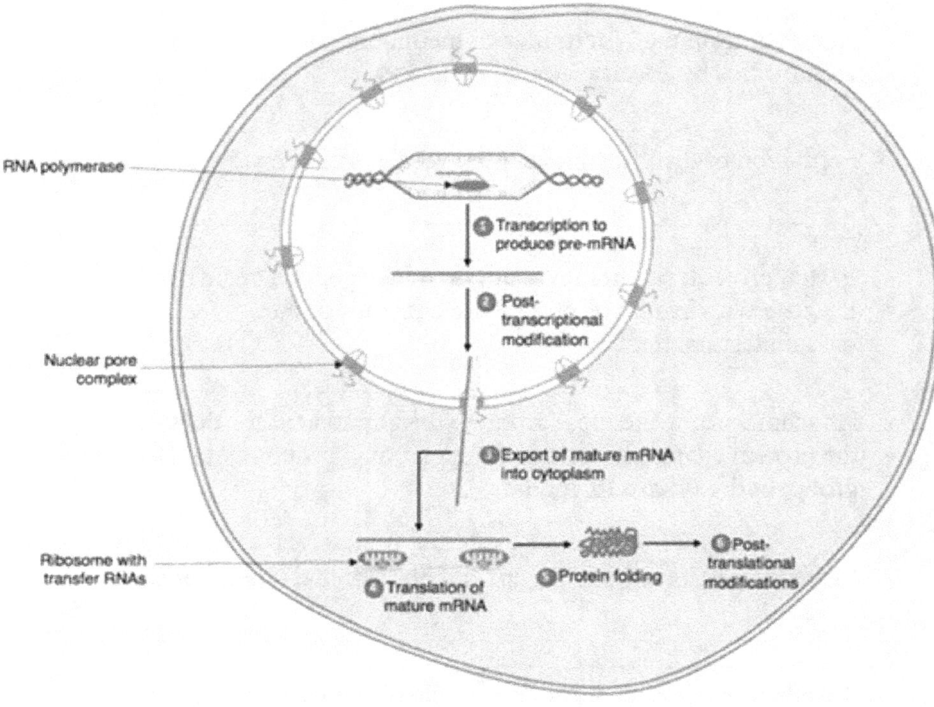

So, even from a brief description, free from many details (for details, see (Urry et al., 2016), it is clear how difficult biogenic synthesis is at the cell level. The "complexity barrier" takes on visible outlines.

THE CELL: METABOLISM AND ENERGY

For biochemical reactions, substances must enter the cell - endocytosis, metabolism in the cell - metabolism, and the removal of metabolic products outside the cell - exocytosis.

Endocytosis. There are several types of endocytosis: passive and active transmembrane transport (Figure 9), as well as pinocytosis (capture of liquid colloidal particles by the cell), and phagocytosis (capture of dense and large particles and even other cells). Pinocytosis and phagocytosis are important in protecting the body (segmented neutrophilic granulocytes, macrophagocytes), bone tissue remodeling (osteoclasts), thyroxine formation by thyroid gland follicles, reabsorption of protein and other macromolecules, etc. (Alberts et al., 2015)

Figure 9. Scheme of passive and active transport up and down the electrochemical gradient (LSumi, 2022)

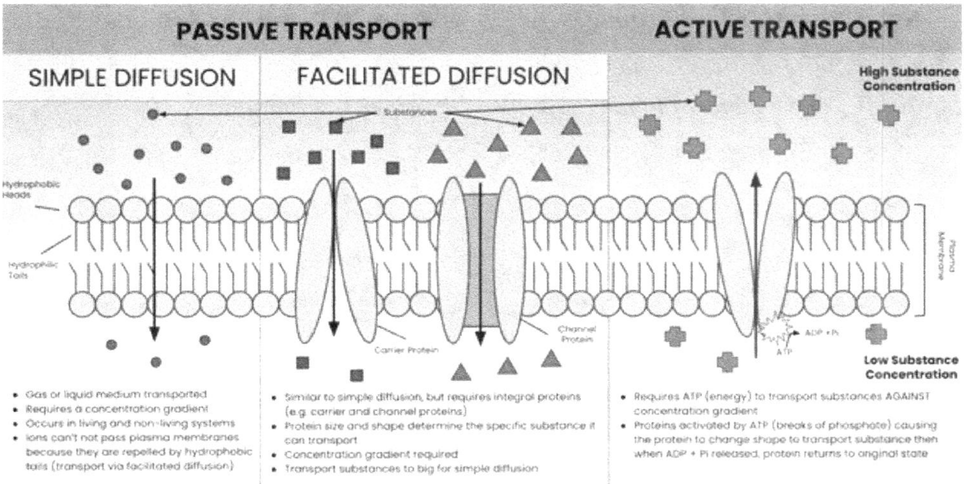

Exocytosis. There are several types of exocytosis: active and passive transport, as well as using transport vesicles (large molecules are segmented in the Golgi complex in the form of transport vesicles; vesicles with the participation of microtubules are sent to the cell surface; the membrane of the vesicle is built into the cytolemma, and its contents are outside the cell; this process can be unregulated or regulated by special signaling molecules).

Metabolism combines assimilation (reactions of biosynthesis of complex biological molecules from simpler ones) and dissimilation (splitting reactions) processes.

Energy processes (Bilich & Kryzhanovsky, 2008). The cell uses the energy of the chemical bonds of amino acids, monosaccharides, and fatty acids, formed from proteins, carbohydrates digestion, and fats, and enters the cell. As a result of their dissimilation, energy is released, which the cell uses to carry out various work, including assimilation.

Many of these reactions are called respiration; if the process occurs with the participation of molecular oxygen, respiration is aerobic, if without - anaerobic. Aerobic respiration is more efficient.

The oxidation of biological molecules is carried out by a chain of reactions catalyzed by enzymes associated with the formation of ATP, which is the main accumulator and carrier of energy; when one phosphate group is cleaved from ATP, adenosine diphosphoric acid and phosphate are formed, and energy is released that is used by the cell for biosynthesis, active transport, movement, and transmission of genetic information. Along with ATP, the cell stores guanosine triphosphate (GTP, used in the synthesis of protein and RNA), uridine triphosphate (UTP, involved in the formation of the cell wall, polysaccharides, RNA), and cytidine triphosphate (CTP, used in the synthesis of lipids and RNA).

Nutrient catabolism (Bilich & Kryzhanovsky, 2008) is shown in Figure 10:

Figure 10. The 3 steps of food catabolism (Bilich & Kryzhanovsky, 2008)

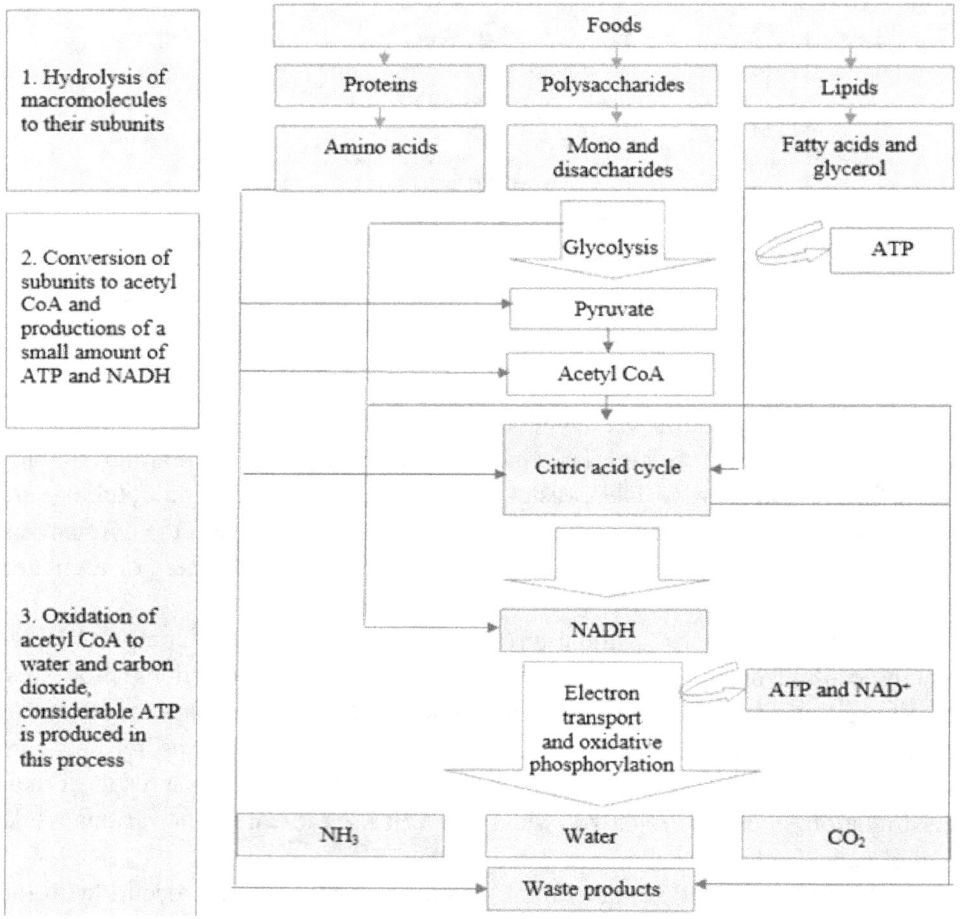

1. First, food substances are broken down into monomers. This stage occurs in the digestive tract under the influence of degrading enzymes in multicellular organisms. Further, the resulting monosaccharides and amino acids are absorbed into the blood and fatty acids into the lymph. In unicellular organisms, cleavage occurs in the digestive vacuoles, with which the primary lysosomes merge.

2. Further, regardless of the nature of the food product, acetyl coenzyme A (acetyl-CoA) is formed. He, as well as other enzymes, including CoA, are key links in many different biochemical reactions.
3. Finally, the acetyl group of acetyl-CoA is completely oxidized to water and carbon dioxide, while many protons and electrons are stored on NADH molecules; further, the energy of the electrons is used to form the proton gradient necessary for the synthesis of ATP.

The citric acid cycle, discovered by G. Krebs, is central to the metabolic process. It consists of 8 stages (Figure 11). The Krebs cycle occurs under aerobic conditions (Figure 12).

Figure 11. The 8 steps of the citric acid cycle (Krebs cycle)

Condensation of acetyl-CoA with oxaloacetate with the formation of citrate and the release of coenzyme A. This reaction is catalyzed by citrate synthase, the rate-regulatory enzyme of the Krebs cycle

↓

The conversion of citrate to isocitrate with the participation of aconitate hydratase through the intermediate stage of cis-aconitate associated with the enzyme

↓

Dehydrogenation (removal of a hydrogen atom from a molecule) of citrate with the formation of α-ketoglutarate and carbon dioxide with the participation of isocitrate dehydrogenase in the presence of Mg_2+ and Mn_2+.

↓

Oxidative decarboxylation of α-ketoglutarate to high-energy succinyl-CoA; the catalyst is α-ketoglutarate dihydrogen complex (Mg_2+)

↓

Transformation of succinyl-CoA under the influence of succinyl-CoA synthase into succinate with removal of CoA, coupled with the formation of GTP from GDP and phosphate

↓

Dehydrogenation of succinate to fumarate catalyzed by succinate dehydrogenase

↓

Reverse hydration of fumarate to form L-malate catalyzed by fumarate hydratase or fumarase

↓

Dehydrogenation of L-malate to form oxaloacetate, catalyzed by NAD-dependent L-malate dehydrogenase and completing the Krebs cycle

Figure 12. Krebs cycle (Kostakis, Tibbetts, 2020)

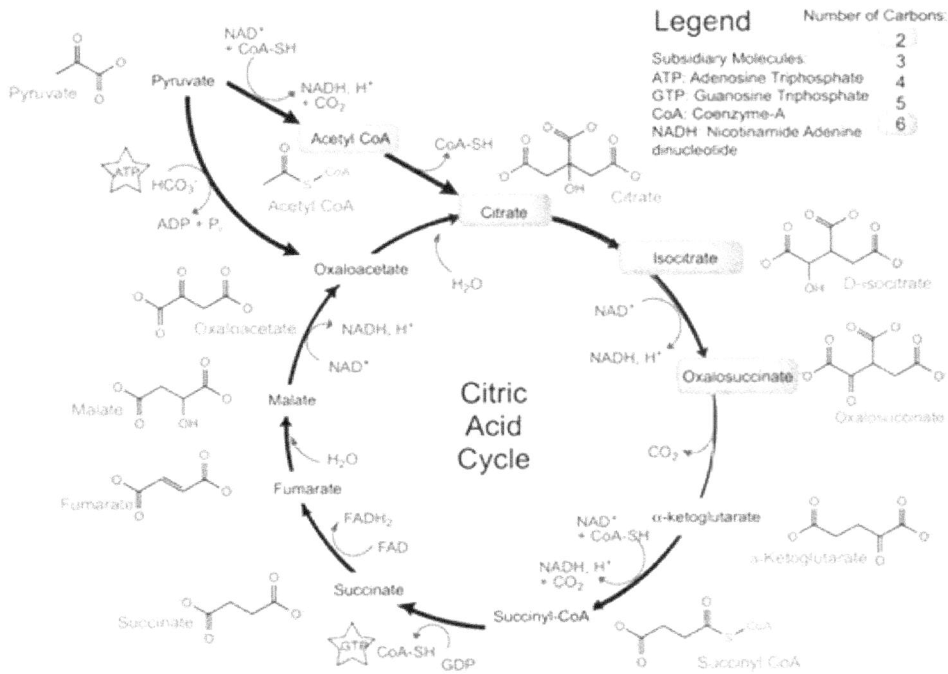

The GTP formed as a result of the fifth reaction can participate in the synthesis of ATP catalyzed by the enzyme ATP synthetase, which occurs already outside the cycle, during oxidative phosphorylation (Mitchell's scheme). The respiratory chain, or the electron transport chain is the main energy conversion system in the cell.

Photosynthesis. At the bottom of the ocean and in other dark places, there are ecosystems based on chemosynthesis, most of which use oxygen as an oxidizing agent and depend on oxygen photosynthesis and sunlight (Pimenova & Pimenov, 2003). However, most modern ecosystems use the energy of the Sun in the form of red-light quanta absorbed by chlorophyll and, through carotenoids, orange and yellow light photons.

As a result of photosynthesis, plants produce simple carbohydrate compounds, but photosynthesis itself is a process with a very complex organization: the spread in the time of various stages of photosynthesis is from 10^{-15} s for the processes of absorption of light quanta, 10^{-10}–10^{-2} s for electron transport and order 10^5 s for processes associated with plant growth, and spatial scales are characterized by values from the molecular level (10^{-27} m^3) to the level of phytocenoses (10^5 m^3) (Polishchuk, 1979).

In higher plants, the organ of photosynthesis is the leaf, and the organelles are chloroplasts. The thylakoid membranes of chloroplasts contain photosynthetic pigments: chlorophylls and carotenoids. Chlorophyll absorbs red and blue-violet light, and reflects green, giving plants a green color. Chlorophyll molecules are organized into photosystems (in plants and blue-green algae - photosystem-1 (PSI) and photosystem-2 (PSII); in photosynthetic bacteria - PSI). Only PSII can decompose water with the oxygen release and take electrons from the water hydrogen.

Photosynthesis reactions are divided into two groups: light-phase and the dark-phase reactions. The light phase occurs only in the presence of light in thylakoid membranes with the participation of chlorophyll, electron carrier proteins, and the enzyme ATP synthetase. Under the action of a quantum of light, chlorophyll electrons are excited, leave the molecule, and enter the thylakoid membrane's outer side, which eventually becomes negatively charged. Oxidized chlorophyll molecules are restored by taking electrons from the water located in the intrathylakoid space. This leads to the decomposition or photolysis of water: H_2O + quantum of light → H+ + OH–.

Hydroxyl ions donate their electrons, turning into reactive radicals •OH: OH— → •OH + e—.

OH radicals combine to form water and free oxygen: 4HO• → $2H_2O + O_2$.

In this case, oxygen is removed from the external environment, and protons accumulate inside the thylakoid. As a result, the thylakoid membrane, on the one hand, is positively charged due to H +, and on the other hand, negatively due to electrons. When the potential difference between the outer and inner sides of the thylakoid membrane reaches 200 mV, protons are pushed through the channels of ATP synthetase, and ADP is phosphorylated to ATP; atomic hydrogen is used to reduce the specific carrier NADP + (nicotinamide adenine dinucleotide phosphate) to NADP H_2: 2H+ + 2e- + NADP → NADP·H_2.

In the light phase, photolysis of water occurs, accompanied by the synthesis of ATP; the formation of NADP·H_2; the formation of oxygen. Oxygen diffuses into the atmosphere. ATP and NADP H_2 are transported into the stroma of the chloroplast and participate in the processes of the dark phase.

The dark phase takes place in the chloroplast stroma (Figure 13). Its reactions do not require light energy, so they occur in the light and dark. The dark phase reactions of the: 1) fixation of carbon dioxide, the acceptor of which is the five-carbon sugar ribulose bisphosphate, and the reaction is catalyzed by the enzyme ribulose bisphosphate carboxylase. As a result of the carboxylation of ribulose bisphosphate, an unstable six-carbon compound is formed, which immediately breaks down into two molecules of phosphoglyceric acid; 2) then the "Calvin cycle" occurs, when phosphoglyceric acid is converted into glucose through a series of intermediate products (Figure 14).

Figure 13. Four major membrane protein complexes in photosynthesis: photosystem I, cytochrome b 6 f, photosystem II, and ATP synthase (Harvey J.M. Hou, n/d)

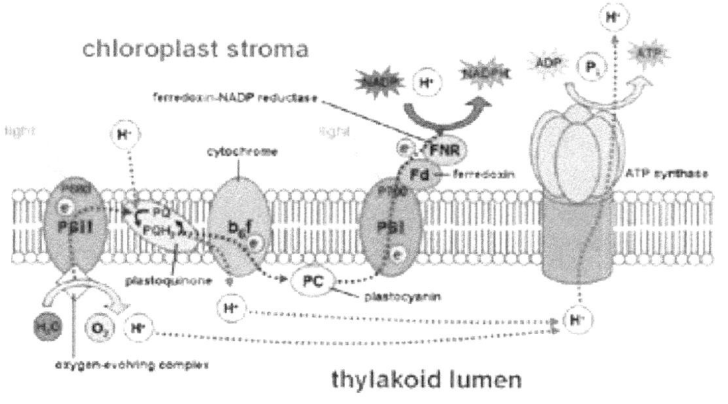

These reactions use the energies of ATP and NADP H_2 formed in the light phase:

$6CO_2 + 24H+ + ATP \rightarrow C_6H_{12}O_6 + 6H_2O$.

In addition to glucose, other complex organic compounds are formed during photosynthesis - amino acids, glycerol, and fatty acids, nucleotides. There are two types of photosynthesis: C3 and C4.

Figure 14. Calvin cycle (Calvin Cycle: Diagram, Definition, and Mechanism. n/d)

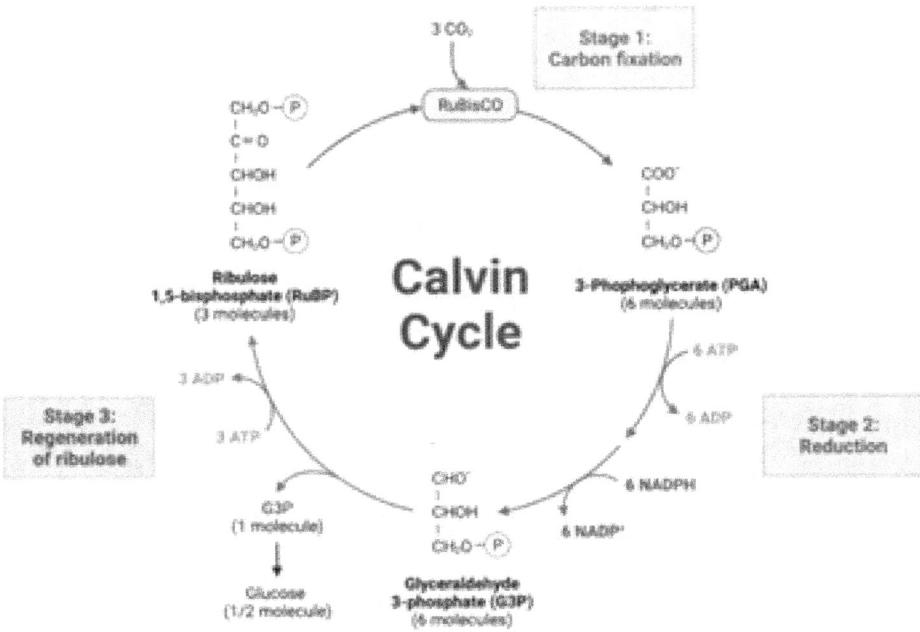

Just as for the processes of the biosynthesis of complex organics, for a better understanding of the complexity of the described processes and their localization, we present the schemes of the structure of cells of modern plants and animals (Figure 15) and, separately, the structure of the membrane (Hanczyc & Monnard, 2017) (Figure 16).

Figure 15. Plant and animal cells (Shields, 2021)

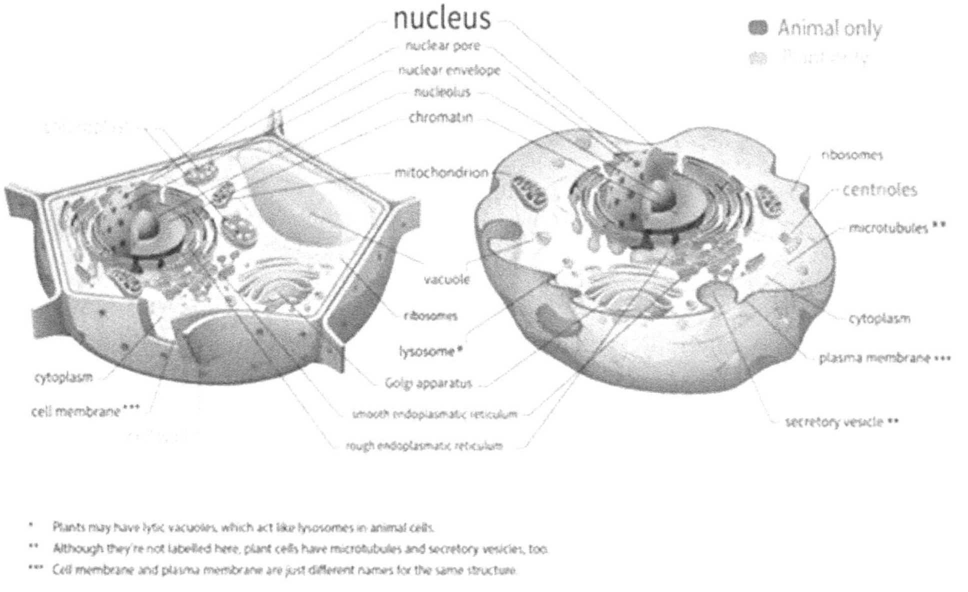

Figure 16. Cell membrane detailed diagram (LadyofHats & Mariana Ruiz, 2007)

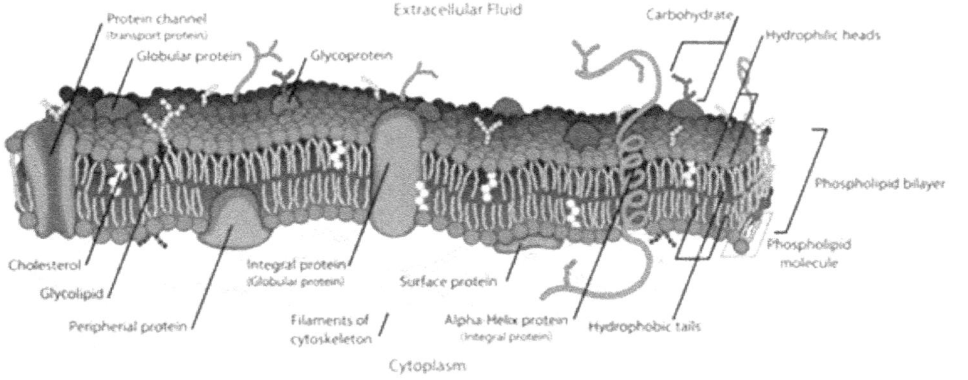

Two circumstances that are important for further modeling and discussion deserve to be noted:

- similar bases for each type of life process and self-healing processes (invariants) are variable for different conditions and different types of living organisms (calibration);
- all these processes are extremely complex and so perfect, that biological immortality takes place[2] for many unicellular and multicellular organisms.

THE FIRST LIVING ORGANISMS AND THE GREAT OXIDATION EVENT

For the nucleic acids and proteins synthesis in cells, complex enzyme complexes are required, which include proteins, and their secondary, tertiary, and quaternary structures (Chapter 1) are necessary for effective catalysis - something like a "vicious circle" is obtained.

There were also specific Ancient Earth conditions. For example, the primordial ocean was a factor preventing the synthesis of polypeptide chains: when building a chain, the addition of each new amino acid unit is accompanied by the release of a water molecule, but in the aquatic environment, the reverse reaction, i.e., hydrolysis of the polymer, is preferred. A simultaneous influx of energy and removal of the formed water is necessary to shift the reaction towards polymerization. In the absence of an ozone layer, under the action of UV rays, the protein would be destroyed even in water - while ultraviolet radiation is just necessary for the synthesis of simple organics.

Yet the complex organics and cells' existence, is an indisputable fact, and several hypotheses have been put forward to overcome the problems described above.

The RNA-world hypothesis (Nikitin, 2018). RNA plays an important role in all major processes of biosynthesis. In 1982, the catalytic activity of some RNA molecules was confirmed; such molecules are called ribozymes. For example, for the considered protein biosynthesis on ribosomes, the catalyst is precisely ribosomal RNA.

The ability of RNA to create a double strand and self-replicate has also been experimentally confirmed (Johnston et al., 2001): small-length RNAs, capable of catalyzing the synthesis of their copies, were obtained. Based on this, the hypothesis of the RNA-world was proposed, suggesting that it was RNA that performed all biologically significant functions in the first living systems, and only then the function of storing hereditary information passed to DNA and the tasks of catalysis, formation of cell structures, and others - to proteins. Over time, the possibility of a simpler system of protein synthesis in the RNA-world was found (Gilbert, 1986; Maizels & Weiner; 1987; Copley et al, 2005; Koonin & Novozhilov, 2009). However, iron ions, alkalis, and high temperatures easily break the RNA chain. Further, cytosine in water gradually loses its amino group, turning it into uracil. Finally, during the formation of hairpins in RNA, catalytic active sites arise that catalyze their cutting. All this limits the size of RNA genomes. Thus, a workable mechanism for protein synthesis in the RNA world has not yet been proposed.

TNK-world? Perhaps RNA was also preceded by simpler compounds capable of self-assembly and replication. Threose nucleic acid has been proposed as such (Yu et al., 2012; Mei, et al., 2018). However, TNC is synthesized artificially and has not been found in nature, and so far, its capabilities and compliance with the conditions of Ancient Earth are not completely clear.

Viruses and the origin of DNA. (Nikitin, 2013k; Di Giulio, 2008). The disadvantages of RNA are eliminated in DNA: it contains deoxyribose, which does not have 2'-hydroxyl groups, from which most hydrolysis reactions begin, does not form self-cutting ribozymes, and instead, uracil contains its methylated analog, thymine, therefore, uracil, formed during the deamination of cytosine, easy to find and fix. But how does the genetic system evolve from an RNA genome to a DNA genome?

A hypothetical scheme for such an evolution was proposed in 1999; the idea was to move from short but replicable strands of RNA to viruses with an RNA genome, then retroviruses with a mixed RNA/DNA genome (the same genome was assumed in the Last Universal Common Ancestor), and then, through viruses, already with DNA -genome, to bacteria, archaea, and eukaryotes. In subsequent publications, this scheme was refined (Leipe et al. 1999; Koonin, 2009).

P. Forter suggested that viruses were the first to change the type of genome, and escape from protection became an advantage: cells use different enzymatic systems that recognize and destroy viral nucleic acids, and the need to bypass them could cause rapid evolution of genomic materials and polymerases to copy them. Therefore, viruses contain a variety of genomic nucleic acids: primitive DNA with uracil in PBS1 and PBS2 phages and DNA with additional nucleotide modifications. Only after working out in viruses, new polymerases were activated by cells.

Since each stage of life emergence depends on the success of the previous one, it is necessary to assess the probability of successful completion of each stage and their sequence as a whole. In the resulting analog of the Drake formula (Drake formula, n/d), one can only speculatively talk about the meaning of each probability coefficient. For example, E. Kunin assessed the events necessary for the start of RNA self-reproduction and evolution as follows:

"At a minimum, the spontaneous appearance of the following is required:

- *Two rRNAs with a total size of at least 1000 nucleotides;*
- *Approximately 10 primitive adapters of 30 nucleotides each, for a total of about 300 nucleotides;*

- *At least one RNA encoding replicase, approximately 500 nucleotides in size (*low bound*).*

Under the notation used here, n = 1,800, resulting in E <$10^{-1,018}$" (Koonin, 2011, p. 435).

From here, in addition to the conclusion about the extreme rarity of life (in our understanding) in the Universe, the question of the time necessary for the emergence of life follows - in comparison with the data of the chronostratigraphic scale (Chapter 1), it may not be enough. Thus, it is necessary to search for specific mechanisms of self-organization - after all, at this stage, natural selection is not yet active.

At the same time, the very possibility of an evolutionary process leading to the emergence of a mechanism for the synthesis of RNA, DNA, and proteins is gradually being confirmed during experiments on artificial evolution (Mizuuchi et al.,2015; Okauchi & Ichihashi, 2021). These are laboratory studies, and the question of their compliance with the conditions of Ancient Earth remains open.

The LUCA. Based on modern reconstructions (Woese et al., 1990; Koonin, 2003; Forterre et al., 2005; Glansdorff et al., 2008; Weiss et al., 2020) performed by methods of comparative genomics, it is believed that the common ancestor had more than 1200 gene families, which corresponds to modern bacteria. This set included genes for metabolic pathways that are now not present in the same cell. Therefore, the common ancestor could constitute an entire ecosystem with closed geochemical cycles. The modern analog of such a system, represented by one species - the bacterium Desulforudis audaxviator - lives at a depth of 2.8 km in a mine in South Africa in hot underground water saturated with sulfates and independently produces all the substances necessary for life from inorganic compounds (Chivian et al., 2008).

The LUCA genome consisted of at least a million base pairs (for comparison: in modern viruses, RNA genomes do not exceed 30 thousand base pairs, while DNA genomes reach 1.2 million; for bacteria, this figure is: in Escherichia coli - about 4 million pairs nucleotides; in parasitic bacteria of the genus Mycoplasma, about 700 thousand base pairs).

However, this genome size is insufficient to encode all LUCA proteins. To resolve this contradiction, and to explain the presence of duplicate biochemical pathways in one organism, ideas have been put forward regarding the non-cellular nature of the common ancestor. E. Kunin (Koonin, 2009) presents LUCA as a community of replicating RNA and DNA molecules that lived on the surface of minerals but had lipid membranes. It is possible that the membranes covered flat accumulations of proteins and nucleic acids on the surface of the mineral, preventing them from dissolving in water, and also limited free-floating bubbles - the first "community organisms" similar to cells. Horizontal gene transfer unified these replicators enough that they could not evolve as separate species. (Nikitin, 2013g).

Prokaryotes (bacteria and archaea) (Nikitin, 2013g). The "zinc world" theory (Chapter 1) explains the divergence of bacterial and archaeal lineages. Separation occurs when a drop in atmospheric pressure no longer allows geothermal vents to bring zinc sulfide ZnS to the surface. Old zinc sulfide deposits melt under sunlight, and the proto-organisms that inhabit them face the prospect of death. There can be two ways out of this situation. One group that lived in the upper, most illuminated layers of zinc sulfide deposits, where even before the crisis they had to create pigment systems to protect membranes and membrane proteins from ultraviolet radiation, left geothermal biotopes and learned to use the energy of light with the help of pigments, from which chlorophyll was subsequently preserved, gave rise to bacteria. Another group did not abandon geothermal chemistry and followed the melting sulfide sources into the depths of the sea, later becoming the Archaean. They had to master chemoautotrophic nutrition.

Both groups faced problems with zinc deficiency and iron ion abundance. There is a lot of sodium and calcium in the seas, compared to geothermal sources. They had to make the membranes impermeable to metal ions to maintain the usual chemical environment in their cytoplasm, as well as systems for transporting zinc, manganese, and other metals into the cell and pumping iron, calcium, and sodium out. To protect against iron, it was necessary to switch from a heteroduplex DNA-RNA genome to double-stranded DNA, after which the multiplasmid genome was assembled by recombination into one circular DNA molecule.

After the bacteria and archaea lines diverged for a long time, they remained isolated. Then the bacteria mastered swimming and populating the open sea. Archaea improved their defense systems and increased the diversity of their metabolisms, and over time were able to move from underwater geothermal sources to other habitats, coming into contact with bacteria and forming a single community. Thus, the prokaryotic biosphere arose on the surface of the Earth and the bottom of the oceans. The bacteria were the main producers that carried out sulfide photosynthesis.

The organic matter, created by bacteria, was consumed by various heterotrophs. Some of them used sulfate breathing; others used various fermentation reactions. Among them were acetic fermenters. In the metabolism of acetic fermenters, along with acetate, CO2, and hydrogen are formed from sugars, which are necessary for methanogenic archaea.

Eukaryotes. Eukaryotic cells are larger than bacterial cells and contain a nucleus that separates the genome and its transcription from the cytoplasm processes; in the cytoplasm are complex systems of organelles. The eukaryotic genome consists of several linear DNA molecules - chromosomes wound around histones for compactness. The genome size can reach tens of billions of base pairs, four orders of magnitude larger than bacteria. The genome is often filled with non-coding sequences of various types. In the life cycle of many eukaryotes, there is a sexual process and meiosis.

The origin of eukaryotes (Nikitin, 2013j) is reconstructed as follows (Figure 17).

This scenario does not explain all the features of eukaryotic cells: in particular, the appearance of sexual reproduction and meiosis; the origin of linear chromosomes; and messenger RNA caps that look strange, like such a part of the eukaryotic translation system. These features of eukaryotes are explained in the viral theory of the origin of the nucleus.

The viral theory of nuclear origin (Bell, 2001; 2006) explains capping and polyadenylation of mRNA, absence of operons, mitosis, and possibly meiosis with a sexual process. It also explains the origin of the membrane systems fusion and the lacing of membrane vesicles. To verify, it is necessary to study archaeal viruses; the caps, or meiotic mechanisms homologs discovery, would be decisive evidence.

The Great Oxidation Event is a global change in the composition of the Earth's atmosphere that occurred at the beginning of the Proterozoic about 2.45 billion years ago. An oxidation event was suggested based on a study of sedimentation patterns (Holland, 2006; Blaustein, 2016; Hodgskiss et al., 2019; Chen et al., 2020; Olejarz et al., 2021). The result was the appearance of free oxygen in the atmosphere and a change in the atmosphere's nature from reducing to oxidizing. It was a long process, including periods of growth, stabilization, and decline in oxygen concentration.

The only significant source of molecular oxygen is photosynthetic organisms. Oxygen photosynthesis originated in cyanobacteria (blue-green algae) 2.7–2.8 billion years ago. The released oxygen was spent on the oxidation of rocks, dissolved compounds, and atmospheric gases. A high concentration was created only locally within the bacterial mats. After the weakening of volcanism and the decrease in the volcanic gases (hydrogen sulfide, sulfur dioxide, methane, and hydrogen) supply and the oxidation of ferrous compounds (Fe_2+ in surface rocks), oxygen began to accumulate in the atmosphere in a free form.

Figure 17. The steps of eukaryote's origin (Nikitin, 2013j)

At the beginning of eukaryogenesis, there are methanogenic archaea living in the lower layers of the bacterial mat in symbiosis with acetogenic fermenters that produce hydrogen.

Archaea acquired the genes necessary for protection from oxygen and a new metabolism - lactic acid fermentation. The absence of a cell wall, an actin cytoskeleton to maintain an elongated and branched cell shape, sterols and the release of the membrane from energy functions contributed to the transition to phagocytosis, which increased the influx of foreign genes into the cell cytoplasm and required the specialization of RNA-polymerases

As a response to the infection of a phagotrophic cell with new types of viruses and mobile elements as a result of the influx of foreign genes, a nucleus appeared to protect the genetic material.

The appearance of a nucleus separating translation and transcription created conditions for the rapid multiplication of introns in protein-coding genes. Mutations of introns led to the appearance of a large number of defective proteins, which led to the formation of an mRNA quality control system and the specialization of some introns to excise damaged introns. Thus, small nuclear RNAs arose, which form the basis of the spliceosome.

New transposable elements activated recombination, which led to the disintegration of circular DNA into linear chromosomes. During replication, their ends protected mobile elements that multiply in parallel with cell division. Telomerase arose on the basis of reverse transcriptase of retrotransposons.

After establishing symbiosis with mitochondria, protoeukaryotes became highly efficient predators. They were able to come to the surface of the mat and begin to phagocytize the cyanobacteria. So, part of the eukaryotes acquired the second type of endosymbionts - chloroplasts. They became algae and went into the ocean, while others remained heterotrophs and continued to feed on bacteria.

Among eukaryotes, predatory ciliates appeared, feeding on other eukaryotes, primarily algae.

Symbiogenesis, gene transfers, and the importance of signaling-regulatory genes show that the process took place in a complex community of prokaryotes.

Methanogenesis became unusable and was replaced due to the advent of oxygenic photosynthesis. The emergence of oxygen photosynthesis in the upper layer of the mat led to the extinction of inhabitants, but some managed to create oxygen defense systems by horizontal gene transfer.

In the middle layer of the mat, among the purple photosynthetic bacteria, aerobic respiration was developed, a valuable substrate for it was organic acids secreted by the fermenters of lower layers.

Adaptations to the environment are aromorphic in nature: aerobic respiration provides an increase in the energy of organisms, and new food

All of them constituted an ecosystem of the open sea, independent of the bacterial mats s.

Since most ancient organisms could not exist at significant oxygen concentrations, anaerobic communities were replaced by aerobic ones and pushed into local ecological niches. As a result of changes in the composition of the atmosphere, its chemical activity changed, the ozone layer was formed, the greenhouse effect sharply decreased, and the planet entered the era of Huron glaciation.

At the end of the Proterozoic, a second oxygen surge occurred, which was possibly caused by the acquisition by plankton of the ability to precipitate organic matter formed during the death of living organisms from the water column to the bottom, removing it from biological cycles: a significant part of the oxygen spent on the oxidation of dead organic matter was released and oxygen began to accumulate. This event is associated with the emergence of multicellular organisms.

We note the following important facts:

- Features of LUCA - possible existence as a cell-free organism or a community of replicating RNA and DNA molecules, in the formation of the genome in which horizontal gene transfer is involved.
- The possible existence of the first ecosystems consisting of organisms of the same species.
- Symbiogenesis is one of the mechanisms providing the fractal property of living organisms.
- complications of organisms and ecosystems during the evolutionary process; especially the emergence of multicellular organisms as a possible way for the existence of a biocenosis within a specific biotope.
- Reproduction and inheritance, because of the need for detailed consideration, are taken out separately in Chapter 6.
- The emergence of parasites and predators, the formation of food chains.
- Great oxygen event, showing that life, having arisen, shows great resistance to changing circumstances. The substances and sources of energy that have appeared become the engine of evolution.

AXIOM CONFORMITY CHECK

1. *Relative homogeneity with the environment.* Homogeneity is expressed in the fact that living organisms include the same substances and fields as the environment. Separation is expressed in chirality, the existence of homeostasis and self-regulation, the instinct of self-preservation, defensive reactions, etc.
2. *Self-organization and structuring*. There is a certain internal potential for self-organization, which can be traced already at the stage of abiogenic synthesis (Chapter 1). Its result is a complication of the body structures (the formation of multicellular creatures, separate systems, organs, cells, organelles, and so on) and ecological systems. However, its actualization requires external initiation.
3. *External factors*. The biotope determines the specific composition of the biocenosis; its changes initiate the evolution of both ecosystems and individual species, whether in the form of aromorphoses, allogenesis, or degeneration.

The direction of such evolution depends on the number of resources and energy; with their excess, parasites, and predators appear, the number of organisms increases, and so on.

4. *Symmetry and self-similarity.* Organisms have substance and field components; one or more symmetry groups characterize them; the anthropic principle testifies to the similarity of man and the universe, children are like parents, and so on. However, symmetry is not absolute; asymmetric structures are also present in organisms. Probably, symmetry is interconnected with isolation and homeostasis (different types of symmetry, according to Noether's theorem, correspond to certain laws of self-preservation), and asymmetry is related to changes and interactions with the environment. This is how the fractal basis of life is manifested, which has several mechanisms of its formation.
5. *Self-regulation and regeneration.* In relatively simple and complex living organisms involving several systems, direct and feedback, self-regulation methods are observed. Complex organic compounds and living beings are characterized by reparation and regeneration. In simple organisms, regeneration is usually more complete than in complex ones.
6. *Replication and translation, reproduction and inheritance.* The isolation of living organisms from the environment sooner or later leads to a struggle for existence and against death, for which evolution has developed various mechanisms. The replication and translation of nucleic acids in cells; reproduction and inheritance of organisms are the most important. These mechanisms are necessary for the survival and development of individual living beings and species.

Now present the axioms themselves.

Axiomatic statements of existence 1. 1 -1. 5 (1-st group) codify the ratio of existence modalities (***A, P***) of space, sets, and elements as well as rules of their mutual transfer depending on the dissolution of U and interaction of C и O, i.e., “from up to down”.

Axiomatic statement 1.1. *Observation transfers S-space in the condition of actual existence:*

$$S°A \qquad (1)$$

Axiomatic statement assigns the existence actuality of *Sp* as a result of S and O interaction when U dissolves. Thus, *Sp* is a primary notion that can't be converted to sets or elements.

Axiom 1.1 postulates the actual existence of S-space as a structure separated from the environment. This axiom determines the possibility of separating *Sp* from the Universe, respectively, allowing modeling the separation of living organisms.

Homeostasis can be compared to the relative stability of the characteristics of solitons. Depending on the situation, solitons can correspond to the material components of living organisms, and waves can correspond to their field components.

Axiomatic statement 1.2. *S-space is a carrier of potentially existing sets that contain potentially existing elements – waves and solitons:*

$$^\circ(\,P(\,\bullet P)) \tag{2}$$

It is stated that S-space is arranged "from up to down" and it is a carrier of potentially existing sets and elements. It means, that space keeping its integrity is not converted to its components. There is no Cantor antinomy here because of the non-set-theoretic nature of relations space – sets – elements.

The relative homogeneity of S-space is secured as follows: 1. Two types of S-elements are relatively homogenous; 2. *Sp* has at least two organization levels that have a different attitude to *Sp,* and react to external influence differently. There are no limits to the number of levels.

Waves and solitons are primary elements of *Sp* and they can't be reduced to the simpler ones. Respectively, we should not compare wave and function (which stipulates reduction) on the element level, but only a wave and a dot. This implies that wave and discrete models are not reduced to space but don't contradict each other. They are abstractions that complement each other and correspond to the wave-particle duality of physical reality.

Axiomatic statement 1.3. *Observation initiates S-space decomposition into actually existing S- sets:*

$$H^\circ P\,A \tag{3}$$

Axiom 1.3 states the role of external influence – observation – as a reason for the Sp evolution which impacts the change of existence modality of S- sets. Thus, *Sp* is an open and dual one.

Axiomatic statement 1.4. *Observation initiates S-set decomposition into actually existing S-elements:*

$$H\,P\,\bullet A \tag{4}$$

Axiom 1.4 defines the role of observation as a reason for the change of existence modality for S- elements.

Axiomatic statement 1.5. *Observation initiates S-elements transfer from waves to solitons and vice versa; its stop – the disappearance of elements:*

$H \bullet A\ (A \bullet A)$ (5)

$\cup \bullet A \rightarrow \cup \bullet P \downarrow$

where – means transfer from waves into solitons and vice versa; ↓ – modality of disappearance.

Axiom 1.5 states that waves are turned into solitons and vice versa within observation; their disappearance makes it possible for convolution to run if there is no observation as a transfer of S- elements in the condition of potential existence rolls back the decomposition. It's impossible to define whether evolution is irreversible, but it derives from this axiomatic statement and the following.

Axioms 1.2-1.5 form the basis for modeling the processes of self-organization and structuring - as we move from axiom 1.2 to axiom 1.5. the structure of *Sp* becomes more complicated. At the same time, the integrity of the *Sp* is preserved.

The initiator of such self-organization is external influences mentioned in the formulations of all axioms of the first group, which also corresponds to the real conditions of self-organization of living beings.

Opposite phases of the wave are symmetrical; by the influence of external factors symmetry may be broken. When waves are formed during self-organization *Sp*, the symmetry order can increase, and the types of symmetries can also differ.

The presence of solitons and waves, their mutual transitions, and the ability to both develop and degenerate and cease to exist (axiom 1.5) are the basis for explaining the existence of both relatively stable and rapidly changing periods in the existence of individual organisms and ecosystems; the initiation of such transitions depends not only on internal modalities but also on external factors.

The nature of the formation of waves and solitons of the current level based on the elements of the previous one shows how the internal resources necessary for self-organization, self-regulation, and regeneration are used and determine the limits of these processes, which cannot exceed the potential of the elements of the previous level. In addition, a basis is created for explaining why the regeneration of simple organisms is more complete than for complex ones - the potential of *Sp* is limited and can be used either to complicate the structure or to restore; in both cases, external resources are also involved.

Axiomatic statements of existence have top priority.

Let us compare the properties of living organisms and the axiom of the 2nd group.

7. Variability. Variability is an immanent property of life, inherent in it to a much greater extent than inorganic matter (trees grow, a person grows in childhood, and he also decreases in old age, adaptation to climate and seasons occurs, biorhythms accelerate and slow down, and so on). Variability is also a natural property of a wave.

It is necessary to distinguish between the modalities of the state of waves and solitons: *expansion* (æ) – emergence and expansion as they extend and absorb other S- elements and within the cooperation with them; *weakening* (ö) – a decrease of length and destruction within interaction; *disappearance* (↓) – transfer from actual to potential existence. Entries of form ∪ö, •ø mean wave (waves) within the condition of weakening and soliton (solitons) within the expansion.

8. Energy exchange, metabolism, endo- and exocytosis. Compared with property 3, we note the influence of external influences on the state of the organism's internal environment. Exchanges with the environment are complex, affecting different types of energy and substances, and involving various levels of the body (systems, organs, cells). They, depending on the conditions, change the state of the body. Direct and reverse links are involved.

9. Reaction to stimuli. A living organism reacts to external stimuli - it moves, changes internal parameters, and so on. Such a reaction implies the existence, along with the bodily, also of a mental component, somehow “evaluating” the harm or benefit of the stimulus. The fact of “symmetrical” components is due to the fundamental differentiation of the external environment at the time of the emergence of life. The set of such reactions may be simple or very complex; simultaneously, it is conditioned by the current state of the organism as a whole: its age, structure, experience, and the like. In turn, the reactions of the organism change the ecosystem.

Axiomatic statement of condition 2.1 (2-nd group) defines indicated special aspects of the behavior of the waves. It makes a global link of *Sp*, its sets, and elements on macro and micro levels as a dependence of condition modality of elements from evolution path of *Sp* and vice versa; dependence of evolution direction of *Sp* from modality condition (æ, ö, ↓) of elements.

Axiomatic statement 2.1. *Stratification progress of S-space defines S- sets structure and condition modality of S- elements – expansion or weakening; condition of elements depicts S- sets the structure and change directivity of Sp:*

°P(A((•A(•ø •ö))) (6)

The priority of the axiomatic statement is lower than the axiomatic statement of the first group but higher than the following ones.

Depending on the modeling situation, the ecosystem, a separate organism, or its part, for example, a cell membrane, can be taken as the S-space. In any case, direct and feedback connections between the system state and its components are postulated.

Adding additional modalities makes it possible to model the variability of organisms and ecosystems accurately.

The influence of external factors on the self-organization of the S-space serves as a basis for modeling metabolism and energy exchange; different levels and sets of *Sp* can be associated with different types of matter and energy, while direct and feedback relationships between their states are preserved.

The S-space's wave nature, sets, and elements facilitate the modeling of the body-psychic duality. The presence of feedback given by the axiom provides a basis for modeling the reactions of the organism and its effects on the external environment.

Let us compare the variety of interactions and axioms of the 3rd group.

10. Periodicity, nonlinearity, threshold effects, and interaction restrictions. The frequency of interactions is associated with the rhythms of nature and biorhythms due to evolutionary adaptation. The nature of interactions and reactions determines nonlinearity. Threshold effects are determined by the sensitivity of receptors and limitations by the lethality of external influences. It is quite obvious that all these features are of a wave nature, and for their modeling, a set of operations is used, determined by the axioms of the third group.

Axiomatic statements of interaction 3.1-3.9 (3rd group). It is necessary to distinguish *parallelism* (||) – the ratio of two or more waves when their interaction doesn't increase the number of components of the resultant wave as compared with initial ones, and *perpendicularity* (⊥) – the ratio of two or more waves when interaction increases the number of components of the resultant wave as compared with initial ones. Entry of the form ∪⊥ means perpendicular waves.

Let's consider the following interactions (operations):

- *Superposition operation* (***S***) – the unification of two or more parallel waves in which components are summed up or compensated.
- *Coincidence operation* (***C***) – the unification of two or more perpendicular S- elements.
- *Interference operation* (***I***) – the interaction of two or more waves and the number of waves, their components, and their condition are changed.
- *Diffraction operation* (***D***) – an interaction of waves and soliton waves and several waves, their condition, and their components are changed (apart from diffraction operation of solitons).

Operations over S- sets, and S-elements are abstractions of different interaction types of physical waves.

Axiomatic statements that describe interaction are applied only with axiomatic statements of existence and condition.

Under superposition operation, parallel components of single or multi-component elements interact. As a result of an operation, their amplitudes are summed up. As axiomatic statement 1.1 is executed, the abovementioned operation doesn't breach conditions of non-binarity and non-additivity.

Axiomatic statement 3.1._Superposition operation of actually existing single component waves turns them into single component waves with linearly changed properties:_

$$S(A\|) \tag{7}$$

Depending on the condition of initial elements the amplitudes are summed up or weakened; in the last case, all interacting waves can transfer from actual to potential existence.

During coincidence operation, perpendicular elements interact, resulting in their unification which contains all perpendicular components of initial elements. These unifications are not binary or additive.

Axiomatic statement 3.2._Coincidence operation of actually existing perpendicular S- elements transfers them into the element whose perpendicular components equal to the total amount of perpendicular components of initial waves and soliton waves:_

$$C(\bullet A) \quad \bullet C \tag{8}$$

There is an interference operation for waves consisting of parallel and perpendicular components.

Axiomatic statement 3.3._Interference operation of actually existing waves turns them into waves with changing a quantity of waves and components and non-linear change of properties:_

$$I(A) \tag{9}$$

Diffraction operation is understood as the interaction of waves and soliton waves. The latter is considered an obstacle that disrupts the linear movements of waves. The result stipulated by the axiomatic statements of the first and second group has the following parameters changed: quantity of waves and components, the condition of waves, and movement direction of solitons. Note the particular case: if only solitons participate in the diffraction operation, they have pushed away and changed movement direction, i.e., behave like particles.

Axiomatic statement 3.4. *Diffraction operation of actually existing waves and soliton waves transforms waves – into waves with changing the number of waves and components and non-linear change of properties; soliton waves – into soliton waves which retain a set of components and changed movement direction:*

$D(\bullet A)$ (10)

Axiomatic statement 3.5. *There is no such consequence of two and more superposition operations that would turn S-elements into their initial condition:*

$CC^{-1}(\| A)$ (11)

Axiomatic statement 3.6. *There is no such consequence of two or more coincidence operations that would transfer S-elements into their initial state:*

$CC^{-1}(A) \quad C$ (12)

Axiomatic statement 3.7. *There is no such consequence of two or more interference operations that would transform waves into their initial state:*

$II^{-1}(A)$ (13)

Axiomatic statement 3.8. *There is no such consequence of two or more diffraction operations that would turn waves into their initial state:*

$DD^{-1}(\bullet A) \quad \bullet$ (14)

Axiomatic statement 3.9. *There is at least one consequence of two diffraction operations that transforms soliton waves into their initial state:*

$DD^{-1}(\bullet A) \bullet A$ (15)

It is necessary to mention that the properties of soliton waves don't change during diffraction.

Diffraction operation is the most general type of operation in S-space; all others are considered special cases, which emerge depending on the properties of components of interacting elements.

Immanent variability of the components of complex systems from Table 7 of Chapter 1 is realized as the impossibility of inverse operations that transfer S-elements to the initial state, as well as waves and solitons, which can be defined as zero.

That's why operations don't make groups except for the special case of diffraction operation of soliton waves (Axioma 3.9).

All characteristics specified in point 9 are realized in axioms 3.1-3.9; based on axiomatic statements in groups 1-3 self-organization theory is built which is given below; they moderate acceptable ways to create S-sets within self-organization of S-space.

In addition, the variety of types of reproduction can also be compared with different wave interactions determined by the axioms of the third group (Chapters 4, 5).

11. Gaining experience. Learnability is another important feature of living beings, necessary for the successful struggle for existence. It is comparable with the axioms of the 4th group.

Axiomatic statements of the measuring 4.1 -4.2 (4th group). It is necessary to introduce the following definitions.

A coordinate system (Cs) is called any set of solitons that belongs to different organization levels of *Sp* (layers); due to their interaction with other waves and soliton waves, it's possible to assess their properties. Let's define a total of such layers as *measurable space. The result* (Pc) of measuring should be interpreted as changes in the coordinate system after interaction with a quantifiable element.

The total number of perpendicular components of all soliton waves Cs should be accepted as *the dimensionality* of the coordinate system. *Sp* is an absolute if compared to totals and elements, therefore Cs will describe only measurable subspace. Components Cs are not related to components of the measurable element and depending on its attribute to subspace, there are different grades of comparison of element and Cs.

Measuring is the observation of interaction results of elements and Cs. Such a procedure is not always suitable. Wave and solitons react in different ways when they are measured: the wave is measured till disappearance (axiomatic statement 3.4.), and repeated measurements can't make results more specific; for soliton waves, only movement direction is changed. Cs is also changed.

Axiomatic statements of measuring prescribe interaction procedures with Cs for waves and solitons. They satisfy groups 1-3's axiomatic statements and the special aspects mentioned above.

Axiomatic statement 4.1.*Observation of diffraction operation in the coordinate system and an actually existing wave of measurable space results in changing of coordinate system – measuring result –and changing of the measured wave:*

HD(Cs A) (Cs Pc) (16)

Axiomatic statement 4.2. *Observation of diffraction operation of the coordinate system and actually existing soliton in measured space leads to changes in the coordinate system – measuring results – and change of direction of soliton:*

HD(Cs •A) (Cs • Pc) (17)

Thus, different forms of gaining experience are compared with the fact that the axioms of interactions (except 3.9) do not form groups, as well as with the difference in the behavior of waves and solitons, described by axioms 4.1 and 4.2.

Two fundamental properties are not mentioned in the considered lists of signs of life.

12. Life cycle. During childhood, everything happens as if "by itself": the size of the body increases, knowledge and skills are acquired, and mental organization becomes more complicated. In old age, the body and mind degrade. The nature of aging remains unclear, but a formal comparison of the life cycle with a wave provides a possible explanation.

13. Death. The finiteness of existence is the irreversible destruction of a living being as a system. It may result from:

- external causes (for example, the extinction of dinosaurs, as a result of a climatic catastrophe after a large cosmic body fell to Earth or degradation of food chains - the displacement of gymnosperms by angiosperms);
- internal factors (weakening and failure of the regeneration mechanism);
- Combinations of external and internal factors (mastery of photosynthesis and changes in volcanic activity led to an oxygen catastrophe, which caused the extinction of many anaerobic species);
- laws of species evolution (the death of old and less perfect individuals creates a "living space" for more perfect and younger ones).

Properties 12 and 13 will be discussed in detail in Chapters 9, 10, and 11.

LIFE IN ITS SYSTEMIC UNDERSTANDING

So, the definitions of life, based on lists of properties have been stated. The manifestations of the most important life processes are also considered, and it is shown that they have a systemic character. Finally, the correspondence of the axioms of the wave model of S-space to the properties of these processes was verified. We have now all the necessary to give a preliminary definition of life in terms of systems theory (clarification will be made in chapter 4):

A living system is a complex, emergent, boundary, open, finite system in a dissipative environment, including immanently changing components that interact nonlinearly with the environment and with each other.

This definition is free of references to carbon or any other basis of life, to forms of reproduction and inheritance, to specific conditions of the Ancient Earth, does not contradict either the lists of signs of life or the 13 generalized properties of life, corresponds to the axioms of the wave model, corresponds to both individual organisms and biotas - and this is its advantage. The quality of emergence separates a living system from additive artificial systems.

The definition is quite abstract - and this may contain both advantages (for example, when modeling viruses, prions, and biotas; all of them, within the framework of the definition, are alive) and disadvantages - if it is necessary to model an organism where specific properties prevail over general ones; also, if artificial emergent systems are ever created, there will be a dispute about whether they should be considered alive.

This definition is an "invariant"; a system that meets it is potentially alive. To become alive, "calibrations" are needed that correspond to a specific situation.

Calibrations include:

- The presence of chemical substances necessary for the formation of vital substances.
- The conditions for the feasibility of a set of corresponding reactions of abiogenic synthesis, including the presence of energy sources, temperature and other environmental parameters, chirality conditions, etc.
- The presence of the necessary time, in other words, all the conditions considered in Chapters 1, and 2.

Note, that for life on silicon or other bases, calibrations will differ from calibrations for carbon life.

Determining calibrations for each type of life base is very important: as data accumulates on the boundaries of environmental parameters, characteristics of sources, etc., as we can already speak about the composition and quantity of vital substances, it will be possible to accurately localize the places of origin of life, determining whether a particular planet or a specific place is promising for its search. But even now, the definition given above allows us to formulate an assumption about the place of origin and features of intermediate forms between inanimate and living matter, and possibly the first living organisms for the conditions of the Ancient Earth:

Intermediate forms appear on the surfaces of crystals in a liquid environment; inherit from crystals methods of self-organization and exchanges with the environment; are liquid biocrystalline systems.

Let's present several arguments in favor of such an assumption.

1. The place of origin of life is debatable: we have competing hypotheses of the "iron-sulfur world", "zinc world", "nuclear geyser system", "primary pizza", "primary mayonnaise", as well as panspermia (Chapter 1). The common problem of these hypotheses is the need for the availability of substances necessary for further syntheses in the required concentrations. Crystals, that during the formation absorb substances from the environment, concentrate ("storage") them, and give them back in the process of dissolution, are open systems and satisfy these conditions. The necessary variety of crystals can be achieved in "black smokers" and other suitable places.
2. It is also necessary to have energy at certain points in time - the processes of crystallization and dissolution can be tied, for example, to periods of nuclear geysers, daily or seasonal changes in temperature and solution concentration; in addition, they occur both with the absorption and release of energy, forming a *dissipative environment*.
3. Intermediate forms are the S-space concept illustration, where the S-space occupies a boundary position in a dissipative environment between S and O (crystals and liquid). Biologically, the "two-dimensional" membrane formations located in this way correspond to the signs of LUCA in the understanding of E. Kunin (Kunin, 2009) - this circumstance allows us to extend the definition of intermediate forms to the first living organisms.
4. The *biocrystalline nature* can be found in the biological structures of different levels of modern organisms, for example, in chloroplasts, muscle and nervous tissues, membranes, and visual receptors (Brown, 1979); it would be natural to extrapolate this characteristic to the distant past - another argument in favor of an expanded understanding of the definition.
5. The boundary position in a dissipative medium corresponds to the *liquid-crystalline nature* of the first organisms, which combines the properties of both crystals and solutions: the influence of crystals is manifested in *self-organization* and *exchanges with the environment*, the influence of the solution is in the plasticity of forms and the ease of *interchanges* already between organisms that become polycrystalline. We also state the openness of living systems-organisms.
6. Phase transitions between liquid and solid crystalline, considering self-organization, give rise to the *emergence* of living systems-organisms, *self-similarity*, and the *fractality of structures* of different levels. They also contribute to the preservation of structures when environmental conditions change, however,

too strong changes lead to the destruction of structures (*finiteness, nonlinearity, threshold effect*).

7. Here are the origins of the first living organisms' *life cycle* origin - from the appearance to the destruction of liquid crystal structures.
8. There is a known morphological similarity between crystals and some biological forms; however, this similarity is not comprehensive and can only be considered an additional argument.

According to the authors, these arguments are a sufficient basis for experimental verification of the assumption. The experiment should reliably reproduce the conditions of Ancient Earth and show the possibility of the formation of a variety of crystals, which is necessary for the formation of the vital substances, fix the appearance of liquid crystal self-organizing structures on their surfaces, and track their exchanges with the environment and with each other. Before the experiment, all these considerations are conjectural. However, the experiment is outside the scope of this study.

The dual nature of liquid crystals, and the presence of their periodic changes, corresponding to periodic influences of external factors and the immanent properties of liquid crystals, make it possible to use soliton-wave abstractions for their modeling.

Now let's conduct a thought experiment.

Let's abstract from the conditions on the Earth's surface and investigate its depths. Here, a kind of analog of the hydrosphere is the asthenosphere - a molten layer of magma, intermediate between the solid rocks of the earth's crust located above and the mantle located below.

Here, as on the Earth's surface, there are conditions for the formation of crystals "washed" by magma; there are pressure and temperature differences, exchange processes occur (i.e., the environment is dissipative), there is a constant source of energy – the internal heat of the Earth. And, if there are the necessary substances and conditions for abiogenic synthesis on a silicon basis, for example, then the emergence of silicon life is theoretically possible. Moreover, the conditions in the depths of the Earth are to a certain extent more favorable than on its surface - there is protection from meteorite bombardment, sudden changes in temperature, etc. Thus, humanity may encounter completely different life forms than ourselves on our home planet. And we can purposefully search for them.

FUTURE RESEARCH DIRECTIONS

All the above determines the program for further research:

- A theory of complex open systems self-organization based on the wave model axioms, and the means of its description and verification, is needed - this will be the subject of Chapters 3 and 4.
- Based on this theory, it will be possible to consider evolutionary theories from the standpoint of a systems approach and build models of certain evolutionary processes (Chapters 5 and 6).
- An understanding of evolutionary processes will enable the modeling of the human being, as well as human-environment relationships (Chapter 7), life cycle (property 12, Chapter 8), and death (property 13, Chapter 9).

CONCLUSION

The basic structures and vital processes of a cell of a living organism are described. The main purpose of this chapter was to check how the axiomatic wave model of S-space is suitable for describing these phenomena.

The definition and signs of life, the features of the synthesis of complex organic matter, metabolism, and energy exchange, the scenarios for the evolution of the first living beings, and their dependence on external and internal factors were summarized in 13 properties, of which 12 and 13 will be discussed in Chapters 9-10. The first 11 properties are compared with objects and operations of S-space axiomatic definitions and the role of external factors in its actualization and existence. The comparison shows the adequacy of the S-space axiomatics and its suitability for modeling living organisms and ecosystems.

Now let's talk about birth, life cycle, death, and their mechanisms.

Along with the external factors noted in Chapter 1, there are immanent factors associated with the wave nature and finiteness of systems. Finiteness is expressed by such phenomena as limiting the number of cell divisions and apoptosis - its self-destruction. But paradoxically, the cell uses not only external but also immanent factors for survival:

- Life's responses to threatening external changes are to ensure the stability of the internal environment (homeostasis), mechanisms of exchange with the environment, including those using the periodicity of external changes (like photosynthesis), the ability to leave dangerous zones (mobility), and others. The response to threatening internal changes is self-repair and reproduction (especially asexual reproduction, which ensures the genetic identity of generations).
- All of them require the organization of complex systems. The cells of living organisms are complex systems - "very complex systems." The question im-

mediately arises as to their origin. Koonin's estimate (Koonin, 2011), which refers not even to the cell, but only to the self-reproduction of RNA, shows an extremely low probability of the accidental emergence of such a complex system. In this regard, it is necessary to consider in more detail the problems of self-organization of systems and the direction of evolution, including for multicellular organisms.

- Openness of the system – not only the need for external resources to build relatively autonomous living organisms is important here, but also their inverse effect on the state of the environment. The Great Oxygen Event is a good illustration of this influence and, among other things, is a historical precedent in the discussion about the impact of living organisms on climate and habitat, as well as the magnitude of such an impact.
- Variability and high adaptability, in addition to immanent factors, are also a consequence of the system's openness. Once this has arisen, life can be adapted to significant environmental changes, and it is undeniable that the life cycle and death are important for this. The life cycle reflects the periodicity of many natural phenomena; death is necessary to release resources and space for more adapted descendants and is one of the mechanisms of evolution - this is one of the points of view on the nature of these phenomena. From these positions, one can also consider a certain "altruism" of parents to rival offspring, which manifests itself in varying degrees - from the absence of cannibalism to prolonged care for them. In this optic, even the death of parents looks like a kind of "self-sacrifice". Of course, such nuances require more detailed consideration in the following chapters.

A preliminary definition of life is given that is compatible with the axioms of the S-space wave model, and an assumption is made about nature and place of formation of intermedial forms or even the first living organisms.

The advantage of this definition is the absence of any links to "terrestrial" life forms and their inherent reproduction methods. The latter is considered the S-space self-organization ways of or a variety of interactions between waves and solitons.

The disadvantage of the definition is its universality - the traditional view of life is perhaps too broad.

REFERENCES

Alberts, B., Johnson, A., Lewis, J., Raff, M., Roberts, K., & Walter, P. (2002). *Molecular biology of the cell* (4th ed.). Garland Science., DOI: 10.1093/aob/mcg023

Aryal, S. (Ed.). (2018). RNA- Properties, Structure, Types and Functions. In *Microbe Notes*. https://microbenotes.com/rna-properties-structure-types-and-functions/

Barton, N. H., Briggs, D. E. G., Eisen, J. A., Goldstein, D. B., & Patel, N. H. (2007). *Evolution*. Cold Spring Harbor Laboratory Press.

Bell, P. J. L. (2001). Viral eukaryogenesis: Was the ancestor of the nucleus a complex DNA virus? *Journal of Molecular Evolution*, 53(3), 251–256. DOI: 10.1007/s002390010215 PMID: 11523012

Bell, P. J. L. (2006). Sex and the eukaryotic cell cycle is consistent with a viral ancestry for the eukaryotic nucleus. *Journal of Theoretical Biology*, 243(1), 54–63. DOI: 10.1016/j.jtbi.2006.05.015 PMID: 16846615

Benner, S. A. (2010). Defining Life. *Astrobiology*, 10(10), 1021–1030. DOI: 10.1089/ast.2010.0524 PMID: 21162682

Biotech Khan. (2014). DNA replication in Prokaryotes. https://biotechkhan.wordpress.com/2014/10/08/dna-replication-in-prokaryotes/

Blaustein, R. (2016). The Great Oxidation Event: Evolving understandings of how oxygenic life on Earth began. *Bioscience*, 66(3), 189–195. DOI: 10.1093/biosci/biv193

Bose, D. (2019). Six Main Cell Functions. *Sciencing*. https://sciencing.com/six-main-cell-functions-6891800.html)

Brown, G. (1979). *Liquid crystals and biological structures*. Elsevier Inc., DOI: 10.1016/B978-0-12-136850-0.X5001-1

Brunk, C. F., & Marshall, C. R. (2021). 'Whole Organism', Systems Biology, and Top-Down Criteria for Evaluating Scenarios for the Origin of Life. *Life (Chicago, Ill.)*, 11(7), 690. DOI: 10.3390/life11070690 PMID: 34357062

Calvin Cycle. (n.d.). Calvin Cycle: Diagram, Definition, and Mechanism. https://researchtweet.com/calvin-cycle-diagram-definition-and-mechanism/

Cell membrane detailed diagram. (2007). Author: LadyofHats Mariana Ruiz. Wikipedia.org. Retrieved August 12, 2024, from https://en.wikipedia.org/wiki/File:Cell_membrane_detailed_diagram_en.svg)

Characteristics of Life. (2021). https://bio.libretexts.org/Bookshelves/Introductory_and_General_Biology/Book%3A_Introductory_Biology_(CK-12)/01%3A_Introduction_to_Biology/1.04%3A_Characteristics_of_Life

Chen, S.-C., Sun, G.-X., Yan, Y., Konstantinidis, K. T., Zhang, S.-Y., Deng, Y., Li, X.-M., Cui, H.-L., Musat, F., Popp, D., Rosen, B. P., & Zhu, Y.-G. (2020). The Great Oxidation Event expanded the genetic repertoire of arsenic metabolism and cycling. *Proceedings of the National Academy of Sciences of the United States of America*, 117(19), 10414–10421. DOI: 10.1073/pnas.2001063117 PMID: 32350143

Chivian, D., Brodie, E. L., Alm, E. J., Culley, D. E., Dehal, P. S., DeSantis, T. Z., Gihring, T. M., Lapidus, A., Lin, L.-H., Lowry, S. R., Moser, D. P., Richardson, P. M., Southam, G., Wanger, G., Pratt, L. M., Andersen, G. L., Hazen, T. C., Brockman, F. J., Arkin, A. P., & Onstott, T. C. (2008). Environmental Genomics Reveals a Single-Species Ecosystem Deep Within Earth. *Science*, 322(5899), 275–278. DOI: 10.1126/science.1155495 PMID: 18845759

Cloverleaf structures of tRNA. (1998). Uploaded by Victor Serebrov. https://www.researchgate.net/figure/Primary-and-cloverleaf-structures-of-tRNA-Phe-from-Ecoli-A-and-its-gene-transcript_fig5_13692313

Copley, S. D., Smith, E., & Morowitz, H. J. (2005). A mechanism for the association of amino acids with their codons and the origin of the genetic code. *Proceedings of the National Academy of Sciences of the United States of America*, 102(12), 4442–4447. DOI: 10.1073/pnas.0501049102 PMID: 15764708

Di Giulio, M. (2008). An extension of the coevolution theory of the origin of the genetic code. *Biology Direct*, 3(1), 37–37. DOI: 10.1186/1745-6150-3-37 PMID: 18775066

Drake equation. (n.d.). Wikipedia.org. Retrieved August 12, 2024, from https://en.wikipedia.org/wiki/Drake_equation

Explain the process of transcription in Eukaryotes. (2021). Upload by Ankush01. https://www.sarthaks.com/1199908/explain-the-process-of-transcription-in-eukaryotes

Forterre, P. (2005). The two ages of the RNA world, and the transition to the DNA world: A story of viruses and cells. *Biochimie*, 87(9-10), 793–803. DOI: 10.1016/j.biochi.2005.03.015 PMID: 16164990

Forterre, P. (2006). Three RNA cells for ribosomal lineages and three DNA viruses to replicate their genomes: A hypothesis for the origin of cellular domain. *Proceedings of the National Academy of Sciences of the United States of America*, 103(10), 3669–3674. DOI: 10.1073/pnas.0510333103 PMID: 16505372

Forterre, P. (2010). Defining Life: The Virus Viewpoint. *Origins of Life and Evolution of the Biosphere*, 40(2), 151–160. DOI: 10.1007/s11084-010-9194-1 PMID: 20198436

Forterre, P., Gribaldo, S., & Brochier, C. (2005). Luca: The last universal common ancestor. *Medecine Sciences*, 21(10), 860–865. DOI: 10.1051/medsci/20052110860 PMID: 16197904

Gennis, R. B. (1989). *Biomembranes: Molecular Structure and Function*. Springer-Verlag. DOI: 10.1007/978-1-4757-2065-5

Gilbert, W. (1986). The RNA world. *Nature*, 319(6055), 618–618. DOI: 10.1038/319618a0

Glansdorff, N., Xu, Y., & Labedan, B. (2008). The Last Universal Common Ancestor: Emergence, constitution and genetic legacy of an elusive forerunner. *Biology Direct*, 3(1), 29. Advance online publication. DOI: 10.1186/1745-6150-3-29 PMID: 18613974

Hames, D., & Hooper, N. (2005). *Biochemistry* (3 ed). Taylor & Francis Group.

Hanczyc, M. M., & Monnard, P.-A. (2017). Primordial membranes: More than simple container boundaries. *Current Opinion in Chemical Biology*, 40, 78–86. DOI: 10.1016/j.cbpa.2017.07.009 PMID: 28802999

Harvey, J. M. (n.d.). Four major membrane protein complexes in photosynthesis: photosystem I, cytochrome b 6 f, photosystem II, and ATP synthase. Researchgate.net. Retrieved August 12, 2024, from https://www.researchgate.net/figure/Four-major-membrane-protein-complexes-in-photosynthesis-photosystem-I-cytochrome-b-6-f_fig2_237046097

Hodgskiss, M. S. W., Crockford, P. W., Peng, Y. A., Wing, B. A., & Horner, T. J. (2019). Productivity collapse to end Earth's Great Oxidation. *Proceedings of the National Academy of Sciences of the United States of America*, 116(35), 17207–17212. DOI: 10.1073/pnas.1900325116 PMID: 31405980

Holland, H. D. (2006). The oxygenation of the atmosphere and oceans. *Philosophical Transactions of the Royal Society of London. Series B, Biological Sciences*, 361(1470), 903–915. DOI: 10.1098/rstb.2006.1838 PMID: 16754606

Ichihashi, N., Usui, K., Kazuta, Y., Sunami, T., Matsuura, T., & Yomo, T. (2013). Darwinian evolution in a translation-coupled RNA replication system within a cell-like compartment. *Nature Communications*, 4(1), 2494. Advance online publication. DOI: 10.1038/ncomms3494 PMID: 24088711

Johnston, W., Unrau, P., Lawrence, M., Glasner, M. E., & Bartel, D. P. (2001). RNA-catalyzed RNA polymerization: Accurate and general RNA-templated primer extension. *Science*, 292(5520), 1319–1325. DOI: 10.1126/science.1060786 PMID: 11358999

Koonin, E. V. (2003). Comparative genomics, minimal gene-sets and the last universal common ancestor. *Nature Reviews. Microbiology*, 1(2), 127–136. DOI: 10.1038/nrmicro751 PMID: 15035042

Koonin, E. V. (2009). On the Origin of Cells and Viruses. Primordial Virus World Scenario. *Annals of the New York Academy of Sciences*, 1178(1), 47–64. Advance online publication. DOI: 10.1111/j.1749-6632.2009.04992.x PMID: 19845627

Koonin, E. V. (2011). *The Logic of Chance: The Nature and Origin of Biological Evolution*. FT Press.

Koonin, E. V., & Novozhilov, A. S. (2009). Origin and evolution of the genetic code: The universal enigma. *IUBMB Life*, 61(2), 99–111. DOI: 10.1002/iub.146 PMID: 19117371

Koshland, D. E.Jr. (2002). The Seven Pillars of Life. *Science*, 295(5563), 2215–2216. DOI: 10.1126/science.1068489 PMID: 11910092

Kostakis, G. E., & Tibbetts, I. (2020). Recent Bio-Advances in Metal-Organic Frameworks. *Molecules (Basel, Switzerland)*, 25(6), 1291. DOI: 10.3390/molecules25061291 PMID: 32178399

Kovalyov, Y., Mkhitaryan, N., & Nitsyn, A. (2020). *Self-organization of the human mind and the transition from paleolithic to behavioral modernity*. IGI Global. DOI: 10.4018/978-1-7998-1706-2

Leipe, D. D., Aravind, L., & Koonin, E. V. (1999). Did DNA replication evolve twice independently? *Nucleic Acids Research*, 27(17), 3389–3401. DOI: 10.1093/nar/27.17.3389 PMID: 10446225

Litwack, G. (2021). *Human Biochemistry*. Elsevier.

Mackenzie, R. J. (2020, December 18). *DNA vs. RNA – 5 key differences and comparison*. Genomics Research from Technology Networks; Technology Networks. https://www.technologynetworks.com/genomics/lists/what-are-the-key-differences-between-dna-and-rna-296719

Maizels, N., & Weiner, A. M. (1987). Peptide-specific Ribosomes, Genomic Tags, and the Origin of the Genetic Code. *Cold Spring Harbor Symposia on Quantitative Biology*, 52(0), 743–749. DOI: 10.1101/SQB.1987.052.01.083 PMID: 3454285

Mei, H., Liao, J.-Y., Jimenez, R. M., Wang, Y., Bala, S., McCloskey, C., Switzer, C., & Chaput, J. C. (2018). Synthesis and Evolution of a Threose Nucleic Acid Aptamer Bearing 7-Deaza-7-Substituted Guanosine Residues. *Journal of the American Chemical Society*, 140(17), 5706–5713. DOI: 10.1021/jacs.7b13031 PMID: 29667819

Mizuuchi, R., Ichihashi, N., Usui, K., Kazuta, Y., & Yomo, T. (2015). Adaptive Evolution of an Artificial RNA Genome to a Reduced Ribosome Environment. *ACS Synthetic Biology*, 4(3), 292–298. DOI: 10.1021/sb5000884 PMID: 24933578

Mulkidjanian, A. Y., Galperin, M. Y., & Koonin, E. V. (2009). Co-evolution of primordial membranes and membrane proteins. *Trends in Biochemical Sciences*, 34(4), 206–215. DOI: 10.1016/j.tibs.2009.01.005 PMID: 19303305

Mulkidjanian, A. Y., Galperin, M. Y., Makarova, K. S., Wolf, Y. I., & Koonin, E. V. (2008). Evolutionary primacy of sodium bioenergetics. *Biology Direct*, 3(13), 13. Advance online publication. DOI: 10.1186/1745-6150-3-13 PMID: 18380897

Nikitin, M. P. (2023). Non-complementary strand commutation as a fundamental alternative for information processing by DNA and gene regulation. *Nature Chemistry*, 15(1), 70–82. DOI: 10.1038/s41557-022-01111-y PMID: 36604607

Okauchi, H., & Ichihashi, N. (2021). Continuous Cell-Free Replication and Evolution of Artificial Genomic DNA in a Compartmentalized Gene Expression System. *ACS Synthetic Biology*, 10(12), 3507–3517. DOI: 10.1021/acssynbio.1c00430 PMID: 34781676

Olejarz, J., Iwasa, Y., Knoll, A. H., & Nowak, M. A. (2021). The Great Oxygenation Event as a consequence of ecological dynamics modulated by planetary change. *Nature Communications*, 12(3985), 3985. Advance online publication. DOI: 10.1038/s41467-021-23286-7 PMID: 34183660

Orgel, L. E. (1968). Evolution of the genetic apparatus. *Journal of Molecular Biology*, 38(3), 381–393. DOI: 10.1016/0022-2836(68)90393-8 PMID: 5718557

Passive vs Active Membrane Transport. (2022). *File:Passive vs active membrane transport.Svg*. Wikimedia.org. Retrieved August 12, 2024, from https://commons.wikimedia.org/wiki/File:Passive_vs_Active_Membrane_Transport.svg

Ray, S., Kassan, A., Busija, A. R., Rangamani, P., & Patel, H. H. (2016). The plasma membrane as a capacitor for energy and metabolism. *American Journal of Physiology. Cell Physiology*, 310(3), C181–C192. DOI: 10.1152/ajpcell.00087.2015 PMID: 26771520

Sakatani, Y., Yomo, T., & Ichihashi, N. (2018). Self-replication of circular DNA by a self-encoded DNA polymerase through rolling-circle replication and recombination. *Scientific Reports*, 8(1), 13089. DOI: 10.1038/s41598-018-31585-1 PMID: 30166584

Shapiro, R. (1987). *Origins: A Skeptic's Guide to the Creation of Life on Earth.* Bantam New Age.

Shields, J. (2021). Here's How Plant and Animal Cells Are Different. [Illustration]. Researchgate.net. Retrieved August 12, 2024, from https://science.howstuffworks.com/life/cellular-microscopic/plant-cells-animal-cells.htm

Tetter, S., Terasaka, N., Steinauer, A., Bingham, R. J., Clark, S., Scot, A. J. P., Patel, N., Leibundgut, M., Wroblewski, E., Ban, N., Stockley, P. G., Twarock, R., & Hilvert, D. (2021). Evolution of a virus-like architecture and packaging mechanism in a repurposed bacterial protein. *Science*, 372(6547), 1220–1224. DOI: 10.1126/science.abg2822 PMID: 34112695

The 7 Unmistakable Characteristics of Life. (n.d.). https://biologywise.com/characteristics-of-life

The Characteristics of Life. (n.d.). https://courses.lumenlearning.com/suny-wmopen-biology1/chapter/the-characteristics-of-life/

The University of Arizona. (2013). Habitability and Biology: What are the Properties of Life? Phoenix Mars Mission. http://phoenix.lpl.arizona.edu/mars141.php

Thomas, J. A., & Rana, F. R. (2007). The influence of environmental conditions, lipid composition, and phase behavior on the origin of cell membranes. *Origins of Life and Evolution of the Biosphere*, 37(3), 267–285. DOI: 10.1007/s11084-007-9065-6 PMID: 17361322

Urry, L., Cain, M., Wasserman, S., Minorsky, P., & Reece, J. (2016). *Campbell biology* (11th Edn.). Pearson Higher Education.

Villarreal, L. P. (2005). *Viruses and the Evolution of Life.* ASM Press. DOI: 10.1128/9781555817626

Wang, D., & Farhana, A. (2021). *Biochemistry, RNA Structure*. StatPearls Publishing LLC.

Weiss, M. C., Sousa, F. L., Mrnjavac, N., Neukirchen, S., Roettger, M., Nelson-Sathi, S., & Martin, W. F. (2016). The physiology and habitat of the last universal common ancestor. *Nature Microbiology*, 1(9), 16116. Advance online publication. DOI: 10.1038/nmicrobiol.2016.116 PMID: 27562259

Woese, C. R., Kandler, O., & Wheelis, M. L. (1990). Towards a natural system of organisms: Proposal for the domains Archaea, Bacteria, and Eucarya. *Proceedings of the National Academy of Sciences of the United States of America*, 87(12), 4576–4579. DOI: 10.1073/pnas.87.12.4576 PMID: 2112744

Yu, H., Zhang, S., & Chaput, J. C. (2012). Darwinian evolution of an alternative genetic system provides support for TNA as an RNA progenitor. *Nature Chemistry*, 4(3), 183–187. DOI: 10.1038/nchem.1241 PMID: 22354431

ADDITIONAL READING

Nikitin, M. A. (2013g). *Myr vyrusov, poslednyj obshhyj predok y proyskhozhdenye DNK*. [The world of viruses, the last common ancestor and the origin of DNA]. *Chemistry and Life. 2013, 8* https://hij.ru/read/2777/

Nikitin, M. A. (2013h). *Proyskhozhdenye membran y membrannaja byoэnerghetyka*. [Origin of membranes and membrane bioenergetics]. *Chemistry and Life, 2013, 9*https://hij.ru/read/2882/

Nikitin, M. A. (2013i). *Zakat «cynkovogho myra», prokaryotycheskaja byosfera y zarozhdenye fotosynteza*. [The decline of the "zinc world", the prokaryotic biosphere and the origin of photosynthesis] *Chemistry and Life, 2013, 10*. https://hij.ru/read/2922/

Nikitin, M. A. (2013j). *Proyskhozhdenye эukaryot*. [Origin of eukaryotes]. *Chemistry and Life, 2013, 11*https://hij.ru/read/3390/

Nikitin, M. A. (2013k). *Rolj vyrusov v proyskhozhdenyy kletochnogho jadra*. [The role of viruses in the origin of the cell nucleus]. *Chemistry and Life, 2013, 12*https://hij.ru/read/4662/

Nikitin, M.A. (2018). *Proyskhozhdenye zhyzny. Ot tumannosty k kletke*. [Origin of life. From nebula to cell]. Alpina dokumentaljnaja lyteratura.

Semak, I. V., Gubich, O. I., Kukulyanskaya, T. A., & Orel, N. M. (2011). *Byokhymyja v skhemakh y tablycakh* [Biochemistry in schemes and tables]. BGU.

KEY TERMS AND DEFINITIONS

Coincidence Operation: (*C*): Unifying two or more perpendicular S-elements.

Coordinate System: (Cs): Any set of solitons that belongs to different organization levels of Sp (layers); due to their interaction with other waves and soliton waves, it's possible to assess their properties. A total of such layers is *measurable space.*

Diffraction Operation: (*D*): An interaction of waves and soliton waves and several waves, their condition, and their components are changed (apart from diffraction operation of solitons).

Dimensionality of Cs: Total number of perpendicular components of all solitons of Cs.

Interference Operation: (*I*): The interaction of two or more waves and the number of waves, their components, and their condition are changed.

Modalities of the State of Waves and Solitons: *Expansion* (æ) – emergence and expansion as they extend and absorb other S- elements and within the cooperation with them; *weakening* (ö) – a decrease of length and destruction within interaction; *disappearance* (↓) – transfer from actual to potential existence. Entries of form ∪ö, •ø mean wave (waves) within the condition of weakening and soliton (solitons) within the expansion.

Parallelism: (||): The ratio of two or more waves when their interaction doesn't increase the number of components of the resultant wave compared to the initial ones.

Perpendicularity: (⊥): The ratio of two or more waves when interaction increases the number of components of the resultant wave compared to the initial ones. Entry of the form ∪⊥ means perpendicular waves.

Result of Measuring (Pc): Changes in the coordinate system after interaction with a measurable element.

Superposition Operation: (*S*): The unification of two or more parallel waves in which components are summed up or compensated.

ENDNOTE

1 In Nikitin (2023), the phenomenon of "circuit switching" is considered - a fundamentally different mechanism for storing and processing DNA/RNA information, based on reversible low-affinity interactions of fundamentally

non-complementary nucleic acids. Such interactions may underlie basic processes in the body, from short-term memory to cancer, aging, and evolution.

Chapter 3
Self-Organization of S-Space:
Description Means and Construction Methods

ABSTRACT

The paradigm of the S-space wave model is completed: After the abstractions and axioms, presented in Chapters 1 and 2, the means of description and research methods are considered. It is established that natural languages describe the objects and relations of S-space better than formal languages. A set of symbols for describing the objects and relations of S-space is supplemented. Intuitive construction is considered the main method of research. It is described in detail and its connections with the epistemological concept of Bergson and Lossky and intuitionistic mathematics are considered. In addition to the study of S-space, an improved apparatus of ternary connectives may be used to work with sources of different epochs.

BACKGROUND

We should note the following before presenting the next two components of the S-space wave model paradigm - description tools and research methods.

The S-space wave model, created by Y. Kovalyov (Kovalyov, 1996), belongs to mathematics' intuitionistic (constructivist, synthetic) branch (Brauwer, 1907; 1908; Heyting, 1930; 1956; Weyl, 1934). The rejection of the axioms of choice and regularity, as well as the law of exclusion of the third, the replacement of the abstraction of actual infinity by the abstraction of potential feasibility means the rejection of logical formalism (Hilbert, 1950/1902; Hilbert & Bernays, 1934;

DOI: 10.4018/978-1-6684-8509-5.ch003

Aleksandrov, 1956; 1987), which inevitably leads to theoretical multiple paradoxes (Buraly-Forti, Kantor, Russell, etc.).

The means of description are needed, firstly, to construct a self-organization theory and its verification (Chapter 4), and secondly, to create a methodology for testing scenarios for the existence of subjective space after the destruction of the soliton-wave system.

In the first case, the question arises about the adequacy of formal and natural languages (Chomsky, 1957; 1975; 1986; 2006; Aho & Ulman., 1972; Losev, 1968; De Saussure, 1978) to the characteristics of S-space. In (Kovalyov & Mezhzherina, 1999; Kovalyov & Grebenyuk, 2006; 2007) show that natural languages better correspond to such characteristics. Therefore, they will be used.

In the second case, description tools are needed to create a ternary connective, where they are settings (projections) for obtaining intuitive knowledge. This approach is known in psychology: when the object of study is very complex or uncertain, the process of cognition itself is modeled (Shchedrovitsky, n/d). Natural language is also better suited for such purposes.

The task is to show what kind of knowledge about the elements and relations of S-space can be expressed using formal and natural languages.

Further, there is an indirect connection between intuitionistic mathematics and philosophical intuitionism as a method of scientific knowledge in the interpretation of N. Lossky (Lossky, 1993/1904). This connection is important here for solving two problems.

The first is the development of the main method for studying S-space: intuitive design (Kovalyov, 1996; Mkhitaryan et al., 2004). This method will be used in Chapter 4 to construct a theory of self-organization of complex systems.

The second is developing a method for interpreting data from various sources used to test some of the predictions of self-organization theory (Chapter 12). This method called the ternary connective (Kovalyov 1996), is also associated with the philosophical intuitionism of N. Lossky.

MEANS OF DESCRIPTION AND VERIFICATION

The mathematical paradigm includes a set of objects and relations, as well as operations, means of description, and verification. When creating a new paradigm, you must redefine all of these components. Objects, relations, and operations are defined in Chapter 2, and we will briefly review the means of description and verification. Their detailed substantiation is given in (Kovalyov, 1996).

The Border Method

Since the qualities of living organisms do not correspond to the properties of the geometric space (Chapter 1), then, by the principle of complementarity of N. Bohr, the study should be carried out based on an additional paradigm, which is determined from the denial of the existing one.

The border method algorithm is defined as follows:

- The components of the original paradigm and its conceptual apparatus are determined, as well as methods for deriving geometric judgments;
- Metaconcepts that generalize the existing ones (implementation of Tarski's approach) are introduced. The characteristics of real open systems determine the direction of generalization. The totality of metaconcepts constitutes a certain concept, which must be implemented in the form of an axiomatic model;
- For metaconcepts and each of the paradigm's components, boundary conditions are studied and explicitly formulated, going beyond which leads to paradoxes. At the same time, the repeatability of conditions for various components of the paradigm is evidence of the integrity of the original paradigm;
- Opposite ones replace boundary conditions; the additionality conditions obtained in this way are the boundary conditions for all new means, defining the limits of use, the structure, and the composition of the new paradigm. Satisfaction of these conditions for all its components means the integrity of the new paradigm if the original one was also integral;
- The positive meaning of the axioms is determined based on the properties of complex systems.

Since proving the completeness and consistency of axiomatic systems is theoretically impossible (Gödel, 1995), boundary conditions and complementarity conditions can only be established for a certain class of problems. It follows that the means of verification are purely practical, if, in the course of solving problems, there is no need for new provisions, means of description, and methods. The boundary conditions are considered to be correctly defined for the class of problems under consideration.

Therefore, the "technology" of the confines method is based on the experience of building axiomatic systems. Thus, the transition from the geometry of Euclid to the geometry of Lobachevsky occurs by replacing the 5th postulate with the opposite one; the same applies, for example, to non-Archimedean geometry; the same principle underlies the classification of geometries according to Kelly-Klein, etc.

Now let's implement the first two points. They are common to all components, so in the future, the properties of the components will be specified, but no new metaconcepts will be introduced.

Metaconcepts

Relative homogeneity is necessary for modeling qualitatively inhomogeneous systems. Objects are relatively homogeneous if some of the ideas about the qualities of one of them are part of the ideas about the other. Relative homogeneity is compared with *heterogeneity* (not one of the ideas about the qualities of one of the objects can be used to describe others) and *homogeneity* (the same representations can be used to describe all objects).

Since we are talking about representations, not the objects themselves, conceptual indistinguishability and the associated distortion of the meaning or meaning of concepts are possible.

Integrity reflects the fact that the properties of the system are not reducible to the properties of the components. Consider the relationship between the whole and the part. The boundary will be called a system of ideas about this relationship that is not immanent to the whole. If the whole is integral, selecting parts is possible only by introducing boundaries from the outside. Then the ratio is characterized

Property 3.1.*The set of parts and boundaries is relatively homogeneous to the whole.*

Property 3.2.*The whole cannot be expressed by particular and boundary concepts.*

Due to immanent boundaries, integrity is opposed to additivity (the properties of the elements determine the properties of the set). In this case, the ratio of the part and the whole is given by

Property 3.3.*The totality of parts and boundaries is homogeneous to the whole.*

Property 3.4.*A set of particular boundary concepts expresses the whole.*

Openness is one of the most important characteristics of complex systems (Chapter 1). Consider the relationship of the parts - the subject, the object - and the Universe. S and O are simultaneously connected with U by a relation of integrity (because they do not exist outside of it) and are separated from U since they are not it. Therefore properties 3.1 and 3.2 hold. However, the boundaries cannot be introduced from the outside, so they must be defined as a consequence of the breakdown of U into S and O in the course of its evolution. They form an essentially isolated component - the boundary S-space - in the S-Sp-O triad; which characterizes him

Property 3.5. *Sp is open space; its existence is possible only as a component of the triad S - Sp - O as a result of the collapse of the universe.*

The openness of the S and-space and its conditionality by the existence of S and O (i.e., the relative homogeneity of U, S, O by property 3.1) is opposed to the non-boundary space (homogeneous to itself - property 3.2), defined independently of external factors.

The following characteristics are due to the above. They also correlate with the properties of complex systems listed in the Chapter 1.

Duality. Consider the mapping S↔O (Figure 1). According to property 3.4, after the decay of U, there are only S, O, and *Sp*, hence *Sp* is the apparatus of this mapping. We will call it a ternary connective (Ts) and denote it

$$U\ (S, Ts, O), \qquad (3.1)$$

emphasizing the conditionality of all components. Let's find out the meaning of Ts.

Since *Sp* is only relatively homogeneous with S and O, they are identical in all respects except those defined by *Sp*. Thus, *Sp* extracts from U and defines S and O, pointing out their differences. Figure 1 shows this symbolically: the one-sidedness of the Klein bottle (identity) is preserved despite a self-intersection line (differences expressed by *Sp*). Consider the mappings S→Sp and O→Sp. Integrity (3.1) is violated: S and O are not connected and are connected with the boundary only projectively. There is a projection schematism of *Sp* concerning S and O. Thus, the TS is not identical to the composition of mappings, which could be illustrated by a cut Klein bottle, when its one-sidedness is violated, which we will treat as a splitting of the Ts. This discussion summarizes

Property 3.6.*Sp is dual, being expressed as the apparatus of Ts under the mapping (3.1) and the projection S or O under the maps S Sp and O Sp.*

Figure 1. Duality. The Klein bottle model is used for illustration (The Klein bottle model. (n /d). Author: pascal62000)

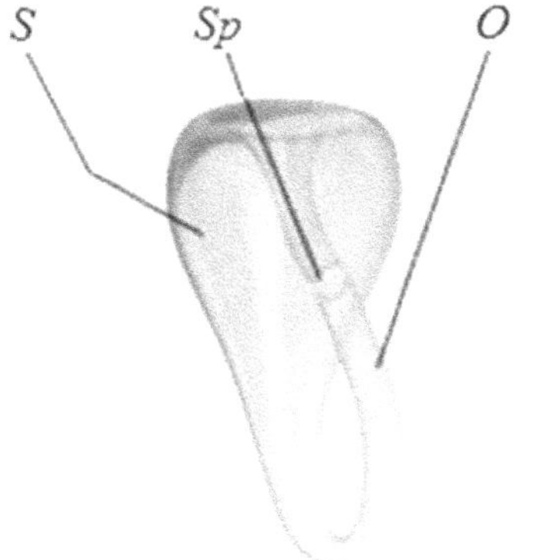

The duality of *Sp* is opposed to separating the concepts of *image* and *projection*[1], and the projection apparatus.

Interdependence. Due to property 3.5, any change in *Sp* is due to the coordinated action of external factors; S and O, as it were, allow or prohibit certain changes. Further, the possibility of implementing changes is due to their absence in the existing structure of the *Sp*; therefore, it depends on the structure and state of *Sp* as a whole. This is possible if the sets and elements of *Sp (S-sets, S-elements)* have the qualities of *variability* and *extension*, and for operations *(S-operations) - non-binarity and non-additivity.*

Property 3.7. *The components of Sp depend on external factors and each other, manifesting itself as the extent and variability of S-sets and S-elements and non-binarity, as well as the non-additivity of S-operations.*

The interdependence of the components of *Sp* is compared with the discreteness of the components of a non-boundary space.

Characteristics of Ts stratifications. We will call the internal stratification (Sb) the subject-boundary correspondence. Takes place

Property 3.8. *The boundary is relatively uniform S; the essence of concepts is subjective.*

Let's call the external stratification (bO) the boundary-object correspondence. He is characterized

Property 3.9. *The boundary is relatively uniform O; the meaning of concepts is objective.*

Property 3.10. *For Sb and bO, the meaning of the concepts may be distorted.*

A discrete boundary model (DGM) is a set of representations in *Sp* with relatively homogeneous boundaries; a continuous boundary model (NGM) is a set of representations or concepts of *Sp*, the boundaries of which are homogeneous. Note that here continuity is not identical to integrity.

The boundary operator is the correspondence between the DGM and the NGM (G-operator; in schemes and diagrams $\xrightarrow{G}$, or $\xleftrightarrow{G}$ when identical). Then the relations within *Sp* are characterized by the following property.

Property 3.11. *Permissible are those and only those correspondences of representations in Cn that are defined by schemes (3.2 3.5):*

(synthesis):

$$\sum_{i=1}^{\infty A}\{DGM_i\} \to NGM \tag{3.2}$$

(analysis):

$$NGM \to \sum_{j=1}^{\infty A}\{DGM_j\} \tag{3.3}$$

(projection):

$$NGM \xrightarrow{G} NGM^*, DGM \xrightarrow{G} DGM^* \tag{3.4}$$

(identification):

$$NGM \xleftrightarrow{G} NGM^*, DGM \xleftrightarrow{G} DGM^*$$

, (3.5)

where ∞_A denotes summation according to the abstraction of actual infinity.

G-operators are comparable to additive and binary transformations since the result depends only on the input data and the matching algorithm.

Ternary connectives and projections will be used as a tool for analyzing artifacts and texts from different eras.

Language Selection

Considering the possibilities of formal, algorithmic, and natural languages is necessary.

Definition of boundary conditions. Properties of formal and algorithmic languages. Since the means of algorithmic languages are more limited than the means of formal M_F languages, we will consider them together. We will follow the construction of M_F on the example of the most frequently used language of the 1st order. It has these levels of organization.

a. Symbols. The language's character set includes designations of subject variables - x, y, z...; functional - f, g, h... and predicate - P, Q, T... symbols. If each variable has its own set of valid values, the language is called multisorted (for example, the language of type theory); more often, the domain of change is common to all variables (single-sorted language).

b. Names. From subject variables and functional symbols, terms are built - names in the theory associated with objects only conventionally.

c. Atomic (elementary) formulas. With predicate symbols, atomic formulas of the form $P(t_1, ..., t_n)$, $n>0$ are formed from terms, meaning that the objects $t_1 ... t_n$ are linked by the relation *P*.

d. Formulas and rules of derivation. Non-elementary formulas are built from atomic formulas with the help of quantifiers and propositional connectives. These become concrete statements when a calculus model is specified (a non-empty area of objects with given predicates). The rules of inference are fixed, which regulate the allowable combinations of quantifiers and propositional connectives in forming new formulas or transforming one formula into another. Most often, these rules are based on the laws of formal or intuitionistic logic, sometimes including efficiency requirements and the possibility of algorithmic resolution, as well as some restrictions, such as the prohibition of recursion.

Hence, the *boundary conditions* of the formal languages of GM_F are expressed as follows:

Condition 3.1.GM_F:

- *Elements, names, formulas – DGM.*
- *Relations – homogeneity, convolution (up to* ∞_P*).*
- *Operations – additive, binary.*
- *Mappings - (3.2-3.3).*
- *Interpretation - regardless of syntax.*

where ∞_P denotes summation according to the abstraction of potential infinity

Properties of natural language. Let us distinguish between the natural language *M* and its rationalistic part *M**. Consider the liar paradox:

"The Cretan claims that all Cretans are liars."

His understanding of it as a paradox is based on the following interpretation:

(statement) ∈ (Cretan) ⊂ (all Cretans),

meaning, by the convolution axiom, the assignment of a particular property of falsity to all Cretans and to all their statements, which is contradictory. However, such an interpretation is not mandatory. If we assume that *M* retains relative homogeneity, then the totality (all Cretans) is integral and cannot be expressed, according to property 3.2., in particular concepts, i.e., no contradiction exists. Thus, *M** expresses the relations of homogeneity and convolution according to properties 3.3., 3.4., while *M* expresses the relations of relative homogeneity and integrity according to properties 3.1., 3.2. this implies

Property 3.12. *Tarski's criterion of truth in formal languages cannot be extended to natural language.*

Thus, it suffices to confine ourselves to defining the boundary conditions for *M* (GM*)*.

The syntactic structure of text segments is specified using a syntactic subordination, described respectively in the form of a subordination tree - graphs of a special type that reflect the binary relations of text segments. Such an apparatus can be compared with the concepts of the category of classes and diagrams of a rather special form. This means *discrete language units, syntactic constructions, and binary and additivity of operations*. Linguistically significant relationships and classifications of language objects are determined and an analytical language model is built. At the same time, syntagmatic and paradigmatic models are distinguished. These relations and classes of language objects are treated from the theoretical set positions. The structures are characterized by *homogeneity and convolution*.

The formal grammar theory refers to the most important aspect of language - processing "meanings" into "texts" and vice versa. It is based on the theory of algorithms and automata and considers binary relations between the processing stages. Binary and additive mappings; refer to (3.2-3.3). The semantic interpretation of the syntactic structure has an intentional (considering the meaning of concepts) or extensional (a complex concept is interpreted only by interpretations of its components) character. Since the second interpretation is accepted in classical and intuitionistic logic, it is precisely this that occurs most often.

T*he relationship subject - M* - object*. It is believed that the signs of the language are a combination of denoting (sounds) and signified (meaning), based on an arbitrary convention - any sign can designate any object. From this, it follows that *C - M * - O* is considered not as Ts, but as *bO: M * O* and Sb: *M * S*. Since *M** is non-homogeneous O, then, in fact, there are relations *M*(uO)* in *Sp*, where *uO* are representations of the object. They are characterized by:

Property 3.13. *(linearization) "Denoted a) has an extension (in time); b) this extension has one dimension - it is a line" (De Saussure, 1977, p. 108):*

$$M^*(uO)\ GM_M(t), \qquad (3.6)$$

where *uO* is a representation (denoted); GM_M - boundary model of the language; *t* - time.

Property 3.14.*M* expresses a discrete idea of an integral object:*

M*(UO)

$\rightarrow \sum_{i=1}^{m} DG\text{M}_\text{M}$

$_i(t_i)$, (3.7)

where DGM_M is a discrete language model; t_i – i-th time interval; *i* run *values* up to *m* or up to ∞_p (potential feasibility abstraction).

Property 3.15.*Means M* combine in one name a set of discrete ideas about an object:*

M*(

$\sum_{i=1}^{m} DGM_i(t_i)) \rightarrow$

NGM (3.8)

According to property 3.14, the integrity of the *uO* in space-time is replaced by successive discretization, which extends up to ∞_p. The integrity of the object in the DGM_M is not preserved. According to property 3.15, the constructive sequence of descriptions is replaced by the actual infinity of the name. From here comes the conversion of concepts and their inconsistency with the *uO* in all cases, except the following:

Property 3.16. *Means M* compares the constructive sequence of representations about the object with the constructive sequence of discrete boundary language models:*

M*(

$\sum_{i=1}^{m}$

UOi(ti))→

$\sum_{i=1}^{n}$

DGMMi(ti) (3.9)

Further, from properties 3.14 3.16 follows:

Property 3.17.M^* *is an instrument of human rational thinking.*

Since logical thinking is "inside" rational, there is

Property 3.18.M^* *contains all the means of logical L and constructive K operators:*

$$L\ M^* \tag{3.10}$$

$$K\ M^* \tag{3.11}$$

Summarizing the above, we define the boundary conditions G_M^*:

Condition 3.2.*Objects for the description of which the* M^* *means are used, should not go beyond the* G_M^**:*

- *Elements and sets – NGM, DGM.*
- *Relations – homogeneity, convolution (up to* ∞_A*).*
- *Operations – binary, additive.*
- *Mappings - (3.2-3.5).*

Determination of additionality conditions. Additional provisions will be those that go beyond the set-theoretic and logistical interpretations of M^* and M_F and deny the sequence of their construction. Based on property 3.13. and conditions 3.1.-3.2., we formulate them as follows:

Condition 3.3.*The means of description must comply with:*

- *Space – open, dual, including interdependent elements.*
- *Sets and elements – immanently extended and changeable.*
- *Relationships – integrity and relative homogeneity.*
- *Operations – non-binary and non-additive.*
- *Mappings – Ts.*

Condition 3.4.*The sequence of constructing the means of description is non-linear and according to the principle "from top to bottom" - language - text - sentence - word - sound.*

Condition 3.5.*The rules for constructing language forms of description means are non-conventional and are determined by the joint action of S and O.*

Condition 3.6.*The means of interpretation are not separated from the syntax.*

Thus, a comparison of conditions 3.3.-3.6. with the properties of M_F and M^* (3.13.-3.18.) shows that their means are insufficient in several positions. So, M^* is linear, while non-linearity is necessary; the type of change of sets and elements does not correspond to name changes, etc. Therefore, to describe *Sp*, it is more expedient to use *M*, the possibilities of which are wider. However, in this case, there are some limitations:

Statement 3.1. The m*eans of M are insufficient to express S and O's integrity.*

Proof: According to property 3.6., TS preserves the identity of the part and the whole, therefore the integrity of S and O as parts of the Universe is preserved. According to property 3.3., particular concepts are not enough to express the whole, and since concepts are determined by means of *M*, they are not enough to express integrity.

It follows from this that using both natural and formal language to describe *S* is possible only within certain limits and up to certain invariants.

With this conclusion in mind, and without going into further clarifications (Kovalyov, 1996), we will continue to use natural language, metaconcepts, and the notation introduced here and in the first chapter to describe the course and results of the study.

Intuitive Design Method

To determine the specific research methods for S-space, first of all, it is necessary to establish the limits of applicability of logical and constructive methods of inference (for which you should use the technique of boundaries), then derive complementarity conditions and determine the essence of the intuitive design method and other applied methods.

Border conditions. Existing methods for deriving and verifying geometric judgments are represented by logical and constructive means (L- and K-operators).

The L-operator includes 15 axioms of the classical predicate calculus and two rules of inference - modus ponens and generalizations (Mathematical Encyclopedia, 1977-85, vol. 4, p. 578). They are formulated as follows.

1. $(\varphi(\psi\supset\varphi))$
2. $((\varphi\supset(\psi\supset\eta))\supset((\varphi\supset\eta)(\supset(\varphi\supset\eta)))$
3. $((\varphi\wedge\psi)\supset\varphi)$
4. $((\varphi\wedge\psi)\supset\psi)$
5. $(\varphi\supset(\psi\supset(\varphi\wedge\psi)$
6. $((\varphi\supset\eta)\supset((\varphi\supset\eta)\supset((\varphi\vee\psi)\supset\eta)))$
7. $(\varphi\supset(\varphi\vee\psi))$
8. $(\psi\supset(\varphi\vee\psi))$

9. $(\neg\varphi\supset(\varphi\supset\psi))$
10. $((\varphi\supset\psi)\supset((\varphi\supset\neg\psi)\supset\neg\varphi))$
11. $(\varphi\vee\neg\varphi)$
12. (∀xφψ⊃(xút))
13. (φ(xút) ⊃∃xφ
14. $(\forall x(\varphi\supset\psi)\supset(\varphi\supset\forall x\varphi))$
15. $(\forall x(\psi\supset\varphi)\supset(\exists x\psi\supset\varphi))$,

where φ, ψ, η - arbitrary formulas of the language, such that the axiomatic scheme 1-15 generates a specific system of calculation with a specific choice of these formulas, for example, when adding non-logical axioms expressing the properties and relations of geometric objects.

In axioms 12 and 13*φ(xút)*- the result of the simultaneous replacement of all free occurrences of *x* by the term *t*, and if *t* is in place of *x* in some formulas, φ, having the form *xψ* or *yx*, where *y* is included in *t*, all free occurrences of *y* in this part should be replaced by the variable, not included in φ, to avoid distortion of the meaning of variables.

Modus ponens rule: *If φ (small premise) and φ ψ (major premise) are deduced, it is allowed to deduce ψ.*

Generalization rule: *If φ is output and x is a variable, it is allowed to output xφ.*

Note that various forms of notation of the calculus are also possible, for example, sequential; differently - in a tree-like or linear form ¬ the conclusion of judgments is also written.

These means describe logical laws, which are "valid for any non-empty area of objects with properties and relations arbitrarily set on it" (Mathematical Encyclopedia, 1977-85, vol. 4, p. 578).

However, the well-known paradoxes (of a liar, etc.) do not contain sophistical conclusions, since the rules of inference are not violated, and the area of objects is also not empty, but a contradiction arises. To avoid antinomies, restrictions on interpretations are used:

- Ban on non-predicative definitions (type theory)
- Restrictions on the means of defining sets - the introduction of classes and class variables.
- Design limitations.

According to Godel's completeness theorem, in classical predicate calculus, all valid formulas are deduced, and only they.

According to Gödel's incompleteness theorem, in formal systems - uninterpreted logical calculi that do not satisfy the requirements of language efficiency and inference - there are non-derivable formulas (Gödel, 1995), and, according to Tarski's results, the classes of true and provable propositions do not coincide (Tarski 1969/1936).

These facts prove the limitations of the objects of logical judgments. To identify limitations, consider the relationship of the L-operator with metaconcepts.

Proposition 3.1.*Concerning the L-operator, the following statements are true:*

- *L-operator is a G-operator.*
- *Scope of the L-operator - NGM and DGM.*
- *The action of the L-operator corresponds to property 3.11.*

Proof. Since the L-operator is contained in M^* (property 3.18), which is a G-operator, it is itself a boundary operator.

Since the scope of the G-operator is NGM and DGM and only they by definition, and the L-operator is a G-operator, then its scope is NGM and DGM.

Satisfiability for the L-operator of property 3.11 follows from the previous assertions.

Natural restrictions on logical objects are the requirements of homogeneity, discreteness or Natural restrictions on logical objects are the requirements of homogeneity, discreteness or ability to discreteness, arising from the concepts, arising from the concepts of DGM and NGM. The following proposition holds.

Proposition 3.2.*No sequence of L-operators is identical to Ts.*

Proof. Since the scope of the L-operator is NGM and DGM, and not the integrity relations S-Sp-O, and the additivity of the composition of operators follows from axioms 1-15, then any sequence of them will have the same scope, i.e., there will be no identical TS that implements the relation of integrity.

This implies the independence of logical operators from the actions S and O. Summing up both propositions, we define the boundary conditions for the L-operator (G_L):

Condition 3.7.*G_L include:*

- *Objects are discrete.*
- *Relations – homogeneity, coagulation (to ∞_A).*
- *Operations – binary, additive.*

These restrictions apply to calculi built based on modal, multivalued, and probabilistic logic, since they, in a different way, determining the possibility of judgment, are based on the same premises as calculi based on classical logic.

CONSTRUCTION METHODS

Constructive means differ from calculus in that it is based on the abstraction of actual infinity and is replaced by the abstraction of potential feasibility or the condition of efficiency of constructions. Axiom 11 (the law of the exclusion of the middle) is also not accepted. Hence, the K-operator has the following boundary conditions (GC):

Condition 3.8.*GC includes:*

- *Constructive objects – discrete.*
- *Relations – homogeneity, convolution (∞_P).*
- *Operations – binary, additive.*
- *Mappings – freely becoming sequences.*
- *Properties (types) – hierarchical.*

Complementarity conditions are obtained by negating the boundary conditions of the operators 3.7. 3.8.

Condition 3.9.*Methods for studying S-space should correspond to:*

- *Sets and elements – not discrete.*
- *Relationships – not coagulation (∞_P and ∞_A), not bulkiness, not uniformity.*
- *Operations - not binary and not additive.*

Since accepting these conditions means going beyond the limits of logical calculations, non-logical inference methods are needed. Consider the available alternatives.

Gnoseological concepts are represented by materialistic, idealistic, intuitionistic, and other doctrines. If we abstract from the specific content of the term's "subject", "object", etc., then the process of cognition can be interpreted using projective schemes.

The cognition scheme binarity is derived from the impossibility of cognition without dividing the universe into cognizing and cognizable components – otherwise, there is nothing to cognize and be cognized. Then cognition is an S-O operation:

$$S\ O \qquad (3.12)$$

Depending on the orientation, S→O or O→S, it is an idealistic or materialistic doctrine. Scheme (3.12) does not express a specific mechanism of cognition; moreover, it is inadequate - if S and O are heterogeneous, then cognition is impossible, and if they are relatively homogeneous, then the scheme does not fix this.

Accounting for boundary operators adds an intra-subject operation of "cognition", which implies binary correspondences between various structures of S. The corresponding scheme looks like this:

$$O\ C(G_c\ G_c*), \qquad (3.13)$$

where G_c and G_c* are structures of the S (memory and reason), and G is an operator that implements (3.2-3.5).

Compared with (3.12), this scheme expresses relative homogeneity and distinguishes between "cognition" (G-operator) and "reflection" (relation S↔O). When (3.2-3.5) is fulfilled, S is identified with reason. This creates a regresio ad infinitum. This is proved as follows.

Let the perception of the external world by the senses evoke a reaction in the recognizing system that needs to be recognized. This can be done by recalling the memory and comparing the recognizable reaction with the characteristic features. The resulting reaction again affects the recognition system, which, turning to the memory, must evaluate this new impact, and so on ad infinitum, which contradicts the finiteness of the recognition time.

This implies the inadequacy of (3.13) and the need to supplement it with a structure different from the mind - the nous (Ra), which will be identified with S, as well as the object model (OM). The scheme will take the form:

$$O \rightarrow G_c \rightarrow G_c*$$

(3.14)

$$OM \leftarrow Ra$$

This is the basic binary scheme. We will call (3.14) the scenario of knowledge of the 1st kind, and its product - *knowledge 1 (K1).*

Note that:

- The means of describing K1 are the means of natural and formal languages.
- The scheme allows deriving any mathematical theory.
- Methods of proof theory or practical checks serve as means of verification, and their possibilities are limited by conditions 3.1-3.7.

Thus, (3.14) is a formal expression of the rationalistic interpretation of knowledge, which underlies modern methods of studying complex systems. This interpretation faces the following difficulties:

- Identification of "I", S, and Ra contradicts the introspection experience of stopping recursive mental activity, which does not lead to the cessation of cognition.
- The same applies to the phenomena of sleep, loss of consciousness, and subconscious activity, which also do not stop cognition.
- When used for the study of *Sp*, (3.14) does not express the boundaries of *Sp*, does not correspond to the extent and variability of S-sets and S-elements, and does not express the integrity relations of Universe, i.e., is not adequate.

The last condition, also present in (3.12) and (3.13), is the primary source of the difficulties experienced by all binary theories.

The ternary scheme of cognition proposed by A. Bergson and elaborated by N. Lossky (Lossky, 1993/1904):

U→(O→S = «I»)

↑ ↓

(3.15)

(RN),

where (*RN*) is the general designation of reason and nous; S = "I" - the identification of "I" and S, characteristic of this theory.

Cognition is *"possible because the world is an organic whole, and the cognizer is S, an individual human self, some extra-temporal and extra-spatial being, closely connected with the world"* (Lossky, 1993/1904).

Such an attitude of S to the world - epistemological coordination - is not yet cognition; for cognition *"The subject must direct to the object a whole series of intentional (target) mental acts - awareness, attention, differentiation, etc."* (Lossky, 1993/1904), which is carried out with the help of the sense organs and the recognition system (only RR; memory is understood as a direct contemplation of one's past). The recognition system is potentially tuned to the knowledge of O due to its effects on the senses, but the actual perception occurs only due to the efforts of S (branch S - (RN) - O). Note that, which is important in setting up the recognition system K1, is not yet true knowledge from the point of view of intuitionism.

Cognition occurs when *"a cognized object, even if it is part of the external world, is included directly by the consciousness of the subject, which cognizes, so to speak, into a person and therefore is understood as existing independently of the act of cognition"* (Lossky, 1993/1904) (branch O - S). Such knowledge of O as a result of its connection with S is called intuitive.

The intuitionist theory, however, is not without flaws:

- Highlighted position of "I", like S out of space and time, complicates cognition - if "I" is heterogeneous (RN), then how can S direct target acts? "I" can only be relatively homogeneous (RN). In general, identifying "I" from S contradicts impersonal cognition.
- Interpreting memory only as a contemplation of the past S contradicts the facts of the impermanence of memory data (forgetting, modification during the transition to long-term memory, etc.).
- The possibility of OM adjusting S for the perception of O is not disclosed (and is not considered).

For these reasons, a correction is necessary, after which (3.15) will take the following form:

U (O, (OM (SMNRW «I»)), S), (3.16)

where (SMNRW← "I") means the sense organs, memory, nous, reason, will, and ego of the individual; (OM↔(SMNRW← "I") are interpreted as *Sp*, separating S from O during the decay of U.

Let's explain it in more detail.

1. The process of cognition is understood as the establishment of the Ts of the impersonal S and O in the integral U, where the G operator acts in (SMNRW ← "I"). This is knowledge of the 2ind kind, the product of which will be
   a. impersonal intuitive knowledge - inclusion of O in S in a holistic U;
   b. personal attunement to a given O in the form of an "imprint" of intuitive knowledge in memory and other personality structures.
2. Let us call this personal attunement (2) *knowledge of the 2nd kind (K2)*, distinguishing it from intuitive non-personal knowledge (1) and knowledge K1, which is obtained by logical deriving.
3. Preliminary adjustment is carried out with the help of the target acts of the "I" (which is possible due to the relative homogeneity of the "I" and the SMNRW) or with the help of object models OM of a special type.

Statement 3.2.*K2 is dual, that is, it is both a projection apparatus and a projection, an attunement to O and a model of O.*

Proof.

1. K2, the setting by definition, limits the subjective perception of U to a specific O, which, according to property 3.6., defines K2 as an apparatus of the Ts.
2. Since *Sp* is a boundary space dividing U into subspaces S and O, it is a "plane" in the Universe's space, and K2 belongs to it as a projection of O onto this plane.
3. Since K2 is tuned to a specific O, it contains boundaries that distinguish O from U, and since K2 is not identical to O, it is its model.

It can also be shown that (3.16) expresses (Kovalyov, 1996.):

- Integrity of U.
- Relative homogeneity of *Sp*, S, and O.
- Openness and interdependence of *Sp*, and, consequently, the corresponding characteristics of elements, sets, operations, and mappings.

Due to the presence of S and O, this theory cannot be interpreted as a sequence of G-operators and be represented as a combination (3.2 3.5).

Consequently, (3.16) *satisfies the necessary conditions and can be used to study S-space directly. The scheme of intuitive design, as the main method of such research, corresponds to (3.16).*

Objectivity and verifiability. First, we consider the existing proof theory to determine the possibilities of known methods. It includes several approaches that are not equally convincing for representatives of different mathematical schools.

1. *The finitistic approach* uses only the abstraction of potential feasibility and extends to constructive objects and functions. Its main advantage is the absence of antinomies and, according to a modern author, *"from a philosophical point of view, it reflects the constructive processes of real nature much more satisfactorily than general set-theoretic mathematics"* (Mathematical Encyclopedia, 1977-85. vol. 2. p. 369). The restrictions correspond to the applicability conditions for K-operators.
2. *The formalization method.* Features and limitations are described above.
3. *Theoretically model methods* are the culmination of formalization; their essence lies in constructing an algebraic system that is a model of a formal theory, provided that all the derived formulas of the latter are true in the semantics generated by the algebraic system.

4. Such modeling is based on the theorems of Gödel (any consistent calculus with classical logic has a model) and Maltsev (if any finite fragment of the calculus has a model, then the whole calculus has a model). In this case, the evidence problem is solved by comparing equivalent models. The restrictions are the same as for the formalization method.
5. *Semi-formal methods.* It is allowed to use an infinite number of constructively generated premises. for inference rules. Boundaries - may be somewhat wider than for L- and K-operators.
6. *Non-classical calculus* is not generally accepted. The proof uses more powerful tools than the L and K operators. Judgments are built for known mathematical objects that meet the usual conditions and restrictions.

How objective are these restrictions? If we understand the K-operator only as the T-operator, then they are objective. But this understanding is not the only possible one. The persuasiveness of a proof carried out with the help of effective constructions is not connected with logical proofs; the existence of a "working" model implies certainty, regardless of how it was obtained. Thus, it is possible to use the K-operator to construct an OM as an element of an extra-logical and non-binary Ts (Figure 2).

Figure 2. Interpretation of the intuitive design scheme (3.16)

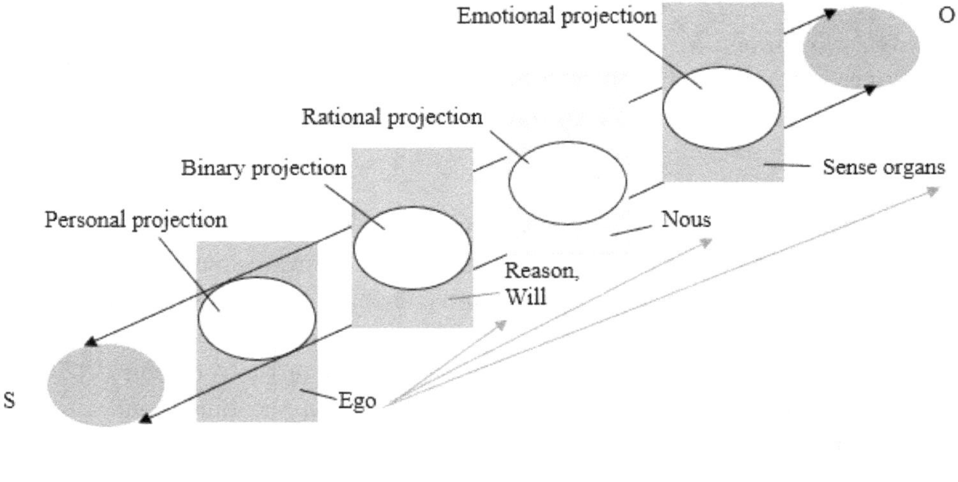

Developing these propositions, we introduce several assertions.

Statement 3.3.*K2 can be expressed in whole or in part using M and OM.*

Proof. Since K2 refers to the structures of the SMNRW← "I", then it is relatively homogeneous M and therefore can be fully or partially expressed by its means. Since K2 is a projection, it can correspond to more than one O; in such a case, O other than the original is the customization object model.

Let us explain the essence of objectivity with the following example. The geometric proof is objective since deduction does not depend on the properties of a particular individual, that is, the impersonal nature of the cognition process ensures objectivity. Fair

Statement 3.4. *K2 is objectively*

Proof. Since intuitive knowledge corresponding to the scheme (3.16) is impersonal, and K2 is its reflection in the SMNRW← "I" structures, obtaining K2 is impersonal, and K2 is objective.

Statement 3.5.*K2 is verifiable.*

Proof. It is enough to indicate the methods of verification - a conclusion on a sufficient basis and OM, which provides practical verification.

The application of the intuitive design method includes the following stages:

- Construction of OM.
- Psychophysiological setting.
- Intuitive perception.
- Fixation and interpretation of the result of perception.

Let's comment on them.

1. The construction of OM is carried out constructively.
2. The need to comply with the boundary conditions gives rise to the following features of the formation of ideas about the object:
    - For tuning, such representations that are not completely rationalistic should be used.
    - View components must be subordinate to the context or composition of the view as a whole.
    - There must be different representations that provide tuning to the same object O.
    - Representations must allow for the impersonal nature of the setting.
3. At the end of perception, it becomes necessary to fix, interpret, and verify K2 when using the resulting OM and M means is possible.

Thus, the intuitive design method, including the construction of constructive OM, the formation of ideas about the object, methods of adjustment and perception, and theoretical and experimental methods for verifying statements, meets the conditions of objectivity and verifiability and can be used for *Sp*. Since the given characteristics do not violate conditions 3.8-3.9. and correspond to statements 3.2-3.5, the final

Statement 3.6. *The intuitive design method, including the building of constructive OM, the formation of ideas about the object, methods of adjustment and perception, and theoretical and experimental methods for verifying statements, meets the requirements of objectivity and verifiability and can be used to study Sp.*

As seen from the above materials, there is a certain unity of the means of description and research. The consequence of this unity is the connection between the speech form and its designation in the case when they are set for the same object. Let us prove a proposition stronger than statement 3.3.

Statement 3.7. *The M means is sufficient to tune in to the perception of any O.*

Proof. Since *M* does not preserve the set-theoretic relation of the whole and the part (the axioms of volume and folding are not satisfied), but the content of the words is determined by the context of the sentence or text, then the following take place:

- Construction of the language on the "top to down" principle.
- Non-binary and non-additivity of both morphology and semantics of the language.
- Inseparability of syntax and semantics.

Further, the language is a boundary, which determines, according to properties 3.8-3.9, the objectivity of the content and the subjectivity of the meaning of concepts, from which it follows that *M* satisfies the complementarity conditions 3.2-3.6, and, therefore, can be used as a means of setting the setting for any object. Then the ratio of K2 and K1 will be determined as follows:

Property 3.17. *K2 is a part of the boundaries of O, which can be expressed using M, and K1 is its conventional DGM or NGM.*

As seen from the above, there is a certain unity of methods of description and research. The consequence of this unity is the connection between the linguistic form and its designation in the case when they are settings for the same O.

The above reasoning may be used when defining signs, symbols, images, and plots and finding invariants when comparing heterogeneous artifacts and texts created in different eras.

FUTURE RESEARCH DIRECTIONS

Directions for further research follow from the above. This is the development of self-organization scenarios for this work, the development of epistemological theory based on this apparatus, their verification, and application.

CONCLUSION

The discussion of the results of this chapter may go in two directions.

First, along with the axioms that reflect the objects and relations of the S-space, the method of intuitive design, the method of boundaries, and the means of description form a new paradigm corresponding to the properties of complex systems.

Secondly, a geometric interpretation of the intuitionistic theory of knowledge has a certain gnoseological significance, making it possible to determine the results of such knowledge concerning formal logical constructions, or, in a narrower sense, to clarify the relationship between intuitionistic mathematics and classical mathematics.

However, both directions interest specialists in the foundations of mathematics and mathematical linguistics. Therefore, let us ask a more general question: how does intuitive construction help to cognize the world and how are the results of such cognition expressed?

To make further reasoning more figurative and acceptable to the reader, let`s use the plot of a fairy tale, known also in many literary works, abstracting from insignificant details.

So, some king, wanting to destroy our hero, or at least get rid of him forever, gives him the following task: "Go there, I don't know where to bring something, I don't know what!".

The hero also has no idea what the king needs and where to look. But he begins to act: he meets mysterious entities, negotiates with them and receives help from them, fights and defeats powerful enemies, gets into more unusual areas, and finally realizes that he has found what he needs. He returns, and what he brings is unexpected and disastrous for the king. The king dies and the hero takes his place.

Let's consider this plot as an example of using intuitive design to build a ternary connective, obtaining intuitive knowledge and knowledge of the 1st and 2nd kinds.

So, in the beginning, our hero is just a servant who does not know his goal and does not know himself. But between him and the object of search already exists some intuitive or "instinctive" connection. Following this unconscious instinct, the hero begins to act ("direct to the object a series of intentional acts"). In the course of his search, solving various "quests", he completely changes himself: he realizes that he is not a servant, but a hero, gains the ability to recognize, the will to win, learns to ask the right questions and understand the answers, control his fear and other emotions. At the right moment, the rational knowledge K1 obtained as a result of questioning and the psyche tuned as a result of the search (knowledge of the 2nd kind, K2) create a ternary connective (Figure 2) and the hero acquires "Something, I don't know what" - intuitive knowledge. With this knowledge of himself - that he is a hero and a potential king and master, not a servant - he returns, finally becoming a real king. Notice that the previous king didn't do anything, he wasn't ready, so he was surprised.

So, we come to a conclusion made in ancient times - "having known yourself, you will know the gods and the Universe", and we supplement it following the theory stated above - "changing yourself with the help of intuitive design, using K1 and K2 as settings, and creating ternary connective, you will gain intuitive knowledge about the gods and the Universe, but you can talk about it using the means of the natural language *M*, passing on the knowledge K1 to the listeners".

REFERENCES

Aho, A. V., & Ullman, J. D. (1972). *The theory of parsing, translation, and compiling*. Prentice-Hall.

Brouwer, L. E. J. (1907). *Over de grondslagen der wiskunde*. Amst.-Lpz.

Brouwer, L. E. J. (1908). *De onbetrouwbaarheid der logische principes* (Vol. 2). Tijdsehz voor Wijsbegeerte.

Chomsky, N. (1957). *Natural language syntax*. Mouton & Co.

Chomsky, N. (1965). *Aspects of the Theory of Syntax*. MIT Press.

Chomsky, N. (1975). *The Logical Structure of Linguistic Theory*. Springer US.

Chomsky, N. (2006). *Language and Mind* (3rd ed.). Cambridge University Press. DOI: 10.1017/CBO9780511791222

Chomsky, N. *Knowledge of Language: Its Nature, Origin, and Use. Greenwood* Publishing Group.

Gödel, K. (1995). Some basic theorems on the foundations of mathematics and their implications. In *Kurt Gödel Collected Works*, v. III, 304-323. Oxford University Press. (Original work published 1951)

Heyting, A. (1930). *Die formalen Regeln der intuitionistischen Logik*. 3 parts, In *Sitzungsberichte der preußischen Akademie der Wissenschaften. phys.-math. Klasse*. 42-65, 57-71, 158–169.

Heyting, A. (1956). *Intuitionism. An introduction*. North-Holland Publishing Co.

Hilbert, D. (1950), *Grundlagen der Geometrie* [The Foundations of Geometry] (2nd ed.). Open Court Publishing. (Original work published 1902)

Hilbert, D., & Bernays, P. (1934). *Grundlagen der mathematic* [The Foundations of Mathematics]. Julius Springer.

Kovalyov, Y., Mkhitaryan, N., & Nitsyn, A. (2020). *Self-organization of the human mind and the transition from paleolithic to behavioral modernity*. IGI Global International Publisher. DOI: 10.4018/978-1-7998-1706-2

Tarski, A. (1969). Truth and Proof. *Scientific American,* 220, 63–77. (Original work published 1936)

ADDITIONAL READING

Aleksandrov, A. D. (1987). *Osnovanyja gheometryy* [Foundations of Geometry]. Nauka.

De Saussure, F. (1977). *Raboty po jazykoznanyju* [Works on linguistics]. Progress.

Kovalyov, Y. M., & Grebenyuk, M. F. (2007). Movy opysu S- prostoru [Description of S-space languages]. *Appl. Geom. and Enj. Graphics*, 77, 29–35.

Kovalyov, Y. M., & Mezhzherina, A. V. (1999). Zastosuvannja teoriji samoorghanizaciji S-prostoru dlja vyvchennja pryrodnykh mov [Application of the theory of self-organization of S-space for the study of natural languages]. *Appl. Geom. and Enj. Graphics*, 66, 113–120.

Kovalyov, Y. N. (1996). *Gheometrycheskoe modelyrovanye erghatycheskykh system: razrabotka apparata* [Geometry Modeling of Man-Machine Systems: The Apparatus Creation]. KMUGA.

Kovalyov, Y. N., & Vereschaga, V. M. (2012). *Prykladna gheometrija: narysna gheometrija, inzhenerna ta komp'juterna ghrafika, suchasni tendenciji* [Applied Geometry: Descriptive Geometry, Engineering and Computer Graphics, Modern Trends]. Omega-L.

Losev, A. F. (1968). *Vvedenye v obshhuju teoryju jazykovykh modelej*. Introduction to the general theory of language models. MGPI.

Lossky, N. O. (1993). *Inuitivism* [Intuitionism]. Progress. (Original work published 1904)

Weyl, G. (1934). *Phylosofyja matematyky: sbornyk nauchnykh trudov* [Philosophy of mathematics: a collection of scientific papers]. State Technical and Theoretical Publishing House.

KEY TERMS AND DEFINITIONS

Ternary Connection (Ts): Connection of the impersonal S and O in the integral U, where the G operator acts in (SMNRW ← “I”) [1] In geometry, a projection is a mapping of an image onto a plane constructed according to certain rules that preserve some of its properties - mapping invariants (Kovalyov & Vereshchaga, 2012).

Chapter 4
S-Space Self-Organization Scenarios

ABSTRACT

Theoretically possible variants of S-space self-organization are considered: stratifications and convolutions, depending on the amount of S and O, and the nature of external influences (decreasing, maintaining, increasing the potential of Sp). The results are presented as 4 tables, 3 conditions, 9 statements, and 15 properties. The second definition of life was achieved within the framework of self-organization theory. The self-organization theory will be used in the following chapters for modeling evolution, reproduction, and inheritance (Chapter 6), human interaction with the environment (Chapter 7), and life cycle (Chapter 8). While studying possible self-organization scenarios, the wave model's axioms (Chapter 2) were tested for their completeness and consistency for this class of problems.

BACKGROUND

The modeling apparatus used in the following chapters should be justified as fully as possible. After verifying the correspondence between abstractions and axioms, it is time to build a theory of self-organization. When constructing this theory, the following should be taken into account.

As noted in Chapter 3, self-organization theory belongs to the intuitionistic area of mathematics. Features of intuitionistic and classical mathematics are described in (Hilbert, 1950/1902; Brouwer, 1907; 1908; Kolmogorov, 1932; Weil, 1934; Hilbert and Bernays, 1934; Heyting, 1936; 1965; Aleksandrov, 1956; 1987; Kleene, 2002/1961; 2009/1950; Kleene and Wesley, 1965). When developing the theory of self-organization, these features were considered (Kovalyov, 1996).

DOI: 10.4018/978-1-6684-8509-5.ch004

When constructing a theory, one should check the completeness and consistency of the axioms. Such verification traditionally comes down to proving the independence and consistency of the consequences of the axioms of one group from the axioms of other groups, as well as clarifying the need to add unfixed assumptions to the axiomatics (Hilbert, 1950/1902). Taking into account the results of K. Gödel on the incompleteness of axiomatic systems (Gödel, 1995/1951), as well as A. Tarsky on the discrepancy between the classes of true and demonstrative judgments (Tarski, 1969/1936), verification should be carried out in a limited form - for certain classes of problems. Until now, this has been done for the problems of modeling human-machine systems, operator activity, and the evolution of human consciousness (Mkhitaryan et al., 2004; Shmelova et al., 2018; Kovalev et al., 2020). Methods of intuitionistic mathematics carried out the verification and will be carried out in the same way now.

The modeling of complex systems is also possible using the methods of classical mathematics in cases where the properties listed in Table 8 of Chapter 1 do not appear or appear so weakly that they can be neglected (Mesarovich et al., 1973; Mesarovich & Takahara, 1978; Thurner et al., 2018, Rzevski & Brebbia, 2017). As a rule, these are technical systems. However, there are studies where classical methods are also applied to systems of natural origin (Urmantsev, 1988; Artyukhov, 2009). In this case, theoretical constructions should be accompanied by experimental verification.

Let us consider whether external influences will lead to some organization of S-space waves and solitons, or whether they will arise and end their existence chaotically.

S-SETS AND S-ELEMENTS FORMATION

Causes and Characteristics of Stratifications

Let us first consider the characteristics of the self-organization of *Sp*, derived from the axioms of the first group, that is, the stratification of S-space under the influence of external factors with the formation of S-sets and S-elements. The axioms of existence qualitatively describe the structure of *Sp* and the sequence of stratifications but do not concretize the causes or introduce quantitative characteristics. Let us establish these causes and characteristics.

Let one S interact with one O. By axioms 1.1.–1.5., *Sp* is an independent boundary space open to external influences, formed due to the Universe's collapse. Its structure expresses both the relation of the identity of S and O in U and their differences. In the first case, *Sp* is an apparatus of the Ts, in the second it is a projection of S or O. Let us consider the consequences of the identity of S and O.

From the existence of the Ts follows the sameness of S and O in all respects, except for those expressed in *Sp*. Therefore, S and O are symmetrical concerning *Sp*, and *Sp* itself, as a projection of both S and O, reflects this fact. So, 1) *S-space is symmetric*. Since the relation S – Sp – O does not change during the stratification, then 2) the symmetry of *Sp is preserved when Sp is stratified*. Since the occurrence of *Sp* does not introduce anything new into U, its "summary potential" equals zero, and 3) the S-space is *skew-symmetric*.

These considerations give a basis for distinguishing the subjective and objective symmetrical halves of Sp. Let us introduce a signature to characterize the membership of sets and elements of *Sp*. Records □⁺, and ∪⁻ will be interpreted as structures related to the subjective and objective parts of *Sp*, respectively.

We can denote:

- *{P}* is the set of admissible stratifications.
- *Pi* is the current stratification from *{P}*.
- $\{x(p_i)\}$ – set of its characteristics.
- x_j is the j-th characteristic from *{x}*.

Takes place

Statement 4.1.*Only such stratifications are admissible, that preserve the skew symmetry Sp:*

$$\{P\}:x_j^+, x_{j-}\{x(pi)\} \quad (4.1)$$

The possibility of identifying S and O requires the feasibility of the process, the reverse of stratification - a decrease in the number of elements and sets due to their disappearance after the termination of external influences - convolution (Co) (axiom 1.5.).

As for Co, it is true:

Statement 4.2.*The convolution process preserves the oblique symmetry Sp:*

$$\{Co\}:x_k^+,x_k^-\{x(c_i)\}, \quad (4.2)$$

where *{Co}* – a set of convolutions; *c* – current convolution.

The connection of c_k with the corresponding *p*, provided that the external influences stopped at the N-the stratification, is determined by the formula:

$$x_j(c_k)\ x_j(p^-_{i+1}), \quad (4.3)$$

where the symbol means a change in the direction of the process.

As a sign of the system's integrity, symmetry is confirmed by many studies (Weil, 1968; Urmantsev, 2017; Artyukhov, 2009; Ovchinnikov, 1966). At the same time, asymmetry also occurs, reflecting the dissipative environment in which living systems exist, regarding the living system's growth and development (Depenchuk, 1963; Jan & Jan, 1999; Penrose, 2005).

Now let's consider the difference in S and O relations. In this case, the ternary connective breaks up. Relations S - Sp and O - Sp are expressed using Sb and bO, respectively. Thus, Sb and bO are interpreted as ***H***, ensuring S-sets and S-elements are actualized.

The stratification reflects the collapse of the U into S and O. This is possible when the characteristics of S-elements and S-sets, which appear at the next stage of stratification, relate to the corresponding attributes of the current layer in the same way as they, in turn, associated with the corresponding characteristics of the previous layer, which occurs while observing the harmonic ratio. Therefore, we have

Statement 4.3.*Admissible are those and only those stratifications whose characteristics of S-elements and S-sets of the current stage are in a harmonic relation with the characteristics of the corresponding elements and sets of the previous stage:*

$$\{P\}:x_j(p_i)/x_j(p_{i-1}) = \Phi, \quad (4.4)$$

where $\Phi = \frac{\sqrt{5}-1}{2} \approx 0{,}618$ – quantitative characteristic of the "golden section"[1].

Since Φ is an irrational number, and the number of sets and elements that are formed during the stratification is an integer, the following are true:

Corollary 1.*The total number of elements and sets that appear in the stratification is not limited.*

Corollary 2.*The ratio of the number of elements and sets of the current stage to the number of elements and sets of the previous stage tends to Φ under an unbounded stratification.*

The Φ is also considered a factor of integrity (Vasyutinskiy, 1990; Korobko, 1997; Soroko, 1984; Stakhov, 1984; Bodnar, 2005).

Sets and Elements Relations

The number of sets and elements of the layer marked with the layer index (analogous to the cardinal numbers), for example, □$^{+5}$, □$^{+5i}$ mean the sets of the subject half, which refer to the layer that contains five sets; the second index indicates the number of elements in this set.

Let us connect the index and the number of layers (stratification stage). According to corollary 2 of statement 4.3., the ratio of indices for the current and previous layers tends to Φ. This pattern corresponds to a series of Fibonacci numbers obtained as follows (Mathematical Encyclopedia, 1977-85):

Let *n* be the current member of the series, and *n– 1, and n– 2* be the previous ones. Then

$$n = n-1 + n-2 \quad (4.5)$$

The substitution gives the following first numbers of the series:

1, 1, 2, 3, 5, 8, 13, 21, 34, 55...

The ratio of the current quantity to the next tends to Φ as the number increases. Then the correspondence between the number and index of the layer will be expressed by the relations:

1 2 3 4 5 6 7 8 9 10... m
1 1 2 3 5 8 13 21 34 55... n,

где

$$n \approx \frac{1}{\sqrt{5}}\left(\left(\Phi + 1\right)^{m} + \left(\Phi + 1\right)^{-m}\right) \quad (4.6)$$

Statement 4.4. *When one S interacts with one O, the stratification-convolution scenario is realized, in which the numbers of elements of all layers, starting from the first, are equal to the numbers of the Fibonacci series.*

The study of integral systems (Vasyutinskiy, 1990; Korobko, 1997; Soroko, 1984; Stakhov, 1984; Bodnar, 2005; Lüttge & Souza, 2019) confirms that the quantities of their elements at various levels of organization gravitate just to the Fibonacci numbers. Figure 1 shows an example of a self-organization scenario; actualized by external influence ***H***.

Figure 1. An example of a self-organization scenario Sp. Here and below, solitons are marked in black, waves are in blue, and contour arrows indicate external influences H

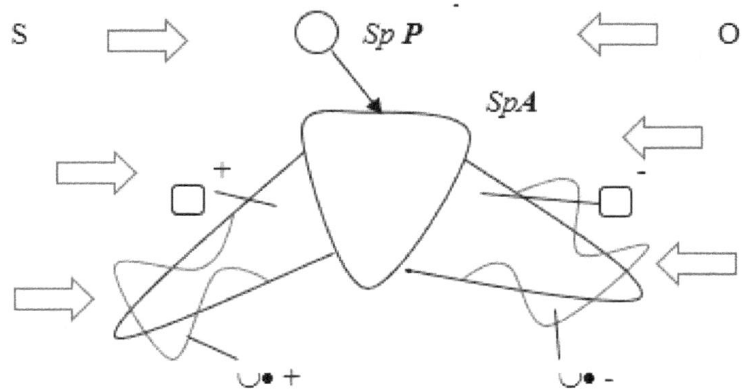

S-sets and S-elements. Axiom 1.4. establishes the existence of S-elements; at the same time, according to Corollary 1 of Statement 4.3, the stratification process is infinite, and there is no such final stage that would contain only elements. Simultaneous fulfillment of both conditions is possible only if the status of some S-sets and S-elements changes in the stratification process. Taking into account Statement 4.4, we have

Property 4.1.*The structures of the current stratification stage are S-elements concerning at least one of the structures of the previous stage. At least one of the structures of the current stage is an S-set for the structures of the next stage.*

Corollary 1.*All manifested layers, except for the last one, contain at least one soliton.*

Property 4.2.*All structures of the current convolution stage are elements relative to at least one structure of the next stage.*

Wave and soliton. According to axiom 1.5., a soliton is formed from a wave when it is observed. Since (statement 4.1) the formation of $\bullet^+$ will necessarily be accompanied by the formation of a skew-symmetric $\bullet^-$, the mechanism of soliton formation can be as follows (Figure 2): 1) H initiates mutual cancellation of two halves of skew-symmetric waves; 2) the halves that remain turn into solitons, since they lose the ability to change according to the law of periodic oscillations.

Figure 2. Soliton formation

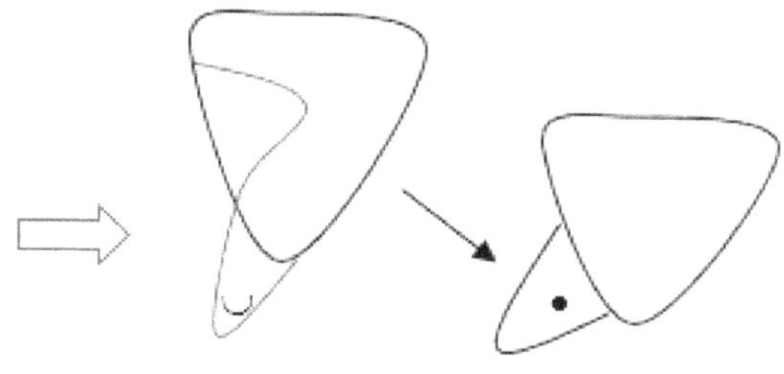

Layering scenarios. Taking into account the role of ***H*** in initiating the stratification of *Sp* and S-sets (axioms 1.3. and 1.4.) and in the transition from a wave to a soliton (axiom 1.5.), as well as the duality of the set and elements established by Statement 4.1, we propose the following stratification scenario:

Under the influence of ***H***, *Sp* is actualized (axiom 1.1.). Then there is a simultaneous (according to property 4.1) transition:

$$\boldsymbol{H}\{\,^{+}_{1}\,\bullet^{+}_{1};\,^{-1}\,\bullet^{-1}\} \qquad (4.7)$$

There is no need to cancel the halves of the waves. The following stratifications of S-sets occur according to the following scheme (Figure 3): 1) the transition of one of the waves of the last layer into a soliton; 2) the formation of waves on the "surface" of the soliton so that their total number is equal to the index of the corresponding layer.

Figure 3. Formation of waves on the surface of a soliton

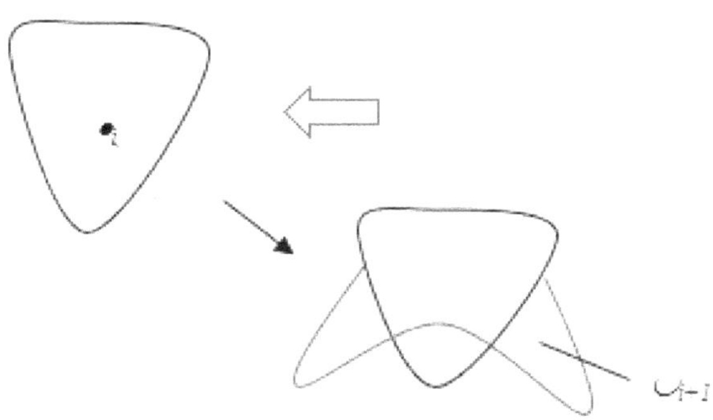

Figures 2 and 3 illustrate the course of the stratification Sp as an analogy for the increase in the length of a fractal curve with a change in scale. This analogy is quite productive, for example, for calculating the dimension of *Sp* and, taking a practical example, for understanding the sequence of pattern recognition by a person (Shmelova et al., 2018, Chapter 5).

Quantitative Characteristics and Qualitative Differences

Potential (π) *Sp* is characteristic of the ability to interact and change, the "captured" part of the "space" U, predetermined by the ratio of decay into S, O, *Sp*. The potential is an intuitive, formally indefinite characteristic, the same as, for example, energy, mass, and charge; it is expressed for each of the S-elements in conventional units relative to the potential *Sp* according to the measurement results (axioms of the 4th group); graphically, it can be represented as an area outlined by a graph showing the change in the amplitude of the wave or soliton and a line showing the time of such changes.

Let us make a digression important for the subsequent research. In the elementary particles' world, there is a corpuscular-wave dualism - any moving particle behaves both as a moving, but localized in space object - a corpuscle, and as a wave. The Louis de Broglie formula determines the length of this wave: $\lambda = h/p$, where $h = 6.6{\cdot}10^{-34}$ J.s is Planck's constant, and p is the momentum of the particle.

In the paradigm of the theory of self-organization, it is logical to interpret the corpuscle as a soliton, and the wave as a wave, as they are understood in the axioms of groups 1-4. Then the wave-particle duality is interpreted as a mutual transition

• under the influence of external influences, which opens up the possibility of a natural interpretation of the Heisenberg[2] uncertainty principle. In abstract constructions, the very principle of connection between a wave and a soliton (an invariant) is important, and the Louis de Broglie relations and the like are gauges that depend on the characteristics of the problem. In the framework of the scenario considered above, the length of the soliton is equal to ½ of the wavelength, and the beginnings of the phases of the wave and the soliton coincide (Figure 4).

Figure 4. The coincidence of the beginnings of the phases of the wave and the soliton; on the right is the variant when the soliton and the wave generate a coinciding pair

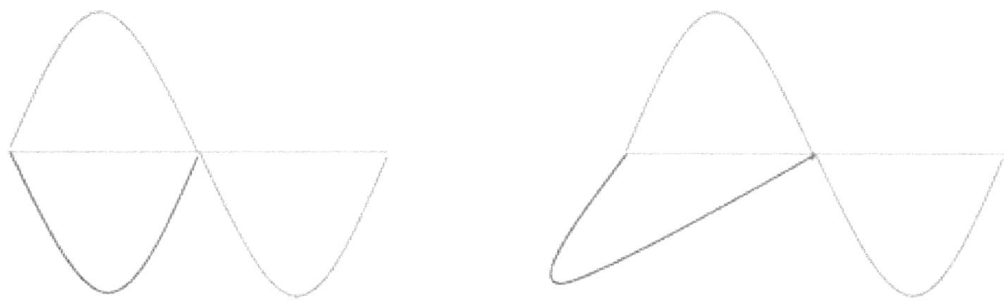

This ratio (including the coinciding pair) makes it possible to simulate the known features of the life cycle of living organisms: birth (beginning of a soliton), death (transition of a wave into antiphase), growth, and increase in vitality in youth, their extinction in old age, and so on. However, some external impulse is needed, which starts the oscillatory process, and then everything happens due to the wave properties of the pair. We will call such a structure a soliton-wave model (SWM).

If we associate a soliton with the human body, and its mental component with a wave, which is discussed in detail in Chapter 8, then, at first glance, both options contradict the obvious circumstance - the body does not disappear immediately after death. But one should also consider external influences that change the potential of the soliton - by the time the wave passes into the antiphase it will not be zero - and the physiological thresholds of the potential, below which systems and organs cease to function. More on this in Chapters 8-10.

Let's go back to considering the potential. The concept of potential makes it possible to interpret the extinction of wave halves during their transition to solitons as a decrease in the potential *Sp* as a whole, its scattering in U. This implies the irreversibility of the evolution of *Sp* in this case; its expression is

Property 4.3. *Let the potential of layer n be equal to π_n. Then $\pi_n > \pi_n{}^*$, where $\pi_n{}^*$ is the potential after { •, $P_{nm}Cv_{mn}$}, m>n.*

However, since the axioms do not determine whether ***H*** compensates for this loss, then the stratification scenarios with decreasing, maintaining, and increasing potential are theoretically possible. When constructing the final classification, this possibility will be taken into account. The presence of several options allows you to simulate a variety of situations.

Let us consider the case when external influences do not compensate for potential losses. To assess the stratification, let us define that part of the potential of the objective or subjective half of *Sp* that passes from the previous stage to the current one. For every S-set,

Property 4.4. *When splitting the S-set with potential π, each element is given the potential π_i:*

$$\pi_i = \frac{\pi\Phi}{i}, \tag{4.8}$$

where i is an index of the current layer.

Thus, the potential Sp *(S-sets) is not equal to the sum of the potentials of the constituent elements.* Consequently, (4.8) formally expresses Sp and S-sets' nonadditivity and relative homogeneity. This implies:

Statement 4.5. *All potential-related characteristics of any layer are different in volume from the characteristics of the next layers and cannot be expressed by them even when summed to ∞_P or ∞_A, that is, they differ qualitatively. Conversely, characteristics of the current layer can be used to characterize subsequent layers.*

It is possible to use S-elements for coding K1. It limits

Corollary 1. *The Shannon estimates of the amount of information encoded using S-elements must be carried out for each layer separately.*

Statement 4.5. is significant for modeling complex systems and should be explained in more detail.

1. The statement allows us to draw a certain analogy between S-space construction and objects, as understood in the object-oriented programming theory. Thus, the ability to access the code of objects - "ancestors", which, as a rule, expresses abstract qualities, by objects - "descendants", where they are concretized (inheritance), also does not have a reverse effect and spreads "from top to bottom". Encapsulation (combining data types and procedures in one object) finds, as will be shown below, a parallel in the conditionality of a set of operations by the properties of elements. The third key concept - polymorphism (giving an action a common name for the entire hierarchy of objects, while implementing it for each of them in a separate way) - is similar to the possibility of obtaining different results from the interaction of the same name, depending on the state and place of the objects that interact. But one should also point out the fundamental difference - for object-oriented programming, these qualities are conventional, while for *Sp* they are a consequence of integrity and non-additivity.
2. Interpreting components of different natures belonging to different layers allows you to combine incomparable characteristics in one model without violating the system's integrity. Their weight will be determined by the magnitude of the potential of the components; the relationship between the conditional value of the potential and physical units is determined from the standards or experimentally.

3. When developing verification methods, Statement 4.5 allows you to establish rules for the illogical inference of judgments that include intuitive concepts of different quality, considering their relationship as a relationship of S-sets and S-elements within the framework of an integral system. One can, for example, trace how the *Sp* potential in different layers is expressed as energy, information, spatial characteristics, etc.

Let's introduce two characteristics that depend on π: the wavelengths for the current and previous stratification stages. Since waves are formed when observing solitons, which are S-sets, and their number is an integer, they must "fit" an integer number of times into a soliton. We have

Property 4.5.*The wavelength l_k of the current stage relative to the wavelength of the previous stage is given by*

$$l_k = \frac{l_{k-1}}{2i}, \tag{4.9}$$

where k is the current stage number.

Let's connect the potential of the wave with the length and amplitude A:

$$\pi = f\,(l/2,\ A), \tag{4.10}$$

where *f* is the proportionality function determined experimentally.

We will consider the potential of its half as an estimate of the wave potential.

Element states. We consider the properties that follow from the general application of the axioms of the first and second groups under the assumption of the interaction of one O with one S.

State characteristics. The modality of the state makes it possible to characterize the behavior of S-elements in the period between stratifications and determine the direction of the evolution of *Sp*. The change of states occurs according to the following scheme:

1. External influence ***H*** on a soliton causes the appearance of waves on its surface (Figure 4). The amplitude of their oscillations increases until

$$\sum_{i=1}^{I} \pi_i < \pi_k, \tag{4.11}$$

where π_i are the layer wave potentials; *I* – layer index; π_k is the total theoretical potential of the layer waves according to (4.8).

This behavior of waves is defined as expansion ø. ø enter into interactions that do not lead to the destruction of their structural independence, while the length of *Sp* increases.

2. When $\sum\pi_i = \pi_k$, the ***H*** is transferred to one of the waves or successively to several waves, leading to the solitons' appearance.
3. Waves not under observation are characterized by a decrease in amplitude and a gradual disappearance - a return to a state of potential existence. This reversal of waves is defined as weakeningö. When interacting, their structure is destroyed.
4. When the ***H*** elements of the last layer are terminated, the stratification stops and the convolution *Sp* begins. This is interpreted as a decrease in the number of features of S and O discrepancies, as opposed to its increase during stratification. Therefore, ø is explained by the simultaneous action of S and O, and ö is explained by the termination of their interaction.

This implies the *limited time of waves and solitons existence* and their conditionality both by interaction with other elements and by the presence of ***H***. Let us characterize the measurement of time and determine the layer's lifetime.

Time measurement. As a unit of t, taking the time equal to one wave period is convenient. In this case, we have

Property 4.6.*Each layer has its unit of time, equal to the wave of this layer period.*

Property 4.7.*The time unit of the current layer t_k relates to the time unit of the previous one in the same way as the wavelengths of these layers:*

$$\frac{t_k}{t_{k-1}} = \frac{l_k}{l_{k-1}}, \tag{4.12}$$

where k is the layer number, and l_k and l_{k-1} correlate according to (4.8).

Property 4.8.*There are such layering scenarios for which the lifetime of each layer, expressed in relative units, is the same:*

$$\forall k, \frac{T_k}{t_k} = const, \tag{4.13}$$

where *Tk* – is the lifetime of the *k-th* layer.

The duration of the stratification. Above, it was pointed out that the stratification is unbounded because no layer would contain elements that are not S-sets for the following, which would mean the end of the process (corollary 1 of property 4.1). There is, however, its time limitation due to the existence of the first *Sp* wave in time.

Property 4.9.*The processes of stratification and convolution have a total duration equal to the time of existence of the first wave Sp.*

Several consequences follow from this:

Corollary 1.*The lifetime of Sp is limited, that is, its evolution is equifinal.*

Corollary 2.*The process of stratification is possible if and only if the lifetime of the current layer is less than the lifetime of the previous one:*

$$k, T_k < T_{k-1} \quad (4.14)$$

Corollary 3.*Let the convolution time be equal to the stratification time. Then the time T_{ok}, which remained by the end of the stratification process at the beginning of the k-th stage, is equal to*

$$T_{0k} = \frac{T_1}{2} - \sum_{j=2}^{k-1} T_j \quad (4.15)$$

Thus, time for S-space is characterized by two qualities:

"The arrow of time" expresses evolution's irreversibility (property 4.9 and corollaries).

A stratification expresses the relative homogeneity of qualitatively different processes (properties 4.3.–4.11). The same, for example, is the number of cell divisions, regardless of their lifetime. The idea of stratification of the temporal flow and of a separate geological, biological, social, and other time is discussed.

The practical meaning of the above properties is estimating the lifetime of individual system components.

Operations. Let us define the characteristics and properties that follow from the general consideration of the axioms of the 1st–3rd groups.

Parallelism and perpendicularity of elements. When C interacts with several {O} (Figure 5), *Sp* is stratified according to scenarios {St} so that St_i O_i (symbol "→" means "conditioned O_i"), St_i *{St}*, O_i *{O}*.

Figure 5. An example of stratification Sp in the interaction of S with {O} (in this case, with 3 {O})

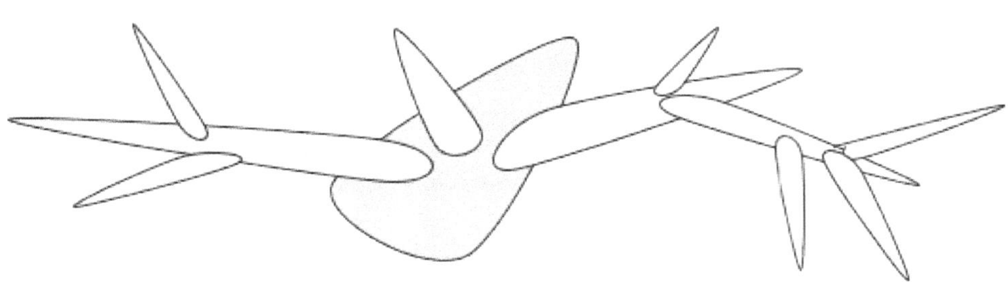

The following conditions are met:

1) qualitative differences $S\,O_i$, $S\,O_j$, O_i, O_j $\{O\}$, must correspond to relatively homogeneous structures of their stratifications.
2) such differences $S\,O_i$, $S\,O_j$, O_i, O_j $\{O\}$, must correspond to homogeneous structures that are formed as a result of P_i, P_j $\{P\}$, which implies a quantitative comparison.

The first condition is interpreted as the interaction

$$\bullet_i \text{ and } \bullet \qquad (4.16)$$

(4.16) is necessary and sufficient, since it covers all operations on relatively homogeneous elements (axioms 3.3.–3.4., 3.6. 3.9.), which requires their perpendicularity.

The second condition is understood as the interaction

$$\cup\bullet_i \text{ and } \cup\bullet_j\text{úú};\ \cup\bullet_i\in\square_k,\ \cup\bullet_j\in\square_k \qquad (4.17)$$

(4.17) is necessary, since the interaction of homogeneous elements - superposition - requires, according to axiom 3.1, the presence of parallel elements. Sufficiency (4.17) implies the absence of a superposition of parallel elements of one stratification, namely:

- elements of one layer;
- elements of different layers.

Let us show that such elements will not interact according to the superposition rule.

a. According to property 4.5., the wavelengths of any layer are such that the soliton, which is the S-set of this layer, forms an integer number of waves. According to Statement 4.1., the symmetry of *Sp* is preserved under its stratification. Therefore, each layer is symmetric and the following holds:

Property 4.10. *Let the layer contain only waves and not contain solitons. Then it has the order of symmetry equal to its index.*

Since the interaction of waves of one layer would break the symmetry, we have

Corollary 1. *Waves of one layer do not interact as long as their symmetry is preserved.*

b. Elements of different layers are also not subject to superposition since they have potentials that are significantly different, and, for example, the potential $_8$ refers to the potential $_3$ as Φ/2*5*3 = 1/30 (4.8), and their lengths - as 1/2*5*2*8= 1/160 (4.9), that is, it differs by more than 10 or 100 of magnitude, and this leads either to the absorption of the "small" wave or to the "big" wave running around it, which leads, according to axioms 3.3. and 3.4., to interference type interaction. Thus condition (4.17) is also necessary and sufficient.

Now we need to consider the layering scenarios that lead to the appearance of parallel and perpendicular elements.

Formation of only parallel structures. In this case, statements 4.4 and property 4.9 hold for each with *Pi*. Only one-component waves are formed, and the number of waves and the order of symmetry increase as the stratification proceeds according to the law of the Fibonacci series. Scenario characterizes

Statement 4.6. *Let one and only one symmetry group S_i act on $^+_{-i}$ (not necessarily S_i=S_n acting on +−n). Then only parallel structures are formed during stratification, and the order of symmetry is m=i.*

Formation of only perpendicular structures. In this case, multicomponent waves and solitons are formed. The scenario of such stratification describes

Statement 4.7. *Let $^+_{-i}$ in the formation of $_{k\ i}$, k=1,...i, be acted upon by its symmetry group S_{ik} (not necessarily S_{ik}= S_{jl}, which acts on $^+_{-j}$ for $_{jl}$). Then only perpendicular structures are formed during delamination, and the number of symmetry groups is n for the layer with index i: n=i.*

Formation of both parallel and perpendicular structures. Statements 4.6 and 4.7 hold. The relationship between them establishes

Statement 4.8. *Let there be M symmetry groups for P_i and S_{ik}=S_{jl} for S_{ik}. Then the formation of perpendicular structures will stop when the number of realized symmetry groups for each layer is L=M, and in the future, only parallel structures can be formed.*

Since the dimension of S-sets is equal to the number of perpendicular components of structures, then we have

Corollary 1. *Under the stratification proceeding according to the scenario described by Statement 4.7, the dimension r of S-sets decreases from M to 1: r_k=r_k−1− i, and, having reached 1, remains constant.*

Various symmetries for S-sets and S-elements make it possible to state such propositions.

1. By analogy with Noether's theorem, it can be assumed that the *presence of certain symmetry groups predetermines the corresponding conservation laws*, for example, the parallel transfer group for a time corresponds to energy conser-

vation, shift, and space reversal - momentum conservation, etc. In this case, the names of the layer characteristics and the symmetry group can be determined based on observing the preservation of the invariants of the system organization.
2. Due to the stratification of space and time (properties 4.5.-4.9), the action of these laws' limits

Statement 4.9.*For +– of any layer, which M symmetry groups characterize, there are M* conservation laws; when you move to the previous or next layer, their action is terminated.*

3. Property 4.9 and statements 4.6.-4.9. for the organization of *Sp* states that the admissibility of interaction and its result are determined not only by the characteristics of the elements but also by the need or lack of need to implement the symmetry group and change the order due to the stratification of *Sp* as a whole (axiom 2.1).

Thus, the role of symmetry in the organization and evolution of *Sp* is adequate to its role in the organization and evolution of natural systems.

Operations Characteristics

Arithmetic operations are reduced to additive binary addition; set-theoretic - to union and intersection. In contrast, the 4 *Sp* operations are special cases of non-additive and non-binary diffraction. We will label their set ***{DIKS}***. We describe each operation according to the following plan:

1. Domain of definition.
2. Role in the formation of structures, stratification, and convolution.
3. Additional remarks.

Note that particular cases of individual operations are similar to the interactions of their physical prototypes. This analogy, however, does not apply to formulas for determining quantitative characteristics, since:

- The number and distribution of waves and solitons *Sp* in complexity exceeds the cases of interaction described in the sources.
- Some coefficients were not determined experimentally.
- Interactions are non-linear.
- The set $\{x(pi_{,j})\}$ will include:
- Potential π.

- Length *l*.
- Amplitude *A*.
- Frequency *o*.
- The number and distribution of waves and solitons *Sp* in complexity exceeds the cases of interaction described in the sources.
- Experimentally determined coefficients.

If the interactions are nonlinear, the set $\{x(pi_{,j})\}$ will include:

- Potential π.
- Length *l*.
- Amplitude *a*.
- Frequency o.
- Phase *f*.
- Period *t*.
- Coefficients:

The measures of non-linearity:

$$e = \frac{A}{M} \leq e^{*},\ d \leq d^{*}, \tag{4.18}$$

where *e* and *d* characterize the nonlinearity in amplitude and motion, respectively; *e** and *d** are the boundary values of *e* and *d*, starting from which the nonlinearity cannot be neglected; *M* is characteristic of the unexcited state of the S-set.

The ratio of lengths:

$$s=l_k/l_m; \tag{4.19}$$

The ratio of potentials:

$$p=\pi_k/\pi_m; \tag{4.20}$$

The modalities ø, ö, ↓ and feature úú and ⊥.

Some of these characteristics are related; recall that not all of them can be defined for all layers.

Special cases of operations were studied (Mkhitaryan, Kovalyov, Badeyan, 2004).

Superposition *S* (axioms 3.1., 3.5.).

1. **Condition 4.1.** Superposition is defined for one-component waves if and only if

$\cup_{ik}, \cup_{lm}$úú, $\cup \rightarrow P_i, P_j$

$l_k < l_m$, $s \in \{\sim 0.2 \div \sim 1'\}$

$$\pi_k < \pi_m, p \in \{\sim 0.2 \div \sim 1'\} \qquad (4.21)$$

$d < d^*$

$e < e^*$,

where the character " ' " means an exception.

The superposition breaks the symmetry of the layer. Takes place

Property 4.11. *When the symmetry of the layer is violated, the following occurs*:

- *Movement of waves over the surface of their solitons.*
- *Interaction of waves of one layer and redistribution of potentials up to the restoration of symmetry with a lower order.*

2. The waves modified with the help of ***S***, being influenced by ***H***, turn into solitons and stratify at the next stage - the influence of ***S*** on St consists in changing the transferred π, or, in case of mutual cancellation of waves, in eliminating scenarios that could lead to stratification. Similar is the influence of ***C*** on Co.
3. It should be remembered that the interactions of the waves are always non-linear; (4.21) simplifies the interaction picture.

Non-binarity is ensured by considering the state caused by ***H*** and the course of this stratification and the possibility of participation in the interaction of more than two waves. The irreversibility is due to the change in the state of the waves under the influence of ***H***. Thus, the complementarity conditions for the superposition are satisfied.

Coincidence *K* (axioms 3.3., 3.6.). The difference between ***K*** and a random set of perpendicular waves and solitons is its integrity, which manifests as a certain consistency of perpendicular components along the length. Their coordination is achieved by redistributing the potentials of the initial waves.

1. Let one S interact with {O*}, which are a set of objects from {O}. Then the coincidence conditions are as follows.

Condition 4.2.*Let the stratifications {St} corresponding to {O*} have an ordered set of waves and solitons:*

$l_1 \perp l_2 \perp l_3 ... \perp l_n,$

$l_1 > l_2 > l_3 ... > l_n,$

where *n* is the number of stratifications in *{P}; l* - lengths •.
Then the necessary and sufficient coincidence conditions for one-component • are:

$l_n/l_1 = m_{11}, l_n/l_2 = m_{12}, ..., l_n/l_{n-1} = m_{1n-1}$

$l_{n-1}/l_1 = m_{21}, ..., l_{n-1}/l_{n-2} = m_{2n-2}$

$l_2/l_1 = m_{n-11}$
(4.22)

$$\forall \Delta \pi_i = \sum_{j=1}^{n} \Delta \pi_j, \ i \neq j$$

{∪ö}∨{∪ø}

d<d*,

where *m, k,* and *l* are integers; *i, j* – potentials of wave changes.
Since *solitons do not change, they are not considered* in (4.22).

2. A multicomponent wave as a whole does not form S-sets; its elements – and • – can form such sets as part of their layers. In this case, coincidence is destroyed, since consistency is violated.
3. The non-binarity of ***K*** is ensured by the participation of several waves and taking into account their states; non-additivity - the impossibility of adding new components, which would violate consistency.

Interference *I* (axioms 3.3., 3.7.).

1. During interference, only one- or multi-component waves interact, which have characteristics that exclude superposition and coincidence. Interference conditions are described as follows:

Condition 4.3.*Let $_kP_{i}$, $_mP_{j}$, P_{i}, P_j {P}, $l_k<l_m$, $\pi_k<\pi_m$. Then the necessary and sufficient conditions for interference are:*

$_{k}$, $_{m}$úú: (e>e*) (d>d*) ((s>>1,~1) (p>>1,~1)) i, j (4.23)

$_{k}$, $_{m}$: (s=M) ($\Delta\pi_K \neq \Delta\pi_m$) ($_k\phi$ $_m$ö), i

=j
(4.24)

$_k u$ $_m$: p=1, (4.25)

for all pairwise parallel components, (4.21) or (4.23) is satisfied,
where *M* is an integer; n_k, n_m are the numbers of components $_k$ and $_m$, respectively.

2. Interference affects the course of stratification or convolution in different ways and may be similar to superposition, similar to coincidence, or does not effect.
3. Non-additivity and non-binarity of interference and the absence of an inverse operation are ensured by the non-linearity of the interaction of multicomponent elements, considering the modalities of states and ***H***.

Diffraction *D* (axioms 3.4., 3.8., 3.9.).

1. The conditions for the implementation of diffraction - operations for single- and multi-component waves and solitons - are as follows:

Condition *4.4.*Let ●$_kP_{i}$, ●$_mP_{j}$, P_{i}, P_j {P}, $l_k<l_m$, $\pi_k<\pi_m$. Then the diffraction ●$_k$ and ●$_m$ takes place in those and only in those cases when:*

∪●$_k$, ∪●$_m$úú: i≠j

●$_k$, ●$_m$: (s≠M) ($\Delta\pi_k \neq \Delta\pi_m$) (●$_k\phi$ ●$_m$ö) i

=j
(4.26)

∪●k, ∪●m: for all pairwise parallel components ∪●$_k$, ∪●$_m$ holds (4.21)

2. Fair

Property 4.12. *Let diffraction take place only for •ϕ, parallel or coinciding in all components, and each of them is characterized by the potential* π_i *and the speed of displacements* v_i. *Then*

$$\sum_{i=1}^{n}\pi_i v_i = \sum_{i=1}^{n}\pi_i v_i^* \tag{4.27}$$

where v_i^* is the velocity of the soliton after diffraction.

Similar to the momentum conservation law for elastic particles in physics, this property follows from the invariance of other soliton characteristics and reflects the soliton-particle analogy.

Corollary 1. *Two symmetry groups act for a layer consisting of solitons: translation and rotation.*

Corollary 2. *Under the same conditions, and if* $\pi_i >> \pi_j$, *separation* $\bullet_j$ *from their S-sets can occur.*

3. The effect of diffraction on the course of stratification and convolutions varies depending on the features and may be similar to interference, identical to coincidence, or does not effect.

The non-binarity and non-additivity of ***D*** and the absence of inverse operations are determined by the participation of several elements of various types, considering the state and ***H***.

Classification of scenarios of self-organization of S-space

In the above study, the main scenario of self-organization *Sp* was used, when space is formed due to the interaction of one S with one O. For real systems, interactions of one or more S with one or more O occur. Therefore, there is a need to consider other options for self-organization. In the previous sections, signs were mentioned that can be classified. We use them to group possible scenarios. For the convenience of the review, we summarize practically significant cases in a separate Table 1; those in bold have been discussed above.

Table 1. Classification of stratification-convolution scenarios

<table>
<tr><th>Classification criterion</th><th colspan="6">Scenarios</th></tr>
<tr><td>{S} and {O} quantity</td><td>1S, 1O</td><td colspan="2">1S, {O}</td><td colspan="2">{S}, 1O</td><td>{S}, {O}</td></tr>
<tr><td>π loss compensation</td><td>missing</td><td colspan="2">incomplete</td><td colspan="2">complete</td><td>redundant</td></tr>
<tr><td>úú or ⊥ of elements</td><td colspan="2">⊥ only</td><td colspan="2">úú only</td><td colspan="2">úúand ⊥</td></tr>
</table>

The table shows that any combination of features is possible.

Let us briefly characterize some of those scenarios that have not yet been considered.

One S interacts with the set {O}. In this case, S-space develops from one center in several directions, the number of which is equal to the number of objects. If the latter have common differences from S, elements with the same characteristics arise, as a result of which the interactions discussed above become possible. As a result, the number of layer structures, their mutual belonging, the distribution of the potential, and the characteristics associated with it change. The "topology" of S-space can be characterized as a tree-like one with separate closed contours.

Interaction of a set {S} with one O. Since S and O contribute equally to the *Sp* stratification, this case has the same features as the previous one.

Interaction of the set {S} with the set {O}. Here S-space has several development centers (by the number of S), each of which has several directions of development (by the number of O). Interactions arise and topological and metric characteristics change. In general terms, the "topology" of S-space is network-like, with separate tree-like structures.

Partial compensation for potential losses. In the process of development according to one of the scenarios described above, the number of elements in the layers increases; the number of "redundant" elements Δn_i can be arbitrary; it is determined according to

$$\Delta n_i = \frac{\Delta \pi}{\pi_{in}}, \tag{4.28}$$

where $\Delta \pi$ – redundant part of the potential; π_{in} – the potential of the element of the *i*-th layer, and rounded to an integer.

Full compensation. In this case, neither the number of elements nor their potentials in the next layers change compared with the current layer.

Overcompensation. The number of elements in the layers decreases; it can be arbitrary and is calculated similarly to (4.28).

Consequently, compensation affects the number of elements in the layers and the direction of the evolution of *Sp*. There is a stratification, or "convolution in reverse", an increase in the total potential of Sp characterizes that.

Measurements in S-space

Consequences of axioms 1–4 of groups are considered.

Measurement procedure. *Dimension r.* The connection of r with the number of perpendicular components, the order, and the number of symmetry groups for *Sp* and S-sets is established by statements 4.6-4.8, property 4.10, and its corollary. Let us formulate some features of measurement for S-space:

1. A one-component wave (soliton) is considered one-dimensional; the direction is measured along the amplitude. this implies

Property 4.13.*S-sets and Sp as a whole are bounded concerning U. The boundary is the surface of their structures.*

2. We will distinguish between the actual dimension *Sp* ra, equal to the total number of distinct perpendicular components that appeared during stratification, and the potential dimension rn, equal to the number of distinct perpendicular components of potentially existing structures. Takes place

Property 4.14.*Sp ra increases with layering and decreases with convolution; rn decreases during stratification and increases during convolution and for any stage of stratification or convolution,*

$$ra+rn=const \qquad (4.29)$$

3. It is natural that the dimension of the resulting S-sets under the bundle *Sp* decreases (Corollary 1 of Assertion 4.1).
4. From (4.9-4.11), assuming a linear dependence of *l*, *A*, and π, it follows

$$A_{i+1} = \frac{A_i \Phi}{n}, \qquad (4.30)$$

where n is an index of $i+1$-th layer.

Then, for the scenario with the formation of only perpendicular elements, it is advisable to determine the lengths of the coordinate segments proportional to the amplitudes. Normalizing them by A_1 and summing up, we get the first 10 numbers

$$\sum_{i=1}^{N} A_i \approx 1 + 0.618 + 0.382 + 0.236 + 0.146 + 0.09 + 0.056 + 0.035 + 0.021 + 0.013$$

$$= 2.597$$

Since, according to Corollary 2 of Statement 4.3, there is a connection between the ratios Φ and Fibonacci numbers through the number of elements n, and from (4.6) it follows that the sum of any two terms of the series is greater than all subsequent ones. The upper bound for N=∞ should be estimated like 2.597+0. 034, i.e., dimension ≤2.631. As we see, the dimension *Sp* can be considered irrational (an analogy with fractal objects).

5. Since at *Sp* stratification annihilation of the halves of the waves of the subjective and objective parts have a place, it is possible to interpret their dimension as having different signs.

Choice of coordinate system. As noted above, any set of solitons can be used as Cs, which determines the measured space. This space is less than the dimension of *Sp* and is different from the dimension of the measured elements. This leads to unrepresentative measurements. To avoid this, we present the following requirements to Cs:

- let the measurements be carried out for the elements of the *k*-th layer. Then $Cs=\{\bullet\phi\}$, where • is the soliton of this layer (Figure 6.1);
- let the measurements be carried out for elements of S-sets of one St. Then $Cs=\{\bullet_n\phi\}$, where *n* is the number of stages that exist, of the given St (Figure 6.2);
- let the measurements be carried out for elements of all or part of S-sets {St}. Then $Cs=\{\bullet_{in}\phi\}$, where *i* are the numbers of all St_i *{St}* (Figure 6.3).

Such *Cs* will be *complete* since they contain the solitons of all layers (corollary 1 of property 4.1); as they separate, they are replenished with new solitons, therefore, the number of their elements and dimension will increase. Other Cs will be called *incomplete*.

Implementation of the measurement. According to axioms 4.1, and 4.2, the measurement procedure is diffraction. In this case, all special cases listed in Table 2 are realized. The possibility of observation means that both the solitons and the resulting waves can act as S-sets in further stratifications.

The result. Let us find out how $\{x(p_i)\}$ is determined. Certain characteristics of S-elements may be obtained: state, l, π; A, F, and o waves, such as those that appear in the interaction of only waves, are not defined here.

Figure 6. Choice of Cs

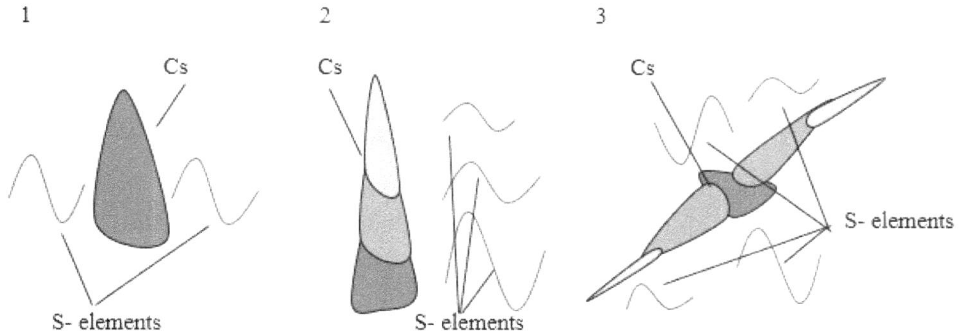

1. The perpendicularity of the measured elements and solitons Cs is established by the absence of the result, which is determined by the absence of changes of Cs and elements after the measurement.
2. The parallelism of the measured soliton and the soliton Cs manifests itself as a change in the velocities of the measured element and the soliton Cs. The soliton potential is determined by the formula following from (4.11):

$$\pi_j = \pi_i \frac{v_i - v_i^*}{v_j - v_j^*}, \tag{4.31}$$

where * means the changed velocities of the i-th soliton Cs and the j-th measured soliton.

3. The parallelism of the wave and the soliton Cs consists of a change in the state or other characteristics of the wave. Special cases are summarized in Table 2.

Table 2. Measurements of the characteristics of waves parallel to solitons Cs

Characteristic	Result	Evaluation
state modality	wave conservation wave disappearance	ø ö
wavelength	reflection from the soliton Cs wave running around a soliton with the formation of a diffraction pattern passage of a soliton through a wave	l∪<<l• l∪≈l• l∪>>l•
wave potential	reflection from the soliton Cs wave running around a soliton with the formation of a diffraction pattern soliton passing through the wave	π∪<<π• π∪≈π• π∪>>π•

4. Coincidence of measured solitons and solitons Cs. In this case, the result is characterized by a change in v_i for all components of the measured soliton. The ratio of potentials, states, and lengths of the components of the measured soliton and solitons Cs is determined from (4.22). Coincidence takes place at incomplete Cs.
5. Coincidence of measured waves and solitons Cs. The result is characterized by the absence of changes in the *{x(pi)}* measured waves. The ratio of potentials, states, and lengths of the components is determined from (4.22). Here coincidence takes place for both complete and incomplete Cs.

Comparisons. Let us define analogies and discrepancies between measurements in Euclidean and S-spaces. Since the measurements are reduced to implementing certain operations groups, let's compare the set ***{DIKS}*** with the group. In general, ***{DIKS}*** does not form a group, since it consists of non-binary operations, the result of which, by changing the modality of the state of elements, is influenced by external factors. Let us establish similarities and differences in terms of associativity, the existence of units, and inverse operations.

Associativity. The set of operations ***{DIKS}*** is not associative. For example, due to changing states, superpositions of waves will give different results, depending on their sequence. Associative in all cases will be only the diffraction of pairs of solitons.

Single operation. As a single operation, one can consider an operation that does not change the state of the elements. With this understanding, a single operation takes place for diffusing pairs of solitons.

The inverse operation, which does not exist for ***{DIKS}*** in integers (according to axioms 3.5. 3.8.), takes place for the diffraction of pairs of solitons (axiom 3.9.). Therefore, we have

Property 4.15.*For****{DIKS}****there is a set of operations that form a group - a set of binary soliton diffractions.*

Without forming a group, {**DIKS**} has a certain closure:

- operations O_i ***{DIKS}*** over • result in only •;
- operations O_i ***{DIKS}*** are commutative.

Denoting "+" the presence, and through "-" the absence of a sign, we summarize the results of comparing the characteristics of operations and measurements in Table. 3.

Table 3. Comparison of a group of operations and **{DIKS}**

Feature	Operation	{DIKS}	Diffraction of solitons
binary	+	-	+
closedness	+	+	+
commutativity	+	+	+
associativity	+	-	+
Existence: single operation inverse operation	+ +	- -	+ +

Now let's compare the characteristics of coordinate systems and measurements. The results are presented in the form of a Table. 4.

Table 4. Comparison of coordinate systems and measurements

Subject of comparison	Euclidean space	S-space
coordinate system	a fixed set of elements of a certain type; r= dimensions of space; no changes during measurements; equal weight of coordinate axes	An evolving set of solitons; r≤ dimensions of space; variability in measurements; unequal axis weights
measurement	same procedure; uniformity of repeated results	two types of procedure; inconsistency in repeated measurements
fulfillment of the metric axioms of symmetry, triangle, identity	performed	performed only in special cases

Sp operations and the measurement procedure described above can characterize the processes occurring inside systems more accurately than classical operations and measurements. However, the practicality of using such a device should be assessed for each specific case: the device's effectiveness is determined by the degree of manifestation of specific "systemic" properties (Table 8, Chapter 1). Let us choose to what extent it was possible to embody the qualities of systems listed in this chapter in the properties of SWM.

1. According to the axioms of the 1st group, *Sp* is a space that:

- Is formed due to the interaction of S and O in U (*openness*).
- Endowed with a structure formed gradually as a result of the influences of S and O (*self-organization*).

  According to the axioms of the 2-nd and 3-rd groups, S-sets and S-elements are characterized by:
- States of excitation, extinction, and disappearance change in the course of interactions; in turn, the result of interactions depends on the state of elements or sets (*variability of elements and processes*).
- Coordinate states and interactions for S-space as a whole *(integrity, self-organization*).

Statements 4.1-4.3 establish that only those bundles and convolutions are admissible that preserve the harmonic relations of the potentials of the layers and certain symmetry groups (*integrity*); It has been shown above that the bundle–convolution scenarios are due only to external factors and is invariant concerning specific implementations (*isomorphism*).

Property 4.4 and Statement 4.5 substantiate the heterogeneity of potential-related characteristics for different *Sp* layers (*nonadditivity*), and Statements 4.6–4.9 substantiate the existence of layering–convolution scenarios that lead to the appearance of perpendicular S-elements (*relative homogeneity of different layers*). axiom 1.2 postulates the existence of two types of S-elements (*heterogeneity of elements*). Properties 4.6–4.8 point to the stratification of time and limit the lifetime of *Sp* (*equifinality*).

2. Interactions within systems are corpuscular or wave and are characterized by energy, information, and other parameters. Because the:
   - Elements are abstractions of corpuscles (solitons) or waves, their amplitude, frequency, phase, and other characteristics, by definition, correspond to their counterparts for real waves and particles.
   - Change of potentials when splitting *Sp* and operations into S-sets and S-elements (interaction axioms, property 4.4) allows energy, information, and the like interpretation.
   - Special cases ***{DIKS}*** correspond to linear, nonlinear, binary, non-binary, additive, and non-additive interactions, that is, the wave model allows expression of the features of real interactions.
3. With the help of measurements, invariant and variable characteristics are determined. The rules established by the axioms of the 4th group allow you to coordinate systems and measurement procedures and present data for situations where one or all of the metric axioms are not feasible. Thus, the wave model has all the properties necessary for modeling complex systems.

Classification of object models by invariants. Practice shows that different tasks require models that have different sets of invariants. Let us put this provision as the basis for the classification of models. Let's distinguish:

Models 1 (M1) - static models that save:

- Membership and sequence of formation of elements, that is, layering-convolution scenarios.
- Number of elements and operations.
- Dimension.
- Groups and orders of symmetry.
- Distribution of relative values π, t_0, etc.

Let`s determine the form of such models. An S-graph is a connected undirected graph whose vertices correspond to potential and actual S-sets and S-elements, and the edges express their membership relations; PC-diagram (Figure 7) is a directed sequence on an S-graph that corresponds to actual S-sets and S-elements, as well as their membership relation. A convolution corresponds to a diagram of the opposite orientation. It is also easy to add modalities, layers, external influences, • designations here.

Figure 7. PC-diagram

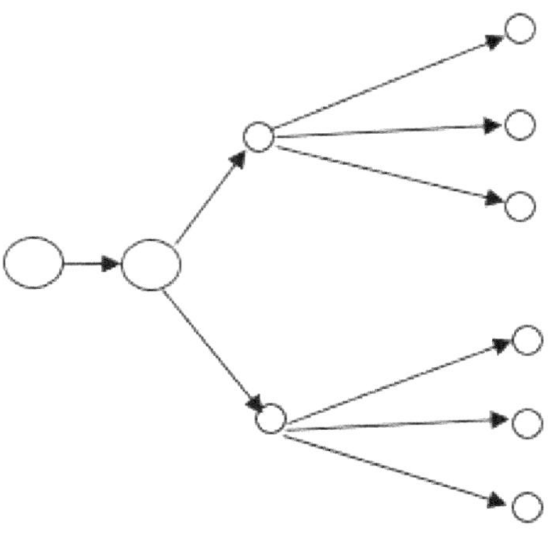

Models 2 (M2) are dynamic models that will retain, in addition to the invariants of M1:

- uniqueness of correspondences of sets of characteristics.
- the dynamics of changes in the structures of *Sp*.

This class includes a variety of analog models that allow you to recreate the dynamics of wave processes by representing them as waves with frequencies, amplitudes, etc., related to the original ones on a certain scale.

Models 3 (M3) - calibration models that partially or entirely retain the absolute values of the system parameters.

These include models that display the values of characteristics in their inherent units, that is, full-scale models. M3, unlike the previous ones, contains qualitatively different components; they are closest to technological systems.

FUTURE RESEARCH DIRECTIONS

Directions for further research follow from the above remarks. This is the application of self-organization scenarios for modeling the evolution of living organisms (Chapters 5 and 6), verification of the above assumption (Chapter 5), the structure and life of a person (Chapters 7 and 8), and the human life cycle (Chapters 9 and 10). In Chapter 9, further clarification will be made regarding this definition.

CONCLUSION

Discussion of the results of this chapter is possible in the following directions.

First, it is the value of the self-organization theory for this work. Various self-organization scenarios develop the wave model's axiomatics (Chapter 2) and will be used to further modelling the stages and processes of the human life cycle.

Secondly, according to intuitionistic mathematics, the means of verifying the proposed self-organization scenarios is their correspondence to various simulated systems and processes. Previously, models of human-machine systems, operator activities, the evolution of human consciousness, and complex organizational and technological systems were considered (Kovalyov, 1996; 1997; Mkhitaryan et al., 2004; Shmelova et al., 2018, Chapter 5; Kovalyov et al., 2020). The following chapters will deal with biological processes.

Therefore, it is advisable to return to the preliminary definition of life given in Chapter 2 and refine it, now within the framework of the theory of self-organization.

1. A living system is an invariant - S-space - calibrated according to environmental conditions, substances and fields used, and the potential magnitude.
2. The living system includes soliton and wave components. This follows not only from the theoretical conditions for implementing any stratification scenario *Sp*, but also from the fundamental foundations of physical reality - the existence of matter and field and wave-particle duality.
3. The boundary condition *Sp* applies to a living system's substance and wave components. For substance, this is understood as the existence on the border of two or more environments (solid, liquid, gaseous, plasm). The wave component is understood as the front of a propagating wave, the collision of waves, resonance, etc.
4. Observation leading to the emergence of *Sp*, for living systems understood as an external impulse necessary for their occurrence. The observation leading to the stratification of Sp is understood as external interactions in a dissipative environment, interactions, first initiating stratification, and second compensating for the decreasing potential (providing various resources for the system's existence).
5. The appearance of intermediate forms, the first ancestors of living organisms, and the division of lines of living organisms is described by the PC-diagram

So, the second version of the definition of life is formulated as follows:

A living system is:

1. (Invariant part): an existing boundary Sp, with potential π and modality A, arousing under the influence of external impulses in a dissipative environment, corresponding to the abstractions and axioms of the S- space model, including soliton and wave components (which modalities are æ, ö or ↓), developing according to Sp self-organization scenarios, and
2. (Calibration part): calibrated by conditions specific to possible bases for the presence of substances and fields necessary for life, as well as conditions for the feasibility of reactions necessary for life, energy sources, and self-organization mechanisms.

In addition to specifying the correspondence to the S-space paradigm, the condition and scenarios of self-organization are added to the definition. They are invariants, while specific mechanisms of evolution are calibrations.

Thus, understanding the driving forces of the evolutionary process that differs from the classical one is possible.

At the prebiological stage, one can point to autocatalysts as a means of self-supporting chemical reactions (Chapter 1), which can be considered a kind of predecessor of the genetic apparatus. In addition, the general course of the early stages of the evolutionary process, according to the data given in Chapters 1 and 2, corresponds to the first stages of the scenario (1S, 1O) and can be shown on the PC-diagram (Figure 8).

Figure 8. PC-diagram (for substance component)[3]

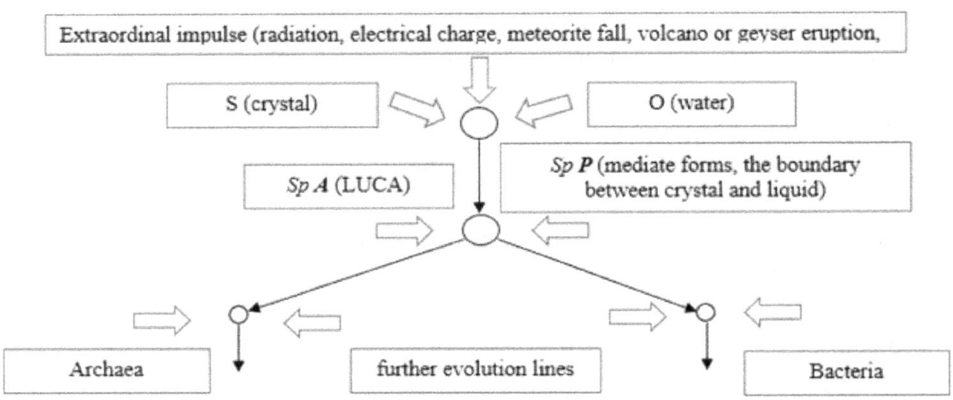

At the biological stage, the calibrations of self-organization are natural selection and the genetic apparatus of inheritance. If so, then the general course of evolution, its characteristics, and forms of inheritance can be represented as relationships between waves and solitons of S-space.

This assumption requires verification, which will be done in Chapters 5 and 6.

REFERENCES

Bar-Yam, Y., & Minai, A. (Eds.). (2002). Unifying Themes in Complex Systems II. In *Proceedings of the 2nd International Conference on Complex Systems Cambridge*. Perseus Pres.

Brouwer, L. E. J. (1907). *Over de grondslagen der wiskunde*. Amst.-Lpz.

Brouwer, L. E. J. (1908). *De onbetrouwbaarheid der logische principes* (Vol. 2). Tijdsehz voor Wijsbegeerte.

Gödel, K. (1995). Some basic theorems on the foundations of mathematics and their implications. In *Feferman, ed., 1995. Kurt Gödel Collected Works*, v. III, 304-323. Oxford University Press. (Original work published 1951)

Heyting, A. (1930). *Die formalen Regeln der intuitionistischen Logik*. 3 parts, In *Sitzungsberichte der preußischen Akademie der Wissenschaften. phys.-math. Klasse*. 42-65, 57-71, 158–169.

Heyting, A. (1956). *Intuitionism. An introduction*. North-Holland Publishing Co.

Hilbert, D. (1950). *Grundlagen der Geometrie* [The Foundations of Geometry] (2nd ed.). (Townsend, E. J., Trans.). Open Court Publishing. (Original work published 1902)

Hilbert, D., & Bernays, P. (1934). *Grundlagen der mathematic* [The Foundations of Mathematics]. Julius Springer.

Jan, Y., & Jan, L. Y. (1999). Asymmetry across species. *Nature Cell Biology*, 1(2), E42–E44. DOI: 10.1038/10036 PMID: 10559895

Kleaney, S. C., & Wesley, R. E. (1965). *Foundations of intuitionistic mathematics*. North-Holland Pub.

Kleene, S. C. (2002). *Mathematical Logic*. Dover Publications. (Original work published 1961)

Kleene, S. C. (2009). *Introduction to metamathematics*. Ishi Press. (Original work published 1950)

Kolmogoroff, A. (1932). Zur Deutung der intuitionistischen Logik. *Mathematische Zeitschrift*, 35(1), 58–65. DOI: 10.1007/BF01186549

Kovalyov, Y., Mkhitaryan, N., & Nitsyn, A. (2020). *Self-organization of the human mind and the transition from paleolithic to behavioral modernity*. IGI Global International Publisher. DOI: 10.4018/978-1-7998-1706-2

Lüttge, U., & Souza, G. M. (2019). The Golden Section and Beauty in Nature: The Perfection of Symmetry and the Charm of Asymmetry. *Progress in Biophysics and Molecular Biology*, 146, 98–103. DOI: 10.1016/j.pbiomolbio.2018.12.008 PMID: 30557534

Mesarovic, M. D., & Takahara, Y. (1975). *General Systems Theory: Mathematical Foundations (Mathematics in Science and Engineering)*. Elsevier.

Mesarović, M. D., Takahara, Y., & Macko, D. (1972). *Theory of Hierarchical, Multilevel, Systems*. Academic Press.

Shmelova, T., Sikirda, Y., Rizun, N., Salem, A.-B. M., & Kovalyov, Y. (2018). *Socio-Technical Decision Support in Air Navigation Systems: Emerging Research and Opportunities*. IGI Global International Publisher of Progressive Information Science and Technology Research. DOI: 10.4018/978-1-5225-3108-1

Tarski, A. (1969). Truth and Proof. *Scientific American,* 220, 63–77. (Original work published 1936)

Thurner, S., Hanel, R., & Klimek, P. (2018). *Introduction to the Theory of Complex Systems*. Oxford University Press. DOI: 10.1093/oso/9780198821939.001.0001

Weyl, G. (1934). *Phylosofyja matematyky: sbornyk nauchnykh trudov* [Philosophy of mathematics: a collection of scientific papers]. State Technical and Theoretical Publishing House.

ADDITIONAL READING

Aleksandrov, A. D. (1987). *Osnovanyja gheometryy* [Foundations of Geometry]. Nauka.

Artyukhov, V. V. (2009). *Obshhaja teoryja system: samoorghanyzacyja, stabyljnostj, raznoobrazye, kryzysy* [General Systems Theory: Self-Organization, Stability, Diversity, Crises]. Librokom.

Bodnar, O. Ya. (2005). *Zolotyj pereriz i neevklidova gheometrija v nauci ta mystectvi*. [Golden section and non-Euclidean geometry in science and art]. Ukrainian technologies.

Kovalyov, Y. N. (1996). *Gheometrycheskoe modelyrovanye erghatycheskykh system: razrabotka apparata* [Geometry Modeling of Man-Machine Systems: The Apparatus Creation]. KMUGA.

Kovalyov, Y. N., & Vereschaga, V. M. (2012). *Prykladna gheometrija: narysna gheometrija, inzhenerna ta komp'juterna ghrafika, suchasni tendenciji* [Applied Geometry: Descriptive Geometry, Engineering and Computer Graphics, Modern Trends]. Omega-L.

Mkhitaryan, N. M., Badeyan, G. V., & Kovalyov, Y. N. (2004). *Erghonomycheskye aspekti slozhnikh system* [Ergonomic Aspects of Complex Systems]. Naukova Dumka.

Soroko, E. M. (1984). *Strukturnaja gharmonyja system* [Structural harmony of systems]. Nauka.

Urmantsev, Y. A. (1988). *Evoljucyonyzm yly obshhaja teoryja razvytyja system pryrodы, obshhestva y myshlenyja* [Evolutionism or the general theory of the development of systems of nature, society, and thinking]. Institute of Plant Physiology of the Academy of Sciences of the USSR.

Vasyutinskiy, N. A. (1990). *Zolotaja proporcyja* [Golden proportion]. Young Guard.

Weil, G. (2003). *Summetriya* [Symmetry]. URSS Publishing House.

ENDNOTES

1 The characteristic 1.618… is also used, equal to $\Phi+1$, or Φ^{-1}; between Φ and the number π, there is such a relationship: $(\pi^2/4)^2 \approx 0.617 \approx \Phi$ (Mathematical Encyclopedia, 1977-85).

2 The exact formulation is: since elementary particles have both corpuscular and wave properties and a pair of conjugate quantities (coordinate and momentum, energy and time, etc.) characterizing a microscopic system, these quantities cannot be measured simultaneously with arbitrarily high accuracy. The Heisenberg relation relates the measurement uncertainties: their product is greater than or equal to the quotient of Planck's constant divided by the particle's mass.

3 Or even vice versa: S - water and O - crystal.

Chapter 5
Evolution Phenomena and Theories

ABSTRACT

The phenomena of the evolutionary process are considered, and 15 patterns are identified. Modern theories of evolution are presented, considering the evolution process at the population, organism, and molecular-genetic levels, and within the framework of ecosystems. It is noted that the theories contradict each other and not all patterns are sufficiently explained. This is especially true for the problem of the rate of evolutionary changes and the hypothesis about the direction of evolution. The theory of the self-organization of the complex system allows us to propose possible solutions. The materials presented in the chapter are the necessary data for modeling evolutionary processes, including reproduction and inheritance, based on the theory of self-organization (Chapter 6).

BACKGROUND

The first task is to present the basic facts relating to the evolutionary process. The second task is to consider theories that explain these facts. The third task is to show difficulties and problems that have not yet been solved. All this together will form the basis for evolution process modeling, using the theory of self-organization (Chapter 6).

The scientific approach to the problem of the origin of species is associated with the names of J. B. Lamarck and J. L. Cuvier, who proposed various ways to solve it. However, the father of the modern understanding of this problem is C. Darwin (Darwin, 1959; 1971), who described many empirical manifestations of the evolutionary process and put forward the ideas of the struggle for existence and natural selection to explain them.

DOI: 10.4018/978-1-6684-8509-5.ch005

Somewhat later, the theory of orthogenesis arose (G. H. T. Eimer, J. W. Haacke), which tried to explain the evolution regularities based on the idea of the initial direction of evolution, independent of natural selection (Bowler, 1979; Grehan & Ainsworth, 1985; Ulett, 2014). The concept of orthogenesis found supporters in the Soviet Union, and several theories were substantiated (Berg, 1922; Lyubishchev & Meyen, 2015).

Naturally, genetics and, in particular, the discovery of the role of DNA and RNA in the process of heredity brought the discussions of the supporters of "classical" evolution and orthogenesis to a new level: on the one hand, we have a synthetic theory of evolution, and on the other, neo-orthogenetic theories.

The creators of the synthetic theory of evolution are J. Huxley, F. Mayr, B. Rensch, J. Stebbins, J. B. S. Haldane, D. Lack, C. Waddington, G. de Beer, E. Baur, V. Zimmerman, W. Ludwig, G. Heberer, I. I. Schmalhausen, G. F. Gause, N. P. Dubinin, A. L. Takhtadzhyan (Huxley, 1942; Fisher, 1958; Rose et al., 2007; Pigliucci & Finkelman 2014; Noble 2011; Laland, et al. 2015; Huneman & Walsh, 2017; Shmalgauzen, (2017/1939). Neoorthogenetic theory is presented in (Levit & Olsson, 2006; Gould, 2002).

The molecular genetic mechanisms of the evolutionary process are known (Bernstein, et al., 1987; Brenner's Encyclopedia of Genetics by S. Maloy, 2013); various models based on the methods of mathematical statistics have been constructed for them (Haldane, 1957; Fisher, 1958; Kimura, 1983; Greenberg & Wu, 2006; Allan & Max, 2010; Liberles, et al., 2020); computer modeling methods are actively used (Tamura et al., 2021); a review is given in (Morozov, 2013). These methods are actively used to research human evolution (Cavalli-Sforza & Feldman, 2003) and are of great practical importance, particularly in medicine (Coleman & Tsongalis, 2018). Several scientific journals have been published, including Molecular Biology & Evolution, Human Molecular Genetics, etc. However, the degree of influence of genetic changes on phenotypic traits has not received an unambiguous assessment.

Another aspect of the evolutionary process is the relationship of living organisms with the environment and with each other within ecological systems (Rosenzweig, 1995; Pianka, 2000; Fox et al., 2001; Mayhew, 2006; Eskov, 2009). This topic is being actively researched; there are several regularly published scientific journals (Ecology and Evolution, Nature Ecology & Evolution, Frontiers in Ecology and Evolution, etc.)

Finally, an important part of the evolutionary process is the evolution of consciousness and behavior in the course of the historical development of mankind. The possibility of its interpretation as self-organization is substantiated by (Kovalyov et al., 2020); W. D. Hamilton explored the relationship between genetic evolution and social behavior. (Hamilton, 1964).

EVOLUTION PHENOMENA

First, we consider the external manifestations of evolution (they can be defined as empirical), and then their theoretical explanations within the framework of existing scientific ideas. For the convenience of analysis, let us number the regularities of micro- and macroevolution.

1. ***The life cycle includes*** several evolutionary events and mechanisms:
   - Inheritance.
   - Heterochrony - differences in the time or duration of the embryo development process in the body compared to its ancestors or other organisms, leading to a change in size, shape, characteristics, and even the presence of certain organs and features.
   - Birth - a living organism enters into a struggle for existence and various resources and is included in ecosystems, food chains, etc.
   - Neoteny - the achievement of sexual maturity and the end of ontogenesis in the early stages of development, which leads to the loss of narrow specialization, more characteristic of the final than the early stages. Neoteny leads to the "rejuvenation" of the species, acquiring the ability for rapid evolution.
   - Allometry - uneven growth of body parts or difference in proportions in organisms with different sizes.
   - Competition and selection.
   - Reproductive phase (mutations are passed on to offspring).
   - Aging and death - the organism ceases to exist, dropping out of competition for food and resources and itself becomes food.

The life cycle is inherent not only to individual beings, but also to populations, species, and other taxa.

2. ***Arogenesis*** - the path of evolution, characterized by an increase in an organization (intensification of the vital activity of the organism; differentiation of parts while maintaining the integrity of the organism; development of active methods of struggle for existence; improvement of the nervous system and sensory organs). The resulting morphological changes are called aromorphoses.

In plants, arogenetic are the transition from haploidy to diploidy, independence from water in the process of fertilization, the transition from external to internal fertilization and the occurrence of double fertilization, the dismemberment of the body into organs, the development of the conducting system, the complication and improvement of tissues, as well as the specialization of pollination with the help of insects and dispersal of seeds and fruits.

The main aromorphoses in the animal's evolution are associated with the emergence of multicellularity; increasing differentiation of all organ systems; the emergence of a strong skeleton; the development of the central nervous system; and social behavior in highly organized animals. And, for example, in mammals it can be noted:

- Division of the heart into right and left halves with differentiation of two circles of blood circulation and an increase in the lungs' working capacity. Effect: improvement of blood oxygen enrichment and organ supply with it, intensifying organ functions.
- Differentiation and specialization of the digestive organs. Effect: complete use of nutrients, increased metabolic processes, increased activity, warm-bloodedness, increased activity of motor organs, and improved design.
- Complication of the structure of the brain and exacerbation of the senses. Effect: development of reactions of behavior, reactions of "reasonable type" to a quick change of scenery.

Aromorphoses are interconnected: thus, perfect organs of movement would be useless without a general increase in activity; the development of the brain and sense organs would become unnecessary without high activity and perfect organs of movement, etc. This complex of aromorphoses allows a mammal to live in various natural conditions.

3. ***Allogenesis*** (idioadaptation) - the development of particular adaptations of organisms in specific environmental conditions. Diverse allogenesis expresses the divergent evolution of a group during its flourishing period. Allogeneic evolution leads to a rapid increase in numbers, species composition diversity, adaptations' completeness, and a high life potential. However, excessive specialization (telogenesis) is an evolutionary dead end. For example, sloths that are not viable outside the forest and have almost lost the ability to move on the ground are doomed to extinction when the forests disappear.

Hypergenesis - an increase in body size or overdevelopment of individual organs. Initially, large forms arise within a group of organisms, which can be beneficial in the struggle for existence, for example, they increase the resistance to predators. Further, gigantism becomes a disadvantage: large organisms require more food, the reserves of which are not unlimited. So, in the course of evolution, all the main classes of terrestrial vertebrates, from amphibians to mammals, gave several giant forms. Then invariably there came a period when all these giants died out, and the dominance of relatively small forms was restored. Therefore, an increase in body size is a special case of telogenesis.

New species are formed not from highly specialized but relatively simple, non-specialized forms. Specialized forms are the least plastic and more prone to extinction under changing conditions. However, they also have a mechanism of evolution - neoteny.

4. ***Degeneration*** (regressive evolution) - simplification of the structure and way of life of the organism during temporary penetration into a simpler environment *(catagenesis)* or constant stay in a simplified environment *(hypogenesis)*.
5. ***Alternation of arogenesis, allogenesis, and degeneration.*** Within a specific group of organisms descending from a common ancestor (monophyletic), the period of arogenesis is always followed by a period of partial allogenesis (the law of A. N. Severtsov, Figure 1).

Arogenesis occurs much less frequently than allogenesis. Each major taxon (phylum, class, order, family) begins with arogenesis, which ensures a new environment and resource development. Then the individual settles in various habitats, and the development process of allogenesis begins, allowing them to survive, even in the vicinity of highly organized and aggressive forms, which determines the emergence of new genera and species. In the presence of particular devices that prevent extermination, aromorphic transformations are not necessary, and such groups retain their main features for a long time. This explains the simultaneous existence of various types of organization - from the simplest to chordates.

Figure 1. The ratio of aromorphosis (a), idioadaptation (b), and degeneration (r) (Severtsov, 1939)

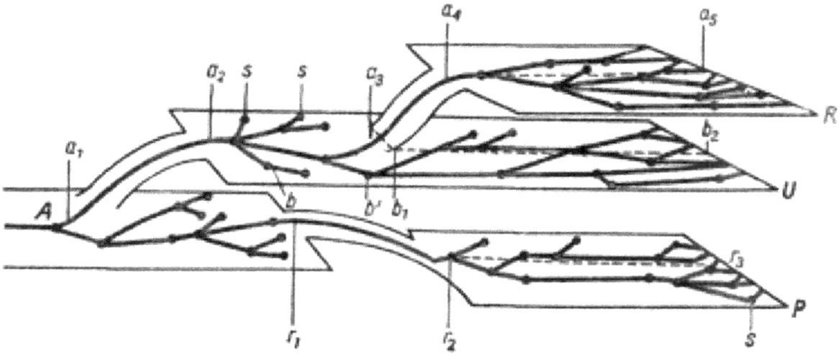

6. ***The biological progress*** of a species is determined by the excess of birth rates over mortality in its populations. If the number of individuals increases, the range expands, the rate of intraspecific differentiation increases, then the species progresses. For example, many groups of rodents are in a state of biological progress.
7. ***Biological regress*** is characterized by: a decrease in numbers due to the excess of mortality over fertility; a decrease in intraspecific diversity; narrowing of the integrity of the range, which breaks up into separate spots; susceptibility due to low numbers to extinctions due to various reasons, for example, epizootics. The muskrat genus, widespread in Europe in the Tertiary, is in a state of biological regression, while only two species have now survived (Paramonov, 1978).
8. ***Divergence*** - the emergence of new species and genera within the group through allogenesis after it emerges due to arogenesis during natural selection. Divergence as a form of species adaptation to environmental conditions is a condition for the existence and a form of development of the species. In the process of development, a species differentiates into descendant species and, as such, ceases to exist. So, divergence from intraspecific turns into intrageneric and further leads to the formation of new genera, families, orders, classes, and so on.
9. ***Convergence*** - the emergence of morphological similarities among representatives of various natural groups when they enter the same environment.
10. ***Parallelism*** - the development of similar adaptations in two species groups that developed divergently after penetrating the same environment.
11. ***The direction of evolution*** can be traced at different levels, for example:
    - The changes in a known direction will tend to change in that direction and beyond.
    - Genetically close species and genera are characterized by similar series of hereditary variability with such accuracy that, knowing the number of forms within one species, one can foresee the presence of parallel forms in other species and genera" (the law of homologous series of N. Vavilov).
12. ***Preservation of integrity*** also takes place at different levels with the involvement of various mechanisms of self-regulation:
    - For living organisms - as restrictions on the development of hypergenesis, heterochrony, allometry; the symmetries and cycles noted above; for example, for a person - consciousness of oneself as a whole, homeostasis, harmonic relations, Fibonacci numbers in the structure of the body.
    - For symbiotic communities, for example, bacteria in the gastrointestinal tract feed at the expense of the "master", but help his digestion, while their reproduction is limited.

- For ecological systems, regulation of numbers and ratios in food chains.
- For behavioral patterns, such as expulsion from packs of overly aggressive individuals in wolves.

13. ***Irreversibility of evolution***. The extinction of old forms makes evolution irreversible: *"If a species once disappears from the face of the earth, we have no reason to think that the same, identical form will ever reappear"* and: *"A group that once emerged never reappears"* (Darwin, 1859).
14. ***The speed of the evolutionary process is different***:
    - In stable environmental conditions (ocean basins, cave lakes) - as a rule, decreases;
    - With rapid changes - as a rule, increases.
    - Widespread species evolve gradually (gradualism is a system of ideas about the evolution of nature, which states that significant changes are achieved due to the slow but continuous processes of minor changes), and small isolates - intermittently and not always adaptively (panadaptationism - evolution always works in the direction of growth fitness).
    - In general, evolution accelerates as more perfect forms are formed. So, if about 2.5 billion years have passed from the appearance of the first living organisms to the appearance of multicellular creatures, and the poor fossil remains of cryptosis lasted about 1 billion years; then over the remaining, approximately 542 million years of the phanerozoic, the first organisms with mineral skeletons, arthropods and chordates, complex forms of plants, dinosaurs, mammals, primates and humans appeared and partially replaced each other at an ever-increasing pace (Figure 4 in Chapter 1).
15. ***Mass extinctions.*** The speed of adaptation has its limits: in the event of catastrophes or too rapid changes, it turns out to be insufficient, and mass extinctions of species occur (there were 5 in total, and the modern period is sometimes characterized as the beginning of the 6th extinction).

The largest of them, the Great Permian extinction, occurred for unclear reasons about 252 million years ago, lasted, according to various estimates, from several thousand to 60 thousand years, and led to the extinction of 96% of all marine species, 73% of terrestrial vertebrate species, about 57% of genera and 83% of insect species (Benton, 2005; Sole & Newman, 2002).

By themselves, mass extinctions do not lead to the emergence of new species "from scratch", as J.L. Cuvier, but contribute to their formation by creating new conditions for adaptation, decreasing or increasing the number of available resources (like the Great Oxygen Event, Chapter 2).

Notes on human evolution. Important acquisitions are upright walking, liberation and improvement of hands; binocular vision, which predominates over other sensory organs, changes in the structure of the larynx, and baldness. An even more important change is the complication of the structure of the brain, an increase in its volume and energy consumption (in an adult, the share of brain metabolism in total energy needs during intellectual work reaches 20-25% with a brain mass of 1.5-2% of body weight). The correlations between anatomical changes and manifestations of intellectual activity have a place. The paradox is that the positive effect manifests itself after many generations. So, upright walking and freeing of hands led to the beginning of the manufacture of tools of the Olduvai type no less than a million years later. Not less than 200 thousand years after the appearance of archanthropes, more advanced than the early Homo, the Olduvian tools improved so much that there were reasons to attribute them to another culture - the Acheulean, etc. The same applies to the symbolic activity: the first Homo sapiens sapiens did not show their superiority over the Neanderthals for at least several tens of thousands of years, and only then did their painting and stonework go far ahead. And just at this time - another paradox - the human brain began to simplify and decrease (from the time of the Upper Paleolithic by 400–450 cm^3).

THEORIES OF EVOLUTION

The observed diversity of species has attracted interest since ancient times. Different views on the origin of species were expressed by J.B. Lamarck (modern species are the result of inherited changes caused by adaptation to the environment) and J.L. Cuvier (species are unchanged; new ones replace old species after their extinction as a result of world cataclysms). The modern stage of the evolution understanding is associated with the name of C. Darwin (Darwin, 1859; 1871), whose works provide substantiated data on the regularities of this process.

Before discussing actual theories, let us show the factors that influence the evolutionary process and evolving structures (Figure 2).

Figure 2. Factors and structures of the evolutionary process. Arrows indicate forward and reverse connections; contour arrows - multiple, mostly direct influences

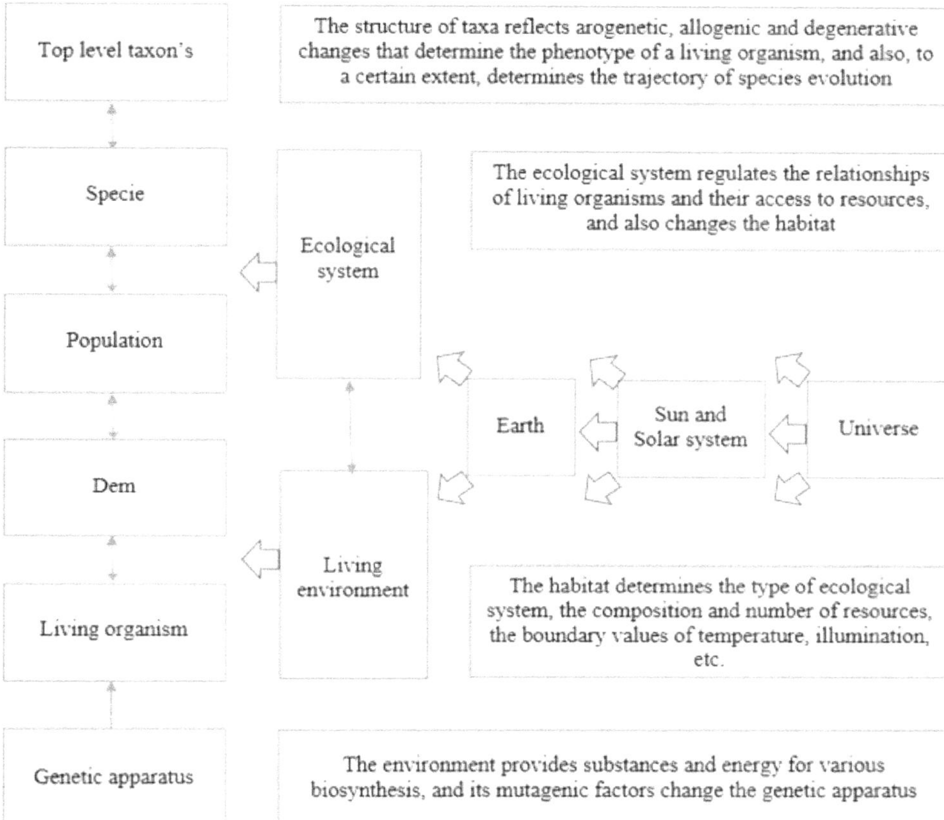

This will help better present the areas of application of the theories and their main ideas.

The Synthetic Theory of Evolution

The development of molecular genetic research and the definition of the role of populations in natural selection stimulated the creation of a general theory called the synthetic theory of evolution (STE), which supplements classical Darwinism with several important refinements. Its main provisions were formulated by J. Huxley (Huxley, 1942); a more formalized representation in the form of 11 provisions contained (Vorontsov, 1999); the development of the current state of thorium is described in (Mayr, 1999/1942; Simpson, 1953; Keilow, 1986; Krasilov, 1986;

Shamalguazyan, 1986; Grant 1985; Reimers, 1994; Modern synthesis (20th century). Wikipedia, n/d). The main theses of the theory are:

- A species is a system of populations reproductively isolated from populations of other species, and each species is ecologically isolated.
- Speciation consists of the emergence of genetic isolating mechanisms and is carried out mainly in conditions of geographic isolation; the achieved reproductive isolation indicates the completion of speciation. The insufficiency of the reproductive criterion of the species is now recognized, but there is no general definition of the species for forms with a sexual process and agamic forms.
- Gradualism and panadaptationism are not universal characteristics of the evolutionary process: speciation can be continuous and linear, discontinuous, continuous and divergent, sharp and convergent.
- The elementary unit of evolution is the local population (or part of the population - dem). Living organisms without a sexual process remain outside the scope of such a definition.
- Material for evolution is mutational and recombination variability.
- Potentially pre-adaptive mutations are widespread in natural populations, which play an important role in macroevolution, especially during abrupt environmental changes.
- Natural selection is considered the main reason for developing adaptations, speciation, and the origin of supraspecific taxa, but it is not the only engine of evolution.
- Species and supraspecific taxa, as a rule, differ in adaptive-neutral features. The main directions of the evolutionary process (arogenesis, allogenesis) are a compromise between adaptability and neutrality. Macroevolution can go both through microevolution and in its ways - it is not excluded that individual macroevolutionary events can also have a sudden character.
- Genetic drift (Wright, 1955) and the founder principle (Mayr, 1999/1942) are the causes of the formation of neutral traits.
- Selection often acts not on individual genes, but on complexes of genes; mutations are neither beneficial nor harmful in themselves, but their selective value depends on the environment.
- The mechanism of action of selection depends on the external and genotypic environment, and the vector of its action on the phenotypic manifestation of mutations.
- Along with monophyly, paraphilia is recognized as widespread.
- Evolution is not always divergent.

- Evolution is not directed to the final goal, that is, it is non-finalists, however, accurate predictions of its events are possible to some extent.
- The main result of evolution at present was the emergence of man, with the emergence of which biological evolution develops into psychosocial (Hamilton, 1964).

Consider patterns 1-15 within the STE.

Mutations, the struggle for existence, and natural selection will explain the life cycle events (1, certain difficulties are associated with heterochrony and allometry) considering environmental conditions. The processes of arogenesis, allogenesis, and degeneration (2-4) also require additional consideration of ecological factors and morphological changes associated with the appearance of symmetries, Fibonacci numbers, etc. - system laws. The alternation of arogenesis, allogenesis, and degeneration (5) can only be understood as the evolution of an open system. Biological progress and regression (6-7) seem more complex than simply the consequences of successful/unsuccessful adaptations to environmental conditions and the number of resources.

So, an important nuance is demonstrated by the experiment "Universe-25" by D. Calhoun - mice, once in ideal conditions (temperature, nutrition, safety), at first rapidly multiplied; when a threshold population density was reached, their social behavior degraded compared to the norm; then they stopped breeding and completely died out, including especially resettled individuals at the stage of destroyed behavior.

It can be noted that such behavior has been mentioned more than once in human history; a classic example is the loss of will and prowess in the late Roman Empire that preceded its fall[1].

Mutations are random, and their adaptive value is determined by natural selection, which sets evolution in a certain direction - this is how Darwinists define this phenomenon. The random nature of mutational variability does not contradict the possibility of the existence of a certain canalization of evolutionary paths that arises as a result of the history of the species. With the complete extinction of a species, the continuity of its development is disrupted; since the conditions of existence are unique, the extinct species is also exceptional, even if its descendants continue to exist.

Divergence (8) is explained quite clearly; as convergence and parallelism (9-10) - with the involvement of the environmental factor.

STE denies the direction of evolution (11) but says that knowing the trends in the evolution of a species in the past, it is possible, with some probability, to predict the nature of changes in the future.

Preserving integrity (12) has a systemic nature - the mechanism of natural selection is one of the means of implementing this general pattern.

The theory of natural selection fully explains the irreversibility of evolution (13); however, this does not rule out other possible interpretations.

The uneven rate of evolution (14) is explained by its coordination with the rate of change in the conditions for the species' existence.

The mass extinctions (15) are explained by the exhaustion of the evolutionary potential of endangered taxa and by the fact that environmental changes occur too quickly to adapt to them.

Discussion about the synthetic theory of evolution. We confine ourselves to scientific arguments (Krasilov, 1986; Reif et al., 2000; Daniel Frías., 2010) and will not consider creationist arguments.

Critics of the theory note the unresolved problem of qualitative certainty, integrity, discreteness, and the species' reality within the synthetic theory of evolution framework. The approach to explaining secondary similarities, that is, close morphological and functional characters that were not inherited, but arose independently in phylogenetically distant branches of the evolution of organisms (10), was criticized. Suppose the genotype and the nature of selection determine the traits of living beings. In that case, the parallelism is explained by the fact that organisms have inherited many identical genes from their recent ancestor, and the origin of convergent traits (9) is entirely attributed to the action of selection. It is known that similarities that develop in fairly distant lineages are often non-adaptive and cannot be plausibly explained either by natural selection or by common inheritance. The independent occurrence of identical genes and their combinations is excluded since mutations and recombination are random processes (Synthetic theory of evolution. Wikipedia, n/d).

According to modern ideas, mutations are random only concerning the environment, but not to the existing organization of the genome: different parts of DNA have different stability; accordingly, some mutations occur more frequently than others. The set of nucleotides is also minimal. Consequently, independent occurrence of the same mutations is possible, up to the synthesis of similar proteins by distant species that could not be obtained from a common ancestor. These factors cause secondary repetitiveness in DNA structure and may explain the origin of non-adaptive similarity. However, this position, is not supported by quantitative estimates of the evolutionary time required for the emergence of parallelism, and its comparison with the time of separate existence of the compared species.

Another point is related to the concept of punctualism ("punctuated equilibrium") based on paleontological data: the duration of stasis is several orders of magnitude longer than the duration of the transition from one phenotypic state to another, which is generally valid for the entire fossil history of multicellular animals (property 14). According to modern ideas, there is no qualitative difference, only a quantitative difference between "gradual" and "intermittent": punctualism and gradualism should be considered complementary concepts. Long-term stasis can be explained by the action of stabilizing selection in a stationary environment, and rapid change can be explained by the theory of S. Wright for small populations under non-stationary conditions of existence or the passage of a species through a "bottleneck" when a small population of it survives, having the necessary adaptations.

However, there is no explanation for the acceleration of evolution noted above, based on the analysis of paleontological data (Chapter 1, Figure 4). Thus, the synthetic theory of evolution faces several difficulties, and to explain some regularities from series 1-15, it is necessary to involve environmental factors.

Let's add a few words. The explanations in paragraphs 11, 12, 14, and 15 are not convincing. Futher, there is a clearly defined direction of morphological changes - the appearance of symmetries of different orders, harmonic ratios, and Fibonacci numbers with the emergence of new taxa of higher orders; at the same time, asymmetry also takes place (Urmantsev, 2017; Ovchinnikov, 1966; Jan & Jan1999; Penrose, 2005; Korobko, 1997; Lüttge & Souza, 2019). As the evolutionary process develops, these manifestations tend more and more to the scenario (1S, 1O), Chapter 3), which, in mathematical terms, is an attractor for evolutionary tracks. While bacteria show very little simmetry, starfish and mollusks such as octopuses show a marked symmetry; reptiles, in addition to different symmetries, demonstrate the similarity of the general plan of the body with a graph expressing the sequence of stratification of S-space; in mammals, there is already a match with such a graph - in various species, only different calibrations are observed; the maximum balance is achieved in humans. Such an observation allows us to suggest that the evolutionary factor in the appearance of new taxa is compliance with the laws of self-organization: a living organism that is more similar to a self-organized system and lives in a world consisting of such systems has a greater chance of survival than an organism that is less self-organized. Living organisms can form complex ecosystems, including both highly organized and more primitive organisms. There, it is already necessary to consider the survival and evolution of not individual species, but systems as a whole. To give credibility to the explanation of the uneven rate of evolution (as well as other characteristics of evolution), quantitative estimates and their correlation with the time of existence of living organisms on Earth are very desirable. The same was noted in Chapters 1 and 2.

Finally, mass extinctions are explained by insufficient evolutionary potential and insufficient speed of adaptation in case of abrupt changes in the environment. But what is evolutionary potential? Let's ask simple questions. Why, for example, as a result of the fall of a large meteorite and a subsequent sharp cooling, dinosaurs died out, but the ancestors of modern lizards and snakes did not die out? Or why did the ichthyosaurs and plesiosaurs living in the seas die out (and large masses of water dampen abrupt changes in conditions), but the ancestors of crocodiles and frogs survived? And those and other species were quite ancient, and their potential should be close to exhaustion?

Let us consider other theories, perhaps better explaining the noted difficulties, based on the processes occurring at other levels compared to STE (Figure 2).

The Orthogenetic Theories

The understanding (and acceptance) of characteristic 11 is the main point of divergence between STE and orthogenetic theories (Levit & Olsson, 2006).

Based on the recognition of the direction of evolution, the theory of orthogenesis arose (G. H. T. Eimer, J. W. Haacke), which tried to give these and other regularities a theoretical justification based on the idea of the initial direction of evolution, independent of natural selection (Bowler, 1979; Grehan & Ainsworth, 1985; Ulett, 2014; Berg, 1922; Lyubishchev & Meyen, 2015), and several theories were substantiated; neoorthogenetic concepts are presented in (Levit a& Olsson, 2006; Gould, 2002).

Recognition or denial of the direction of evolution is not the only discrepancy between STE and orthogenetic theories. For example, for nomogenesis - one of the orthogenetic theories - such a list of differences was compiled by L. S. Berg (Table 1).

Table 1. Differences between Darwinism and nomogenesis

Darwinism	Nomogenesis
1. All organisms have evolved from one or a few primary forms, i.e., monophyletically or oligophyletically.	1. Organisms have evolved from thousands of primary forms, i.e., polyphyletically.
2. Further development was divergent.	2. Further development was predominantly convergent (partially divergent).
3. based on random variations to which individual individuals are subjected, using slow, hardly perceptible continuous changes	3. based on patterns that capture huge masses of individuals, in leaps, outbreaks, and mutations over a vast territory.
4. There are a lot of hereditary variations, and they go in all directions	4. There are a limited number of hereditary variations, and they go in certain directions.
5. The factor of progress is the struggle for existence and natural selection.	5. The struggle for existence and natural selection are not factors of progress; moreover, being conservative figures, they protect the norm.
6. Species, by their origin, are connected by transitions with each other using divergence.	6. Species, by their mutational origin, are sharply demarcated from one another.
7. The process of evolution consists entirely of forming new features.	7. Evolution is to a great extent the unfolding of pre-existing inclinations.
8. The extinction of organisms comes from external causes: from the struggle for existence and the survival of the fittest.	8. Extinction is a consequence of internal (autonomic) causes and external (choronomic) ones.

(Berg, 1977)

Molecular Genetics and The Neutral Theory of Molecular Evolution

They are necessary to explain some subtle evolutionary processes and to substantiate quantitative estimates. Without aiming to give a detailed overview of the methods and results (we refer those interested to (Taper & Case, 1985; Brown & Pavlovic, 1992; Allan & Max, 2010; Greenberg & Wu, 2006; Fisher, 2011; Morozov, 2013Liberles et al., 2020), we will give only some illustrations of the achievements and problems. Thus, it is believed that specific genetic mechanisms (hybridization, polyploidy, chromosomal aberrations) underlie discontinuous speciation; the synthesis of genetics with the concept of recapitulation (the Haeckel-Muller law) and the interpretation of the relationship between ontogenesis and phylogenesis (Takhtadzhyan, 1943) underlies the epigenetic processes of evolution, and so on. In some cases, it is possible to give quantitative estimates.

So, in 1931, S. Wright proposed the concept of ***random gene drift*** (the Wright effect, Sewall Wright effect - random changes in the frequencies of alleles and genotypes that occur in a small polymorphic population during generational change). These changes lead first to fluctuations in the allele frequency (different forms of the same gene located in the same regions of homologous chromosomes, which determine the direction of development of a particular trait from generation to generation), and then to the fixation or elimination of this allele. Thus, the deme gene pool (a small, up to several dozen, an intraspecific grouping of individuals living in a limited area and interbreeding), considered a small sample from the gene pool of the entire population, is formed by chance. Genetic drift has proven to be useful in explaining the origin of maladaptive differences between taxa. Here, the question arises: When will selection prevails in the evolutionary process, and when will drift prevail?

This problem belongs to mathematical statistics, and R. Fisher found its solution (Fisher, 2011); his theorem, which states that the intensity of selection, and the rate of evolution through selection, is proportional to the value of genetic dispersion in the adaptability of the evolving population. So, if Ne is the effective population size (only individuals that can produce viable offspring); F_a and F_A - the value of the adaptability of each of the two alleles of the gene; s is a selection coefficient reflecting the effect of a mutation on fitness: $s>0$ is a beneficial mutation, $s<0$ is a harmful mutation, then with the ratio

$1/Ne >> \backslash s \backslash$ – there is an almost neutral evolution controlled by gene drift; $1/Ne << \backslash s \backslash$ – selection-driven evolution takes place; $1/Ne \approx \backslash s \backslash$ - has a combined evolution with a comparable contribution of drift and selection.

Thus, if differences in mutation efficiency are small and the population is small, the evolution towards lower fitness is possible. Selection will be the determining factor for a sufficiently large population, and in the course of evolution, guided by natural selection, the population's average fitness cannot decrease.

In 1957 J.B.S. Haldane showed that more than 12 genes in a population cannot be replaced simultaneously by more adapted alleles without its reproductive falling to zero (Haldane, 1957).

In the 1960s, M. Kimura, studying the rates of amino acid substitutions in proteins, drew attention to the discrepancy between the data he obtained and earlier by Haldane: the rate of substitutions per genome per generation for mammals exceeded Haldane's estimate by several hundred times.

As a result, he formulated the ***neutral theory of molecular evolution*** (Kimura, 1983; 1991). The theory is based on the statement that not all changes in the DNA nucleotide series lead to a change in the amino acid sequence in the corresponding protein molecule. Those amino acid substitutions that have taken place do not necessarily cause a change in the shape of the protein molecule. When such a change does occur, it does not necessarily change the nature of the protein's activity. Therefore, many mutant genes perform the same functions as normal, so selection behaves neutrally towards them. For this reason, the disappearance and fixation of mutations in the gene pool depend on chance: most of them disappear soon after their appearance, and a minority remains and can exist for quite a long time. As a consequence, much of the intraspecific variability (especially in small populations) is due not to the action of selection, but to the random drift of mutant alleles that are neutral or nearly neutral. As a result, a selection that evaluates phenotypes *"essentially makes no difference what genetic mechanisms determine the development of a given form and corresponding function, the nature of molecular evolution is completely different from that of phenotypic evolution"* (Kimura, 1985).

The discussion about the ***neutral theory of molecular evolution***. Like the STE, the neutral theory has generated an active discussion (Nazarov, 2007; Duret, 2008; Nei et al., 2010; O'Dwyer & Chisholm, 2013). The theory agrees that mutations' fixation rate is constant at the molecular level, which makes it possible to estimate the time of species divergence. The theory does not dispute the decisive role of natural selection in the development of life on Earth, and the proportion of mutations that have adaptive significance remains debatable.

However, the subsequent discussion also showed the limitations of the theory: the evolutionary process would require huge periods and prohibitive amounts of DNA. Therefore, the theory's success within the population genetics and molecular biology framework has not spread to disciplines studying macroevolutionary transformations. The concept of genetic drift is evaluated similarly.

The contradiction indicated in Table 8 of Chapter 1 should also be noted between the properties of living organisms as complex open systems and the axioms of classical mathematics. Therefore, it is necessary to apply its methods correctly and not hope they will solve all problems.

The Ecological Point of View

The main branches of evolutionary ecology are the evolution of life, sociobiology, the evolution of interspecies relationships, and the evolution of biodiversity and communities (Rosenzweig, 1995; Pianka, 2000; Fox et al, 2001; Mayhew, 2006; Eskov, 2009). Here are the main provisions of the ecology.

The Impact of Environmental Factors(Figure 3)

Figure 3. The impact of environmental factors on the life processes of organisms

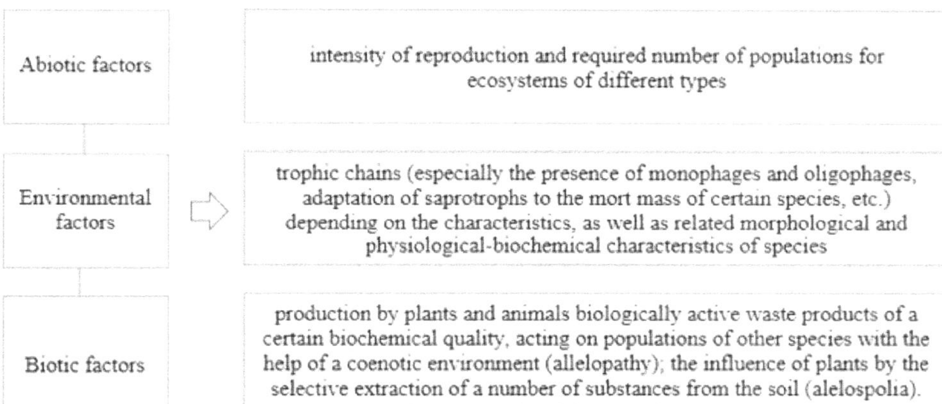

Let's consider abiotic (physicochemical influences that can cause a response of the living organism) and biotic (communities of living organisms, including humans, exerting their vital activity on other beings) factors in detail (Eskov, 2009), Figures 4, 5.

Figure 4. The types of abiotic factors and their influence

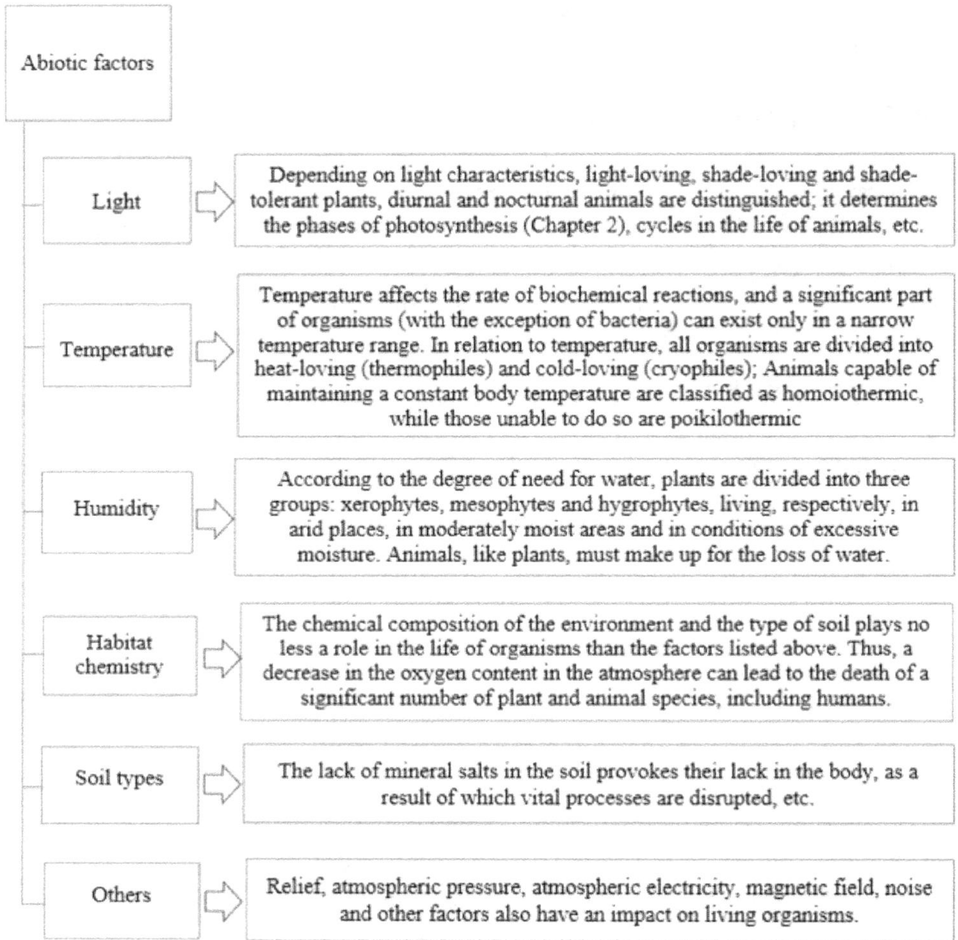

Living organisms form certain communities, and various interactions between living organisms occur (Figure 5).

Habitat of organisms. An organism has direct and feedback connections with the environment, not only experiencing its influence, but also changing the course of its life activity (for example, the already mentioned Great Oxygen Event is associated with the activity of photosynthetic bacteria, and then plants). In various types of environments, supraorganismal systems related to their specificity are formed.

An ecosystem (biogeocenosis) is a set of living organisms that closely interact with each other and their environment. The properties of biogeocenosis are integrity, openness, self-regulation, and self-reproduction - all the "standard" characteristics of open complex systems.

Figure 5. Biotic factors as interactions between living organisms

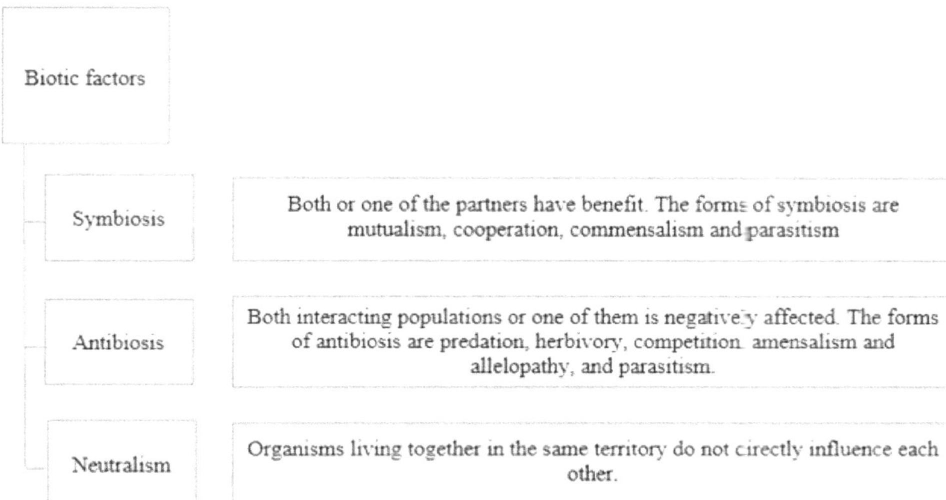

Figure 6. Habitat and living organism types

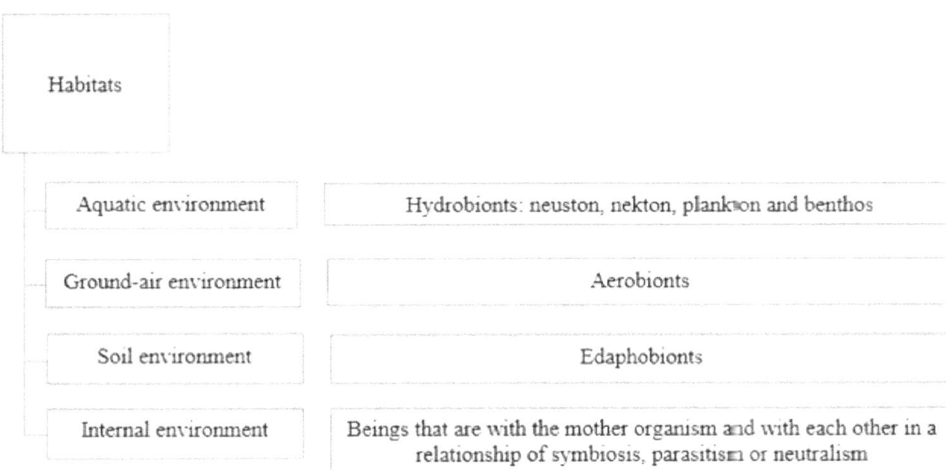

Biotic and abiotic components are distinguished in the ecosystem - biocenosis and biotope. As parts of biogeocenosis, three types of organisms are distinguished:

- Producers (autotrophic organisms that synthesize organic substances from inorganic substances).
- Consuments (consume substances synthesized by autotrophs; these include herbivorous and carnivorous animals and mushrooms).
- Decomposers (decompose organic substances into inorganic ones, for example, bacteria, fungi, earthworms, and so on).

Due to these three groups of organisms in biogeocenosis, substances are circulated, while most of the energy is dissipated. Species and spatial structure characterize each ecosystem.

Food chain (trophic chain). Several interrelated species, each of which serves as food for the next, is called the food chain, which ensures the transfer of energy in organic substances from producers through several organisms by eating some species by others. When transferring, 80-90% of the energy is dissipated as heat, so most food chains contain 3-5 links. The length of food chains is also limited by other factors - the availability of preferred food, territorial behavior, etc. There are chains of grazing (beginning with producers and going to consumers of various orders) and decomposition or detrital chains (small animals, microorganisms, feed on plant and animal remains excrement). Food chains (we can also talk about networks) within the biogeocenosis have a pronounced structure, consisting of components, levels, and relationships. Their relationship can be depicted as ecological numbers, biomass, and energy pyramids.

Producers can store significant amounts of energy in the chemical bonds of organic substances, but they partially expend it on respiration processes. Larger or smaller amounts of energy (usually 80–90%) are lost by consumers at each subsequent trophic level, retaining only about 10%, and ultimately a stable biogeocenosis spends it on its functioning almost completely. Based on this pattern, the rule of the ecological pyramid, or the 10% rule, is formulated: in each subsequent link in the food chain, the amount of energy decreases by 10 times.

Accounting for environmental factors - systemic in nature - clarifies many patterns of evolution that were not understood before. For example, "altruistic" behavior, some aspects of mass extinctions, the ratio of species in ecosystems and food chains, etc. It should be remembered that the ecology of populations and communities entered the evolutionary theory due to the synthesis of Gause's law (two species cannot stably exist in a limited space if the growth in the number of both is limited by one vital resource, the quantity and or availability of which is limited). The concept of an ecological niche is now considered (along with reproductive isolation) as the most important criterion for a species. This approach is more general than the genetic one since it applies to species that do not have a sexual process.

Stability range, optimum law, minimum law. Living organisms, as well as super-organisms systems, have limited reserves of adaptation, therefore, for an organism, population, and ecosystem, there is a range of stability (survival), within which the life of objects occurs. Outside this range, the living system either dies immediately, produces seeds, spores, etc., or passes into a temporary state of rest (bulbs, tubers, and other storage organs of plants, anabiosis in animals, etc.). Within the stability range, organisms' growth and development rate is not the same. The values of the ecological factor at which organisms and populations achieve the best development and maximum productivity are called optimal, and deviations from this optimum cause inhibition of vital processes. Accordingly, the law of optimum determines that any environmental factor has certain limits of positive impact on organisms. The law of the minimum (Liebig's law): the factor that is most deficient at the moment has the greatest influence on the growth and development of the organism.

Note that environmental factors do not act separately, but in interaction with each other, and these interactions are nonlinear and have threshold effects.

Natural and biological rhythms. Rhythmic fluctuations of environmental factors force living organisms to coordinate their vital activity with the periods of action of the most favorable values of these factors. Such periodic changes in the intensity and direction of biological processes are called biological rhythms (Zidermane, 1986).

Distinguish daily rhythms characteristic of most species of plants and animals; circadian (circadian) rhythms with a period of 20 to 28 hours; seasonal (near-seasonal), and annual rhythms due to the rotation of the Earth around the Sun.

Photoperiodism is the reaction of organisms to the ratio of the length of day and night, which is expressed in a change in the intensity of growth and development processes. Photoperiodism is inherent in both animals and plants.

Discussion about the role of environmental factors in the evolutionary process. The importance of ecological factors in explaining the laws of evolution is generally recognized; terms such as “evolutionary ecology” have long entered the scientific community (Pianka, 2000; Fox et al., 2001; Mayhew, 2006; Albuquerque et al., 2015; Cavender-Bares et al., 2018). Three main topics are relevant: the relative influence of individual environmental factors; mechanisms of such influence; integration of the ecological paradigm and various theories of evolution (articles in Methods in Ecology & Evolution).

In each of these areas, the authors have certain disagreements (Pease et a., 1989; Holt & Gaines, 1992; Brown & Pavlovic, 1992). However, the most important question is the compatibility of ecological ideas and neo-Darwinian theories. Such integration has occurred partially, but the lack of a unified apparatus makes it difficult to achieve final success. There have been and continue to be attempts to create such an apparatus based on one system theory or another, but they are based on the classical understanding of set-theoretic relations that are incompatible with the properties of biological systems (Table 8, Chapter 1) (Urmantcev, 2017; Cameron et al., 2022)

To determine the further tasks of this study, it is important to state the need for a conceptual theory and recognition of the systemic nature of evolution as a whole.

FUTURE RESEARCH DIRECTIONS

Now the immediate task is to test the possibility of modeling evolutionary зphenomena (1-15) using the scenarios of the self-organization theory, considered in Chapter 4. At the same time, the processes of reproduction and inheritance, which were not considered here, will be considered.

CONCLUSION

The modern understanding of evolution, enriched by the ideas of ecology and data from molecular genetic studies, is much deeper than even a few decades ago. The synthetic theory of evolution gives quite satisfactory explanations when explaining individual empirical patterns of the evolutionary process from the list (1-15). Accounting for environmental factors, subtleties of molecular genetic interactions, mathematical modeling data, and experimental attempts to organize artificial evolution improve its modern understanding.

And yet many questions remain. If we talk about the main thing, and not dwell on the discussion of individual phenomena (1-15), then the following should be noted.

1. The discrepancy between the estimated time required for the formation and consolidation of certain changes, as well as overcoming the "complexity barrier" in the transition from abiogenic synthesis to living organisms with the actual duration of evolution (Kimura's neutral theory of molecular evolution, Kunin's calculations (Chapter 2), geochronological scale (Chapter 1) respectively).
2. Uneven course of evolution at the molecular genetics, organismal, species, and other levels, as well as for ecological systems (Figure 2).
3. The appearance of changes before living beings "adapt" to them and begin to use their potential advantages, an example of which is anthropogenesis.
4. At the stage of abiogenic synthesis, natural selection does not act; its impact on psychosocial change is debatable—nevertheless, evolution occurs here and there.

All this allows us to make two assumptions.

First: evolution has a systemic character and should be considered within the framework of a systemic approach. The idea is not new, it was expressed, for example, by Y. A. Urmantsev (1988), and the systemic approach itself was proposed by L. von Bertalanffy primarily for the study of biological systems. However, the correct choice of mathematical apparatus is important here. Living systems are non-additive: properties of structures of higher levels cannot be reduced to properties of elements; the same applies to interactions and external influences (Chapter 1, Table 8). Therefore, using the wave model and the theory of self-organization as a research apparatus is expedient.

Second: evolution is a self-organization that implements the scenarios described in Chapter 4. This explains its "directed" nature, "accelerated" pace, unevenness at different levels, and the need for time to "adjust" to changes, complicate the structures of substances at the stage of abiogenic synthesis. Evolution at the level of an organism or an ecological system naturally cannot be reduced to genetic mutations only.

Chapter 6 will be devoted to testing these assumptions.

REFERENCES

Alberts, B., Johnson, A., Lewis, J., Raff, M., Roberts, K., & Walter, P. (2002). *Molecular biology of the cell* (4th ed.). Garland Science., DOI: 10.1093/aob/mcg023

Albuquerque, U. P., De Medeiros, P. M., & Casas, A. (2015). *Evolutionary Ethnobiology*. Springer. DOI: 10.1007/978-3-319-19917-7

Allan, G. J., & Max, T. L. (2010) Molecular Genetic Techniques and Markers for Ecological Research. *Nature Education Knowledge, 3*(10).

Berkemer, S. J., & McGlynn, S. E. A. (2020). New Analysis of Archaea–Bacteria Domain Separation: Variable Phylogenetic Distance and the Tempo of Early Evolution. *Molecular Biology and Evolution*, 37(8), 2332–2340. DOI: 10.1093/molbev/msaa089 PMID: 32316034

Bernstein, H., Hopf, F. A., & Michod, R. E. (1987). The molecular basis of the evolution of sex. In *Advances in Genetics* (pp. 323–370). Elsevier. DOI: 10.1016/S0065-2660(08)60012-7

Bowler, P. J. (1979). Theodor Eimer and Orthogenesis: Evolution by 'Definitely Directed Variation'. *Journal of the History of Medicine and Allied Sciences 34*(1), 40–73._DOI: 10.1093/jhmas/XXXIV.1.40

Brown, J. S., & Pavlovic, N. B. (1992). Evolution in heterogeneous environments: Effects of migration on habitat specialization. *Evolutionary Ecology*, 6(5), 360–382. DOI: 10.1007/BF02270698

Cameron, D., Hartig, F., Minnuno, F. J., Oberpriller, J., Reineking, B., Van Oijen, M., & Dietze, M. (2022). Issues in calibrating models with multiple unbalanced constraints: The significance of the systematic model and data errors. *Methods in Ecology and Evolution*, 13(12), 2757–2770. DOI: 10.1111/2041-210X.14002

Cavalli-Sforza, L. L., & Feldman, M. W. (2003). The Application of Molecular Genetic Approaches to the Study of Human Evolution. *Nature Genetics*, 33(S3), 266–275. DOI: 10.1038/ng1113 PMID: 12610536

Cavender-Bares, J., Kothari, S. J., & Pearse, W. (2018). *Evolutionary Ecology of Communities. (2019). Evolutionary Biology*. Oxford University Press., DOI: 10.1093/obo/9780199941728-0111

Coleman, W. B. & Tsongalis, G. J. (Eds.). (2018). *Molecular Pathology: The Molecular Basis of Human Disease*. Academic Press. DOI: 10.1016/C2014-0-02163-6

Darwin, C. (1859). *On the Origin of Species by Means of Natural Selection, or the Preservation of Favoured Races in the Struggle for Life* (1st ed.). John Murray. DOI: 10.5962/bhl.title.82303

Darwin, C. (1871). *The Descent of Man, and Selection in Relation to Sex* (1st ed.). John Murray.

Duret, L. (2008) Neutral theory: The null hypothesis of molecular evolution. *Nature Education, 1*(1), 218. https://onlinelibrary.wiley.com/journal/20457758

Fisher, R. A. (1958). *The genetic theory of natural selection* (2nd ed.). Dover Publications.

Fox, C. W., Roff, D. A., & Fairbairn, D. J. (2001). *Evolutionary Ecology: Concepts and Case Studies*. Oxford University Press. DOI: 10.1093/oso/9780195131543.001.0001

Frías, L. D. (2010). Omissions in the synthetic theory of evolution. *Biological Research*, 43(3). Advance online publication. DOI: 10.4067/S0716-97602010000300006 PMID: 21249301

Grant, V. (1985). *The Evolutionary Process. A Critical Review of Evolutionary Theory*. Columbia University Press.

Greenberg, A. J., & Wu, C.-I. (2006). Molecular Genetics of Natural Populations. *Molecular Biology and Evolution*, 23(5), 883–886. DOI: 10.1093/molbev/msj077 PMID: 16361304

Grehan, J. R., & Ainsworth, R. (1985). Orthogenesis and Evolution. *Systematic Zoology*, 34(2), 174–192. DOI: 10.2307/2413325

Haldane, J. B. S. (1957). The cost of natural selection. *Journal of Genetics*, 55(3), 551–524. DOI: 10.1007/BF02984069

Hamilton, W. D. (1964). The Genetical Evolution of Social Behaviour. *Journal of Theoretical Biology*, 7(1), 1–52. DOI: 10.1016/0022-5193(64)90038-4 PMID: 5875341

Hughes, J. F., Skaletsky, H., Pyntikova, T., Minx, P. J., Graves, T., Rozen, S., Wilson, R. K., & Page, D. C. (2005). Conservation of Y-linked genes during human evolution revealed by comparative sequencing in chimpanzees. *Nature*, 437(7055), 100–103. DOI: 10.1038/nature04101 PMID: 16136134

Huneman, P., & Walsh, D. M. (2017). *Challenging the Modern Synthesis: Adaptation, Development, and Inheritance*. Oxford University Press. DOI: 10.1093/oso/9780199377176.001.0001

Huxley, J. (1942). *Evolution: the Modern Synthesis*. George Allen and Unwin Ltd.

Jan, Y. N., & Jan, L. (1999). Asymmetry across species. *Nature Cell Biology*, 1(2), E42–E44. DOI: 10.1038/10036 PMID: 10559895

Kimura, M. (1983). *The Neutral Theory of Molecular Evolution*. Cambridge University Press. DOI: 10.1017/CBO9780511623486

Kimura, M. (1991). The neutral theory of molecular evolution: A review of recent evidence. *Japanese Journal of Genetics*, 66(4), 367–386. DOI: 10.1266/jjg.66.367 PMID: 1954033

Kovalyov, Y., Mkhitaryan, N., & Nitsyn, A. (2020). *Self-organization of the Human Mind and the Transition from Paleolithic to Behavioral Modernity*. IGI Global. DOI: 10.4018/978-1-7998-1706-2

Laland, K. N., Uller, T., Feldman, M. W., Sterelny, K., Müller, G. B., Moczek, A., Jablonka, E., & Odling-Smee. (2015). The extended evolutionary synthesis: its structure, assumptions and predictions. *Proceedings. Biological Sciences, 282*(1813), 20151019. DOI: 10.1098/rspb.2015.1019

Lee, T. I., & Young, R. A. (2013). Transcriptional regulation and its misregulation in disease. *Cell*, 152(6), 1237–1251. DOI: 10.1016/j.cell.2013.02.014 PMID: 23498934

Levit, G. S., & Olsson, L. (2006). 'Evolution on Rails': Mechanisms and Levels of Orthogenesis. *Annals of the History and Philosophy of Biology*, 11, 99–138.

Liberles, D. A., Chang, B., Geiler-Samerotte, K., Goldman, A., Hey, J., Kaçar, B., Meyer, M., Murphy, W., Posada, D., & Storfer, A. (2020). Emerging Frontiers in the Study of Molecular Evolution. *Journal of Molecular Evolution*, 88(3), 211–226. DOI: 10.1007/s00239-020-09932-6 PMID: 32060574

Lüttge, U., & Souza, G. M. (2019). The Golden Section and Beauty in Nature: The Perfection of Symmetry and the Charm of Asymmetry. *Progress in Biophysics and Molecular Biology*, 146, 98–103. DOI: 10.1016/j.pbiomolbio.2018.12.008 PMID: 30557534

Maloy, S., & Hughes, K. (Eds.). (2013). *Brenner's Encyclopedia of Genetics* (2nd ed.). Academic Press.

Mayhew, P. J. (2006). *Discovering Evolutionary Ecology: Bringing Together Ecology and Evolution*. Oxford University Press. DOI: 10.1093/oso/9780198570608.001.0001

Mayr, E. (1999). *Systematics and the Origin of Species from the Viewpoint of a Zoologist*. Harvard University Press. (Original work published 1942)

Morozov, A. (2013). Modelling biological evolution: Recent progress, current challenges and future direction. *Interface Focus*, 3(6), 20130054. DOI: 10.1098/rsfs.2013.0054

Nei, M., Suzuki, Y., & Nozawa, M. (2010). The Neutral Theory of Molecular Evolution in the Genomic Era. *Annual Review of Genomics and Human Genetics*, 11(1), 265–289. DOI: 10.1146/annurev-genom-082908-150129 PMID: 20565254

Noble, D. (2011). Neo-Darwinism, the Modern Synthesis and selfish genes: Are they of Use in Physiology? *The Journal of Physiology*, 589(5), 1007–1015. DOI: 10.1113/jphysiol.2010.201384 PMID: 21135048

O'Dwyer, J., & Chisholm, R. (2013). Neutral theory and beyond. In *Encyclopedia of Biodiversity* (pp. 510–518). Elsevier. DOI: 10.1016/B978-0-12-384719-5.00429-9

Pease, C. M., Lande, R., & Bull, J. J. (1989). A model of population growth, dispersal, and evolution in a changing environment. *Ecology*, 70(6), 1657–1664. DOI: 10.2307/1938100

Penrose, O. (2005). An asymmetric world. *Nature*, 438(7070), 919–919. DOI: 10.1038/438919a PMID: 16355197

Pianka, E. R. (2000). *Evolutionary Ecology*. Benjamin-Cummings.

Pigliucci, M., & Finkelman, L. (2014). The Extended (Evolutionary) Synthesis Debate: Where Science Meets Philosophy. *Bioscience*, 64(6), 511–516. DOI: 10.1093/biosci/biu062

Reif, W.-E., Junker, T., & Hossfeld, U. (2000). The synthetic theory of evolution: General problems and the German contribution to the synthesis. *Theory in Biosciences*, 119(1), 41–91. DOI: 10.1007/s12064-000-0004-6

Ridley, M. (2003). *Evolution* (3rd ed.). Wiley-Blackwell.

Rose, M. R., & Oakley, T. H. (2007). The new biology: Beyond the Modern Synthesis. *Biology Direct*, 2(1), 30. DOI: 10.1186/1745-6150-2-30 PMID: 18036242

Rosenzweig, M. (1995). *Species diversity in space and time*. Cambridge University Press., DOI: 10.1017/CBO9780511623387

Simpson, G. G. (1953). *The major features of evolution* (3rd ed.). Columbia Univ. Press. DOI: 10.7312/simp93764

Sole, R. V., & Newman, M. (2002). Extinctions and Biodiversity in the Fossil Record. In *Encyclopedia of Global Environmental Change*. John Wiley & Sons.

Tamura, K., Stecher, G., & Kumar, S. (2021). MEGA11: Molecular evolutionary genetics analysis version 11. *Molecular Biology and Evolution*, 38(7), 3022–3027. DOI: 10.1093/molbev/msab120 PMID: 33892491

Taper, M. L., & Case, T. J. (1985). Quantitative genetic models for the coevolution of character displacement. *Ecology*, 66(2), 355–371. DOI: 10.2307/1940385

Ulett, M. A. (2014). Making the case for orthogenesis: The popularization of definitely directed evolution (1890–1926). *Studies in History and Philosophy of Science Part C Studies in History and Philosophy of Biological and Biomedical Sciences*, 45, 124–132. DOI: 10.1016/j.shpsc.2013.11.009 PMID: 24368232

Wright, S. (1955). Classification of the factors of evolution. *Cold Spring Harbor Symposia on Quantitative Biology*, 20(0), 16–24. DOI: 10.1101/SQB.1955.020.01.004 PMID: 13433551

Wright, W. (1977). *Evolution and Genetics of Populations*. v. 3. Experimental Results and Evolutionary Deductions. University of Chicago Press.

ADDITIONAL READING

Berg, L. S. (1922). *Nomoghenez, yly Evoljucyja na osnove zakonomernostej* [Nomogenesis, or evolution based on patterns]. State publishing house.

Nazarov V. I. (2007). *Evoljucyja ne po Darvynu: Yzmenenye evoljucyonnoj modely.* [Evolution not according to Darwin: Changing the evolutionary model]. Publishing house LKI

Ovchinnikov, N. F. (1966). *Pryncypy sokhranenyja* [Preservation principles]. Nauka.

Reimers N.F. (1994*). Ekologhyja. Teoryy, zakony, pravyla, pryncypy y ghypotezy.* Ecology. Theories, laws, rules, principles, and hypotheses. Zhurnal «Rossyja Molodaja».

Severtsov, A. N. (1939). *Morfologhycheskye zakonomernosty evoljucyy* [Morphological patterns of evolution]. Publishing House of the Academy of Sciences.

Shmalgauzen, I. I. (2017). *Puty y zakonomernosty evoljucyonnogho processa* [Ways and laws of the evolutionary process]. 2nd ed.). URSS. (Original work published 1939)

Takhtadzhyan, A. L. (1943). *Vzaymosvjazj ontogheneza y fylogheneza u vysshykh rastenyj (yssledovanyja po эvoljucyonnoj morfologhyy)* [Relationships of ontogenesis and phylogenesis in higher plants (studies in evolutionary morphology)]. *Scientific works of the Yerevan State University 22*, 71-167

Urmantsev, Y. A. (1988). *Evoljucyonyzm yly obshhaja teoryja razvytyja system pryrodы, obshhestva y myshlenyja* [Evolutionism or the general theory of developing systems of nature, society, and thinking]. Institute of Plant Physiology of the Academy of Sciences of the USSR.

Urmantsev, Y. A. (2017). *Symmetryja pryrody y pryroda symmetryy: Fylosofskye y nauchnye aspektы OTSU* [Symmetry of nature and the nature of symmetry: Philosophical and scientific aspects of GTSU.

Vorontsov, N. N. (1999). *Razvytye evoljucyonnykh ydej v byologhyy* [Development of evolutionary ideas in biology]. Progress -Traditsia.

Zidermane, A. A. (1998). *Nekotorye voprosy khronobyologhyy y khronomedycyny: obzor lyteratury* [Some issues of chronobiology and chronomedicine: a literature review]. Zinatne.

ENDNOTE

1 This classical explanation turns out to be insufficient: works have recently been published in which the fall of Rome is explained by climatic and epidemiological circumstances.

Chapter 6
Evolution as a Self-Organization

ABSTRACT

The regularities and course of evolution are modeled based on the abstractions and axioms of the wave model as self-organization scenarios. The modeling showed that the evolutionary process corresponds to theoretical scenarios and can be considered the consequence of the general laws of the self-organization of the open complex systems. The methods of reproduction and inheritance are considered separately. It is shown that the known methods of reproduction and inheritance can be modeled based on the soliton-wave model within the framework of the theory of self-organization. All this forms the basis for modeling the human-environment system (Chapter 7) and the human life cycle (Chapter 9).

BACKGROUND

Let us make some remarks about the objects and features of modeling.

First, it is an evolutionary process as a whole. To understand this process, particularly the 15 regularities noted in Chapter 5, self-organization scenarios discussed in Chapter 4 will be used.

Secondly, these are the mechanisms of reproduction and inheritance, which are deliberately included in this chapter. Here, the emphasis is again on the conceptual, without unnecessary details, understanding of these processes as relationships within the framework of the soliton-wave model.

However, the description of reproduction and inheritance, as was done in previous chapters, will be quite detailed.

Reproduction, as a necessary condition for evolution, can be considered at two levels - cellular and organismic.

DOI: 10.4018/978-1-6684-8509-5.ch006

The reproduction process at the cellular level is well known by now (Molinari 2000; Mitchison 2003); our task will be to compare it with the appropriate scenario of self-organization.

Reproduction at the level of multicellular organisms is more interesting because, for both plants and animals, it takes a wide variety of forms (Lovett & Lesley, 1988; Lodé, 2001; Ridley, 2003; Fasoula et al., 2020), which requires consideration of different self-organization scenarios. We also note the involvement of humans in the reproduction of cultivated plants and the breeding of domestic animal breeds using both traditional and modernб in particular, genetic engineering technologies, well as the discussion about the safety of new breeding methods (Breseghello & Coelho, 2013; Hartung & Schiemann, 2014).

Molecular and genetic changes are another key condition for evolution (Bernstein et al., 1987).

Finally, it is necessary to consider the system-ecological level of evolutionary processes (Rosenzweig, 1995; Mayhew, 2006). It is expedient to consider self-organization scenarios as constrained multi-criteria optimization (Mkhitaryan et al., 2004).

The preceding defines the means of modeling: these are scenarios, diagrams, graphs, i.e., static models of M1, as well as formulas describing self-organization scenarios. M2 and M3 (Chapter 4), which are more powerful, will not be used here.

Thus, a pragmatic task - to show the adequacy of the theory of self-organization before using it to model the life cycle - is solved.

MODELING OF THE EVOLUTION PROCESS

It is necessary to show the possibility of using abstractions and the apparatus of the wave model and the theory of self-organization. Hence the requirements for accuracy-sufficiency for understanding the essence without undue attention to detail, that is, the images of waves and solitons, graphs, and PC-diagrams will be detailed only as much as necessary to solve the main problem.

Based on the description of the evolutionary process in Chapter 5, we state that the most general case of an organization takes place, that is, *Sp* is formed as a result of the interaction of {C} and {O}, and the integrity conditions take the form:

$$H(\{S\}_o, Sp, \{O\}_p)\ PC\ PC*\ PC**, \qquad (6.1)$$

and

$$PC\ \ n(\{S\}_{i=1,\dots,n}, {}_{xy}, \{O\}_{i=1,\dots,n}) \qquad (6.2)$$

$$PC * \ (m\text{-}n)(\{S\}_{i=n+1,\ldots,m}, {}_{ixy}, \{O\}_{m\text{-}n})$$
(6.3)

$$PC ** (o\text{-}m)(\{S\}_{o\text{-}m}, {}_{i=o\text{-}m,\ldots,o,j=p\text{-}m,\ldots,pxy}, \{O\}_{p\text{-}m}), \quad (6.4)$$

where *PC* - diagrams display the type of scenario; *n, m, o, p, o>p* denotes the amounts of *S* and *O*; *i, j* - serial numbers of stratifications *Sp*, corresponding to the interactions *S* and *O*; *x, y* - numbers of layers and *S*-elements and *S*-sets belonging to them.

This means that within the framework of the general scheme (6.1), the stratification occurs in three ways: *n* times according to scenario (6.2), *m-n* times according to scenario (6.3), and *o-m* - according to (6.4), which is not reducible to the previous ones.

The stratification (6.2) is modeled by a tree-like S-graph, which displays the general principles of system construction, division into levels and elements, and the number of independent characteristics.

The stratification (6.3) corresponds to the scenarios when one S interacts with the set {O} or the interaction of the set {S} with one O takes place. In this case, the S-space develops from one center in several directions, the number of which is equal to the number of O (or S in the second case). If they have common differences from S (or O), elements with the same characteristics arise, and various interactions, described by the axioms of the 3rd group, become possible. As a result, the number of layer structures, their mutual affiliation, the distribution of the potential, and the characteristics associated with it change. The topology of the S-space can be characterized as a tree-like one with separate closed contours.

The stratification (6.4) corresponds to the interaction of the set {S} with the set {O}. Here the S-space has several development centers (according to the number of S), for each of which several directions of development are distinguished (according to the number of O). As in the previous cases, interactions arise and topological and metric characteristics change. In general terms, the “topology” of S-space is network-like, with separate tree-like structures.

In all cases, partial compensation of potential losses, complete compensation, or excessive compensation are possible, which affect the number of elements in the layers and the direction of *Sp* evolution. There is a stratification, or “convolution in reverse”, that is characterized by an increase in the total potential *Sp*, as described in Chapter 4.

The modeling of regularities (1-15) mainly uses scenario (6.2), and, if necessary, also (6.3, 6.4).

1. ***Life cycle.*** Human life cycle events will be discussed in detail in Chapter 6; because of the importance of understanding evolution, the forms of reproduction and inheritance are below in a separate part. For now, let's note the correspondence: the life cycle is ∪ø•∪ö↓.

2. ***Arogenesis*** is modeled as follows (Figure 1).

Figure 1. The appearance of arogenetic changes: 1 – soliton in a stable state; 2 - influences of the environment - generation of changes in the potential of the soliton; 3 - fast changes - the appearance of waves (two waves are taken for illustration) on the surface of the soliton; 4 - stabilization of changes - 2 waves turn into 2 solitons, partly due to the resources of the medium, partly due to the initial soliton, the potential of which decreases

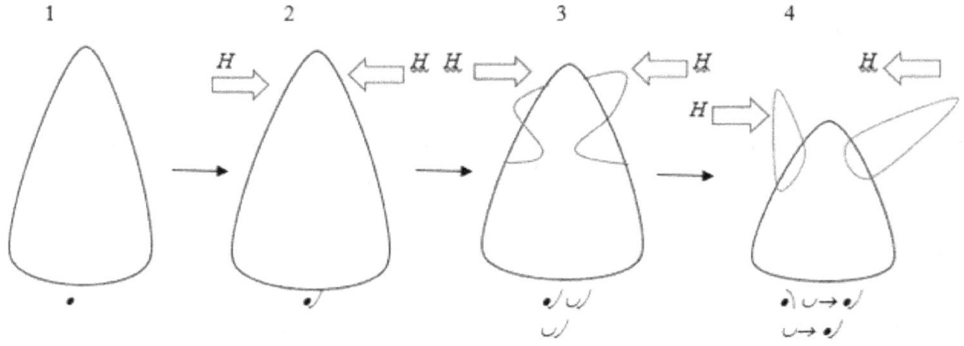

Let us comment on Figure 1. Stage 1 is interpreted as the existence of a population in stable external conditions. Genetic drift occurs (potential conditions for changes are created), in general, natural selection acts as a conservative factor that maintains the norm. Stage 2 shows changes in external conditions: the environment ceases to be stationary, the soliton acquires an excitation modality, its potential increases or decreases due to the resources of external influences (Figure 1 shows an increase in potential): the population size changes, ecological niches appear. Stage 3 - potential changes turn into actual ones, which manifests as rapid changes in the number of demes, the "bottleneck" effect, the ratios of certain mutations, and increased competition. Some of the mutations turn out to be useful in changed conditions (they make it possible to use new resources), and their carriers leave offspring, others turn out to be harmful - and their carriers die without including in the reproductive cycle. We also note another circumstance. The time of existence in the form of waves is much less than in the form of solitons, while the number of changes for waves is much larger and the evolution is faster. This may explain the problem of "intermediate links" - the number of remains will also be much less than for the stages of stable existence. Stage 4 - surviving carriers of beneficial mutations use new resources - is another level of self-organization (the formation of higher taxa). The potential of the initial soliton decreases - the initial population, less corresponding to the new environmental conditions, reduces its number up to complete extinction (a consequence of the Gause law).

Note that at all stages there is a tendency to preserve the integrity of the system (symmetries, Fibonacci numbers, ratios associated with Φ, and so on) - such conditions act as restrictions on certain changes in the future, although, at a particular moment, they may not be fulfilled (for example, the solitons in Figure 1.4 are asymmetric).

3. ***Allogenesis***. Features of allogenesis are illustrated in Figure 2.

Figure 2. The appearance of allogenetic changes: 1 – initial state corresponding to stage 4 of Figure 1; 2 - allogenetic changes, unlike arogenetic ones, occur at the same level (new species appear, but higher taxa do not form)

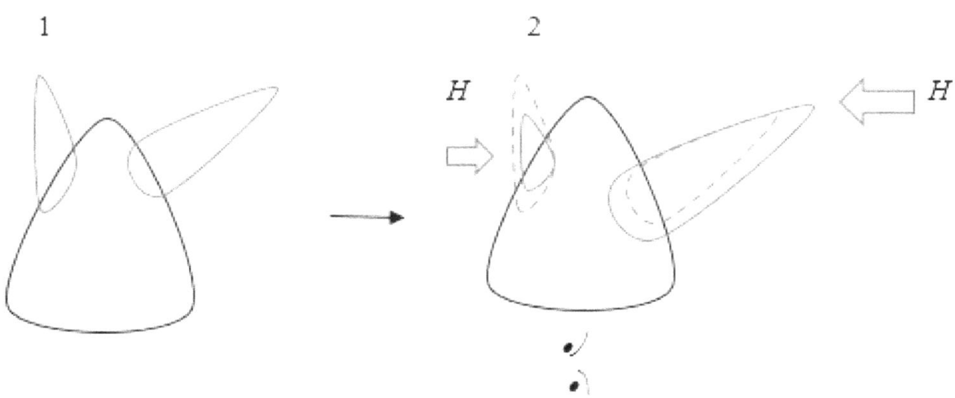

The fate of new species, depending on external conditions, can develop in different ways: if at some point in time, there are many resources, the species will progress •ø (regularity 6, right side of Figure 2.2 2), regress •ö (regularity 7, left part of Figure 2.2), or will be in a stable state if the environmental impacts become stationary. The mechanisms of telogenesis ("chase" for a specialized resource) and hypergenesis ("redundancy" of a non-specialized resource) are also clear. When external conditions change, these evolutionary "strategies" turn out to be losing. The tendency to preserve integrity is also present here.

4. ***Degeneration***. For modeling, let's slightly modify Figure 2 and give it a different interpretation (calibration): let the individuals of the population (black color) have inherited organs formed during evolution in the primary environment (blue color). Their functioning requires external resources and also involves the potential (as, for example, in scenarios 1S, 1O, or 1S, {O}, Chapter 4, table 1). If in a new environment, these organs are useless for obtaining resources, the population will have the opportunity to increase its potential if its individuals do not support unnecessary organs (Figure 3).

Figure 3. The appearance of degenerations: 1 – initial state; 2 - intermediate state: in the new environment, the effects of H_1 no longer act and the organs (blue) adapted to the original environment are useless; 3 - the result of the changes - an increase in the potential that can be used to adapt to the environment with the effects of H_2.

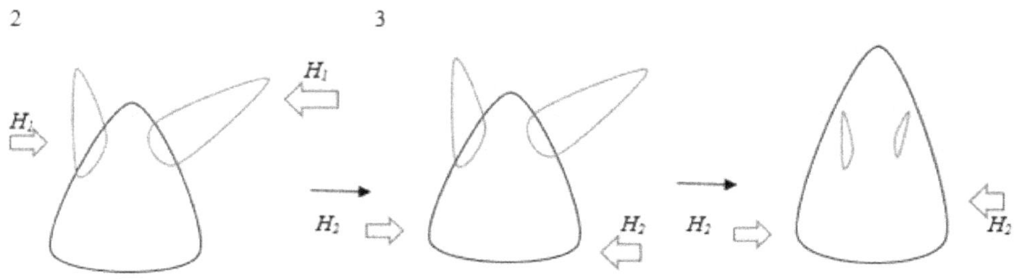

This potential will provide additional opportunities for adaptation to the new environment: for example, protection from toxins, increased reproduction, and so on. Naturally, these potentialities must be actualized at the population level during natural selection. Thus, it is impossible to speak only about the structure simplification in the course of degeneration - adaptations to the new environment must also appear.

In the case of a temporary stay in a new environment, catagenetic changes occur with a tendency to rudimentary unnecessary organs; in the case of permanent residence in such an environment, the tendency will be more pronounced (hypogenesis). However, it should be noted that the mechanisms of inheritance and conditions of integrity will prevent the complete disappearance of unnecessary organs.

5. ***Alternation of arogenesis, allogenesis, and degeneration***. Stage 3 in Figure 3 is very similar to Stage 2 in Figure 1. If the population returns to an environment requiring arogenetic or allogenetic changes, and the available potential is sufficient, then the pattern will repeat.

6. ***Biological progress*** and 7. ***Biological regress*** is illustrated in Figures 1-3.

8. ***Divergence*** in the Darwinian version, as a derivation from one ancestor, corresponds to the scenario *(1S, 1O)*, that is, (6.2), and can be expressed by a PC diagram showing kinship relationships.

Figure 4. Classical divergence scenario. Dots indicate related species, and arrows indicate the sequence of their formation

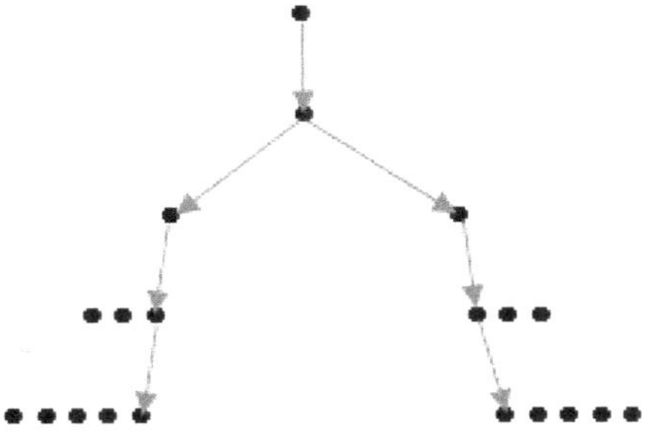

It should be noted that the number of descendant species in Figure 4 reflects the development of the Fibonacci series and is a system characteristic; in practice, it is determined by the available resources of the environment and the adequacy or inadequacy of the adaptations of specific species to its conditions.

Choosing scenarios *(1S, {O})* or *(1O, {S})* from a group (6.3), we get the possibility of modeling relations of paraphilic origin (Figure 5).

Figure 5. Paraphilic origin. Populations of one species give rise to new species either independently or as a result of crossing

The scheme is illustrative, as well as quantitative ratios.

Note that the scenario ({S}, {O}), i.e. (6.4) makes it possible to model a polyphyletic origin.

9. ***Convergence***. Imagine two solitons under the same influence. They will mutate similarly if their potential is sufficient (Figure 6). We interpret them as populations of different species. Adaptations occur according to the patterns of araogenesis, allogenesis, or degeneration shown in Figures 1-3.

Figure 6. Convergence - the similarity of external influences determines the similarity of changes: 1 - 2 populations belonging to different species enter a new environment with different influences; 2 - the traits of individuals in both populations change similarly (blue color), "old" organs that have become unnecessary can become rudimentary (black color)

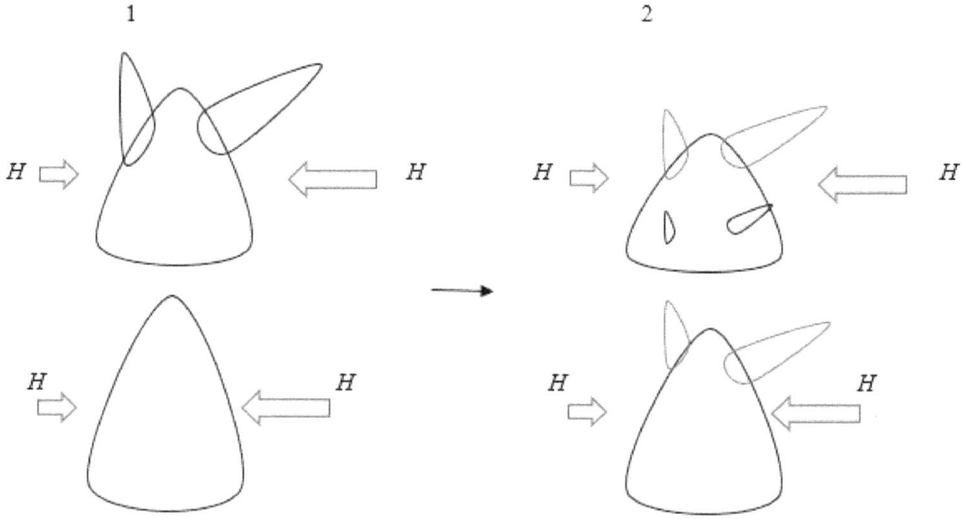

10. ***Parallelism*** modeling as shown in Figure 7: species descended from common ancestors, in the same environment, acquire similar adaptations. The similarity here, in addition to the general influences of the environment, is also due to the inheritance.

Figure 7. Parallelism: 1 – initial stage – two populations of species descended from common ancestors and having developed different allogeneic changes based on common arogenesis, enter the environment with the same impacts; 2 - new similar allogeneic adaptations develop (blue color)

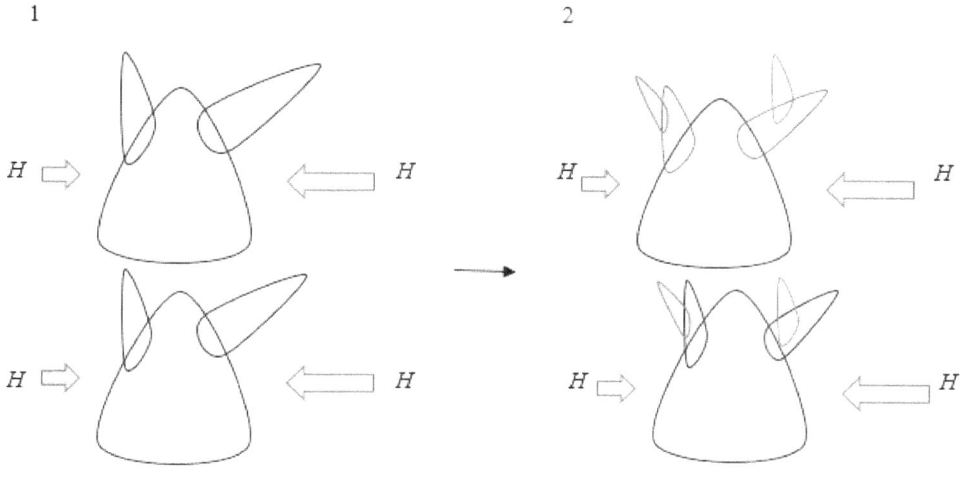

11. ***The direction of evolution*** can be shown in Figures 2 and 7: in the first case, this corresponds to the development of allogeneic changes over time, and in the second, to the law of homologous series.

12. ***Preservation of integrity underlies all scenarios of self-organization***. There are various natural mechanisms for achieving this effect. The preservation of integrity is a tendency: in an open dissipative environment and experiencing variable influences, integrity can be violated at a certain time.

13. ***The irreversibility of evolution*** follows from property 4.9, given in Chapter 4. We also note the absence of groups for most of the operations of wave interactions (Axioms of the third group, Chapter 2).

14. ***The speed of the evolutionary process*** is different, which manifests itself in three ways.

Firstly, for early levels of organization: molecular-genetic, cellular, individual living organism, species, and other taxa, ecological system - this follows from the qualitative difference between the components belonging to different levels of the S-space structure, which, among other things, have and their time scales (Properties 4.6-4.9). This can, in particular, explain the inconsistency in the duration of the evolutionary process according to the neutral theory of molecular evolution of its actual duration.

Secondly, the observed acceleration of the evolutionary process, despite numerous exceptions. The ratios of the duration of important evolutionary periods are quite close to the "golden section": the time elapsed since the emergence of life (≈4 billion years), the time elapsed from the appearance of the first living organisms to the appearance of multicellular creatures (≈2.5 billion years), the duration of the Cryptozoic (≈1 billion years), the duration of the Phanerozoic (≈542 billion years) correlate with each other as follows: 4/2.5 = 1.6; 2.5/1.5542 = 1.62; 1.542/1 = 1.542; 1/0.542 = 1.85 (Phanerozoic is not finished yet, and this ratio may change).

Thirdly, the very fact of the numerous exceptions follows the general "plasticity" of the S-space, the ability of waves and solitons to change modalities and mutually transition into each other when environmental influences change, compensate for potential losses, the presence of different types of interactions for own levels of an organization, etc.

15. ***Mass extinctions*** can have two mechanisms.

First, it is the death of the ecological system as a whole, accompanied by the destruction of food chains and other relationships described in Chapter 5. Global catastrophes are an example of this way.

Secondly, there is one more mechanism due to Corollary 3 of Property 4.9 and relation (4.12): the existence of the current layer cannot exceed the duration of the existence of the previous one. This can be illustrated in Figure 1: if the existence of the initial soliton ends, then all derivative solitons will also die. But, if these solitons exist autonomously, and lost potential is compensated by the environment, then such a mechanism does not work.

Notes about human evolution. Features of human evolution (Chapter 5) look paradoxical. But in the theory of self-organization, these paradoxes are regularities.

A human, systems, and organs within the framework of this theory are waves and solitons that have a certain potential, passing into each other under the influence of extra-systemic factors that exist in the boundary environment for a limited time and are characterized by modalities of existence and state.

Thus, the potential possibility of bipedal locomotion and freeing of the hands emerged as a result of genetic mutation quickly (actualization of the wave, $\cup \boldsymbol{A} \emptyset$) and became fixed (the wave turned into a soliton $\cup \rightarrow \bullet$). At the same time, the potential possibility of using hands to make tools appeared - and after a long time, after another external impulse that changed the system potential - it became relevant, and the potential was sufficient to make Olduvai tools. Similar external impulses preceded other changes, both anatomical (binocularity of vision, its predominance among other senses, changes in the structure of the larynx, loss of hair) and technological (transition to the Acheulean, Mousterian, and Upper Paleolithic stone processing techniques).

The same applies to the improvement and then degradation of the human brain - an external impulse led to a genetic mutation and continued to act further, as a result of which, from generation to generation, the brain gradually grew and became more complicated, the body adapted to "this parasite", and only after for a long time, the brain of a Cro-Magnon (to dwell only on this example) was able to show its usefulness. And then the momentum dried up, and it was time for degradation (•ö).

REPRODUCTION AND INHERITANCE

Reproduction and evolution. The general direction of the reproduction methods evolution is from asexual to sexual forms, from isogamy to anisogamy, from the participation of all cells to the division of cells into somatic and sexual ones, and from external fertilization to internal with intrauterine development. (Reproduction. Wikipedia, n/a; Lewin, 2004).

The reproduction rate, the number of offspring, and the frequency of generational change, along with other factors, determine the species' adaptation rate to environmental conditions. There are two approaches to the reproduction of a species: producing a huge number of eggs with a minimum expenditure of energy for each of them (R-strategy); or the production of a small number of eggs (babies), requiring a lot of energy for each of them (K-strategy) (Johanson & Wong, 2009). In the evolution of vertebrates, from fish to warm-blooded ones, there is a trend towards a decrease in the number of offspring, and an increase in their survival, including due to parental care, that is, a gradual transition to the K-strategy. However, the extreme example of the K-strategy - large anthropoids (gorilla, chimpanzee, and orangutan), giving birth to a cub every five or six years, is an example of its inefficiency.

Cell division underlies both asexual and sexual reproduction. Reproduction by division is carried out by dividing the cell in two; sometimes (in malarial plasmodia, trypanosomes) there is a division into a larger number of cells (schizogony). In this case, the resulting cells are identical to the original. There are several types of cell division.

The division of prokaryotic (non-nuclear) cells is the process of the formation of daughter cells from the mother. A distinctive feature of prokaryotic cell fission is the direct participation of replicated DNA in the division process (Lewin, 2004). The prokaryotic cells divide into two daughter cells (binary fission, Figure 8).

Figure 8. Prokaryotes binary division: 1. The bacterium before binary fission has the DNA tightly coiled. 2. The DNA of the bacterium has replicated. 3. The DNA is pulled to the separate poles of the bacterium as it increases in size to prepare for splitting. 4. The growth of a new cell wall begins the separation of the bacterium. 5. The new cell wall fully develops, resulting in the complete split of the bacterium. 6. The daughter cell has tightly coiled DNA, ribosomes, and plasmids (Ecoddington14, 2014)

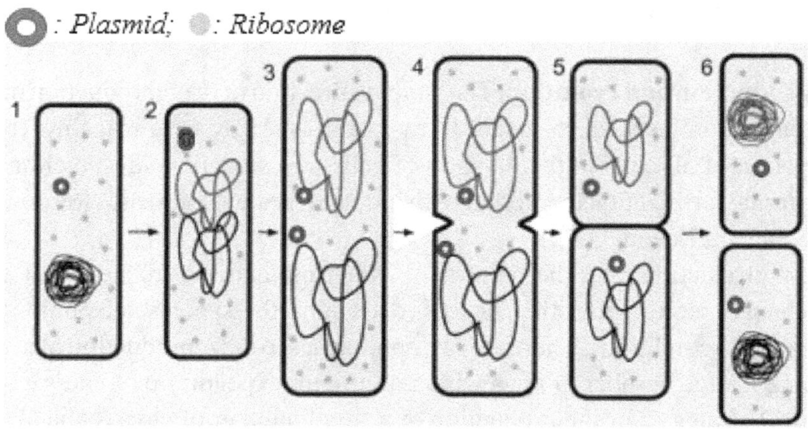

Mitosis is an indirect division of a eukaryotic (nuclear) cell, consisting of a strictly identical distribution of chromosomes between daughter nuclei, which ensures the formation of genetically ident daughter cells to the parent cell and maintains continuity in many cell generations (Alberts et al., 2008; Mitosis. Wikipedia, n / e). The following stages of mitosis are distinguished: prophase, prometaphase, metaphase, anaphase, and telophase. The main function of mitosis is the copying of genetic information. In addition, mitosis provides:

- Growth and development by increasing the number of cells in the body, starting from a single cell - the zygote.
- Cell replacement – in the skin, digestive tract, and other organs, as well as in the blood, cells are constantly replaced; new cells are formed by mitosis and are exact copies of their predecessors.
- Regeneration – due to mitosis, some organisms restore lost body parts.
- Asexual reproduction – due to mitosis, some organisms produce genetically identical offspring, for example, during budding, during vegetative propagation of plants, and so on.

Mitosis of animal cells, as a rule, lasts 30-60 minutes, and plants - 2-3 hours.

Meiosis (reduction division) is the fission of the nucleus of a eukaryotic cell with a halving of the number of chromosomes (4 haploid (n) cells are obtained from one diploid (2n) cell). Occurs in two stages (reduction and equational) (Babynin, 2007; Bogdanov, 2003).

During the reduction stage, during cell division, its ploidy decreases ("maternal" and "paternal" homologous chromosomes diverge into two different daughter cells). In the prophase of the reduction stage, pairwise fusion (conjugation) of homologous chromosomes occurs, therefore, the correct course of meiosis is possible only in diploid cells or in tetra-, hexaploidy, and other "even" cells. Chromosome rearrangements (large-scale deletions, duplications, inversions, or translocations) impose restrictions on chromosome conjugation.

The equational stage is similar to mitosis. The ploidy does not change because of the second division: the "sister" chromatids - copies of DNA - diverging. There is no DNA replication between two meiotic divisions. With a decrease in the number of chromosomes because of meiosis, a transition from the diploid phase to the haploid phase occurs in the life cycle.

Meiosis in animals is observed during the formation of gametes. Meiosis in plants and fungi usually occurs with the formation of haploid spores. In unicellular eukaryotes, meiosis can be observed at different life cycle stages. To restore diploidy in a cycle, the fusion of haploid cells is necessary.

The germ cells of the parents, formed by meiosis, have a haploid set (n) of chromosomes. In a zygote, when two sets are combined, the number of chromosomes becomes diploid (2n). The formation of a new organism occurs by mitotic divisions of the zygote, and each new cell contains a diploid (2n) set of chromosomes. Each pair of homologous chromosomes contains one paternal and one maternal chromosome. Meiosis is the basis of combinative variability due to crossing over (prophase I) and independent segregation of homologous chromosomes (anaphase I and II). Due to the decrease in the number of chromosomes in gametes, a constant (2n) set of chromosomes is maintained in new organisms. In sexually reproducing organisms, the doubling of chromosomes the number in each generation is prevented, since during the meiosis of germ cells, a reduction in the chromosome number occurs.

Amitosis[1], or direct cell division by simply dividing the nucleus in two, is observed in aging or pathologically altered cells, often doomed to death (cells of the embryonic membranes of mammals, tumor cells, etc.). During amitosis, the interphase state of the nucleus is morphologically preserved, DNA replication is absent, chromatin spiralization does not occur, and chromosomes are not detected. Only the nucleus divides, and without the formation of a fission spindle; hereditary material is distributed randomly. With repeated amitoses, two- and multinuclear cells can form (Gilyarov, 1986).

The number of cell fissions is limited by the Hayflick limit (Hayflick, 1961). The maximum number of cell divisions depends on its type and the organism to which it belongs. For most human cells, the Hayflick limit is 52 divisions. During human life, about 10^{14} cell divisions are carried out in the human body (Buldakov & Kalistratova, 2003).

Asexual reproduction (agamogenesis) is characterized by the fact that the next generation develops from somatic cells without the participation of reproductive cells (Gilbert, 1995; Alberts et al., 2008).

Asexual reproduction is the most ancient and simplest; it is widespread in unicellular organisms, such as bacteria, blue-green algae, chlorella, amoeba, and ciliates. In addition, this method is used by many multicellular organisms, in particular, almost all plants and fungi, as well as lower forms of animals.

The advantages include minimal energy and time requirements, as well as the absence of the need for a sexual partner, which allows for high reproduction rates. For example, the quantity of bacteria can double every 30-60 minutes under favorable conditions. Thus, the R-strategy is implemented. The disadvantages include the risk of fixing harmful mutations and limiting the ability to adapt, as a result, under adverse conditions, since all the descendants are identical to one parent, they all die. There is also a decrease in the rate of evolution, which has a particularly unfavorable effect on the increase in the organization of organisms. Consider the main varieties of asexual reproduction (Figure 9).

Figure 9. Methods of asexual reproduction: 1 - division; 2 - schizogony; 3 - budding; 4 - fragmentation; 5 - vegetative reproduction; 6 - spore formation (Reproduction of organisms, n /d)

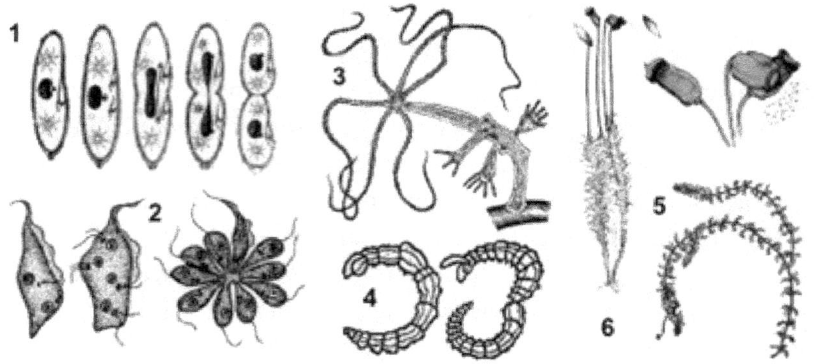

Budding is a type of vegetative reproduction of animals and plants, in which daughter individuals are formed from outgrowths of the body of the mother organism (kidneys).

Budding is based on the mechanism of mitotic cell division. One of the formed nuclei moves into the emerging protrusion of the mother cell, then protrusion buds. The daughter cell is significantly smaller than the mother cell, and it takes time for it to grow and complete its structures, after which it becomes a mature organism.

Budding is characteristic of many fungi, liver mosses, and such animals as sponges, some worms and flagellates, tunicates, hydras, and sporozoa. Note that in animals, budding does not always reach the end, young individuals remain connected to the mother's body, sometimes this leads to the formation of colonies.

Fragmentation is the reproduction by dividing the body into several parts; from each part, a full-fledged organism grows, similar in everything to the parent individual (flat and annelids, echinoderms). Fragmentation is like regeneration.

Vegetative reproduction is the formation of a new individual from the multicellular part of the body of the parent individual. In algae and fungi, it occurs by separating the thallus non-specialized sections or through the specialized section's formation. In higher plants, it is happening as the disintegration of the maternal individual into two or more daughter individuals (when creeping shoots or rhizomes die off, the separation of root offspring) or as the separation of the rudiments of daughter ones from the maternal individual (for example, tubers, bulbs, brood buds). In some plants, shoots separated from the mother plant may take root. In animals, vegetative reproduction is carried out by division or by budding.

Vegetative propagation is based on processes similar to those of regeneration; usually, in the absence of the ability to regenerate, vegetative reproduction is also absent, and in the presence of developed regeneration (annelids, hydroids, flatworms, echinoderms), vegetative reproduction is present.

Reproduction by spores is the formation of cells with reduced metabolism, surrounded by multilayered membranes, resistant to adverse conditions (spores), preceding the asexual reproduction of bacteria. Spores are usually formed by mitosis, sometimes (especially in fungi) in huge numbers. They reproduce the mother organism when germinating. Asexual reproduction with the help of unicellular spores is characteristic of fungi and algae.

Sporulation serves to survive bacteria in adverse conditions and to resettle: once in a suitable environment, the spore turns into a dividing cell.

Sexual reproduction is associated with the fusion of sex cells (gametes) of two complementary male and female organisms. Initially, the formation of gametes with a haploid set of chromosomes occurs. Animals have a double set of chromosomes in ordinary cells, so gamete formation happens during meiosis. In many algae and all higher plants, gametes develop in the gametophyte, which already has a single set of chromosomes, and is obtained by simple mitotic division.

There are several types of gamete formation: isogamy (gametes of the same size and structure, with flagella); anisogamy (gametes of different sizes, but similar in structure, with flagella); oogamy (gametes of various sizes and structures). The small, flagellar male gametes are called spermatozoa, while the large, flagellaless female gametes are called eggs. When two gametes merge (in the case of oogamy, a fusion of different types of gametes is necessary), a zygote with a double set of chromosomes is formed. A zygote develops into a daughter organism, which cells contain parents' genetic information. There are some phenomena and special cases:

- hermaphroditism in animals and monoeciousness in plants (both male and female gonads);
- parthenogenesis in animals and apomixis in plants (a new organism develops from an unfertilized egg and there is no exchange of genetic information, as in asexual reproduction);
- Progenesis (gametogenesis at the larval stage in the form of neoteny, paedomorphosis, and paedogenesis).
- Metagenesis (alternation of generations, reproducing, respectively, sexually and asexually).
- Heterogony (alternation of different sexual generations, for example, alternation of separate sex generations with hermaphroditic, or breeding parthenogenetically).

Human sexual reproduction. (Jones & Lopez, 2013). Gametogenesis – the male and female germ cell formation – is shown in (gametogenesis spermatogenesis and oogenesis, Figure 10).

Figure 10. Gametogenesis (BrainKart. 2020)

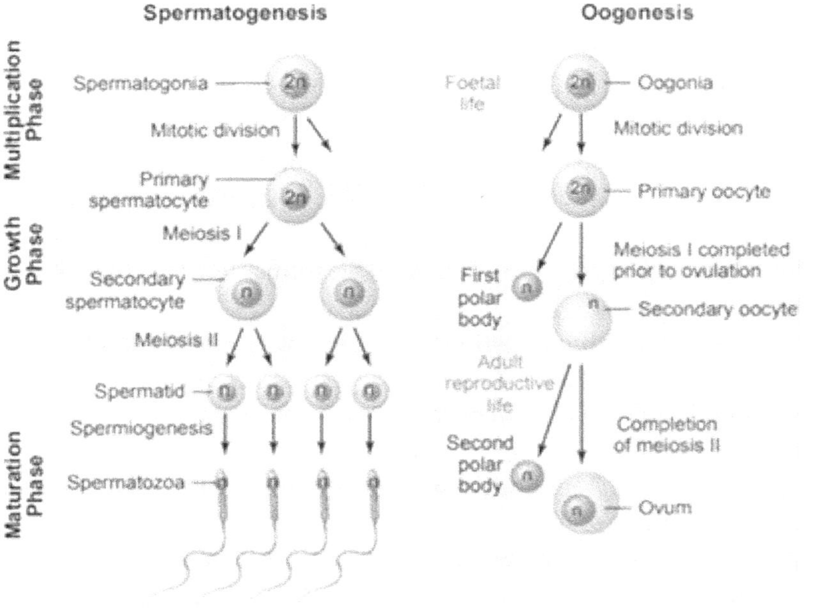

Spermatogenesis occurs in the spermatogenic epithelium of the convoluted tubules of the testicles. Among the spermatogenic cells are Sertoli cells - non-spermatogenic cells of the spermatogenic epithelium - whose functions are: support and support for developing spermatozoa; nutrition of developing spermatozoa; protection of developing spermatozoa from toxins and pathogens; phagocytosis of excess cytoplasm of spermatids and degenerating germ cells; conversion of testosterone to estrogens; secretion of fluid for the transport of spermatozoa in the seminiferous tubules; synthesis of some sex hormones. The developing spermatids and spermatozoa are immersed in the Sertoli cells. Spermatogenesis begins with the onset of puberty. The duration of human spermatogenesis is approximately 70 days. 1 cm^3 of semen contains about 100 million spermatozoa. Spermatogenesis includes reproduction (formation of spermatogonia by mitosis); growth (a slight increase in the size of spermatocytes of the first order); maturation of spermatocytes of the second order and the formation of spermatids by meiosis; formation of spermatozoa from spermatids.

In the process of formation of spermatozoa, specific features appear in spermatids: a tail (an organelle of movement), a mitochondrial clutch, and an acrosome. Each sperm cell has a haploid set of chromosomes - 23 chromosomes. Of these, 22 are autosomes and one is the sex, X or Y chromosome. The emerging spermatozoa are separated from the Sertoli cells and enter the lumen of the tubule, where there is a fluid produced by the Sertoli cells. The movement of the sperm in the female genital tract towards the egg is due to chemotaxis. Before the sperm meets the egg, it moves through the female genital tract for several hours. At the same time, factors of the female body (pH, mucus, etc.) affect the spermatozoon, supporting the ability to migrate, fertilize, and activate them. The spermatozoon membrane becomes permeable to Ca_2+ ions, which, upon entering the spermatozoon, increases tail motility and causes the acrosome and spermatozoon membrane to fuse.

Ovogenesis includes reproduction (formation of oogonia from stem cells using multiple mitoses); growth (formation of oocytes of the first order through synthesis and accumulation of substances); saturation: the formation of oocytes of the II order and the egg by meiosis.

The maturation of follicles and the transition of oocytes of the first order into oocytes of the second order, and then into eggs, is accompanied by meiotic divisions. Around the oocyte of the second order, a transparent membrane is formed that separates the oocyte from the cells of the follicular epithelium. Numerous oocyte microvilli and cytoplasmic processes of follicular cells are introduced into the transparent zone. Around the transparent zone is one layer of follicular cells, forming a radiant crown. Thus, female gametes during oogenesis are protected from harmful effects by a barrier formed by a thick basement membrane, follicular cells, and a transparent membrane. After maturation, the mature follicle, located directly under the integumentary epithelium of the ovary, breaks. The egg, surrounded by a brilliant membrane and follicular cells, enters the peritoneal cavity (ovulation), where enters the fallopian tube. The egg cell has a haploid set of chromosomes (23 chromosomes: 22 X-autosomes, 1 X-sex chromosome). Its cytoplasm contains many mitochondria, elements of the granular endoplasmic reticulum, free ribosomes, lysosomes, and nutrients (yolk, glycogen, and lipid droplets).

Fertilization is the interaction of male and female haploid gametes, leading to the fusion of their nuclei, the unification of chromosomes, and the emergence of a diploid zygote of a new organism. For fertilization, the sperm must consistently overcome three barriers: a radiant crown, consisting of several layers of follicular cells; a transparent shell; and the plasma membrane of the egg. As a result of the acrosomal reaction, a narrow channel is formed in the transparent membrane, through which the spermatozoon passes; The plasma membranes of the sperm and egg come into contact and fuse. Soon, the fused membranes are destroyed, and the head of the spermatozoon is immersed in the egg cytoplasm. Gamete nuclei migrate to the center of the egg and approach each other. Their nuclear membranes disappear, maternal and paternal chromosomes are mixed, and a diploid zygote appears.

MODELING OF REPRODUCTION AND INHERITANCE

It is necessary to show the possibility of using wave abstractions and self-organization scenarios to display the presence of a chromosome-cell pair (or, more abstractly, an information-substance pair), a short division time compared to the lifetime of cells, and the main types of reproduction. The details of sexual reproduction described above, the processes of synthesis of complex organics (Chapter 2), and even the differences between prokaryotic and eukaryotic cells will not be affected here.

The Cell Division Modeling

It is possible to use as images of chromosomes and a cell: 1) an abstraction of a perpendicular soliton (wave) of two components, one of which will display the information (blue color), and the other - the real component (black color); 2) modalities of potential existence (information component) and actual existence (material component). We will apply the first option, reserving the second for some subsequent tasks.

Then ***the binary division of prokaryotic cells and mitosis of eukaryotic cells*** will be expressed as shown in Figure 11. Schizogony is carried out according to the same scheme, but in this case, the wave stage will include several periods. We note the dual role of external influences: the trigger of transitions •ø → Uø→ •, and also the source of resources for cell growth.

Figure 11. Binary division: 1 – initial stable state; 2 - DNA replication and cell growth; 3 - cell division into two identical cells

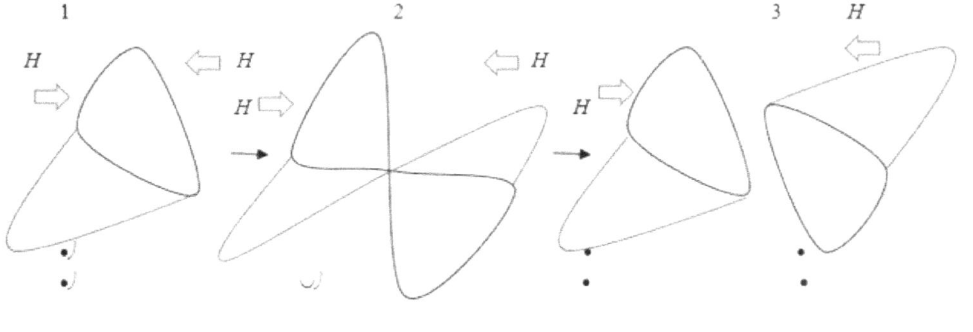

The stages of meiosis are illustrated in Figure 12.

Figure 12. Mitosis: 1 – initial state (one diploid (2n) cell); 2 - reduction stage (2 haploid cells); 3 - equational stage (4 haploids (n) cells)

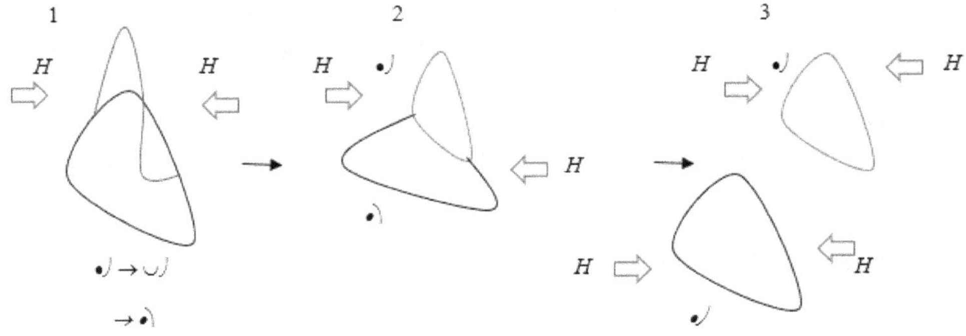

Mutual transitions of waves and solitons, similar to those shown in Figure 11, are not shown.

Amitosis. As already noted, some biologists don't recognize the existence of amitosis. However, it can also be illustrated (Figure 13); since there is no DNA replication and no chromosomes appear, the information component is not shown in the figure.

Figure 13. Amitosis: 1 - initial state; 2 - intermediate stage; 3 - separated cells

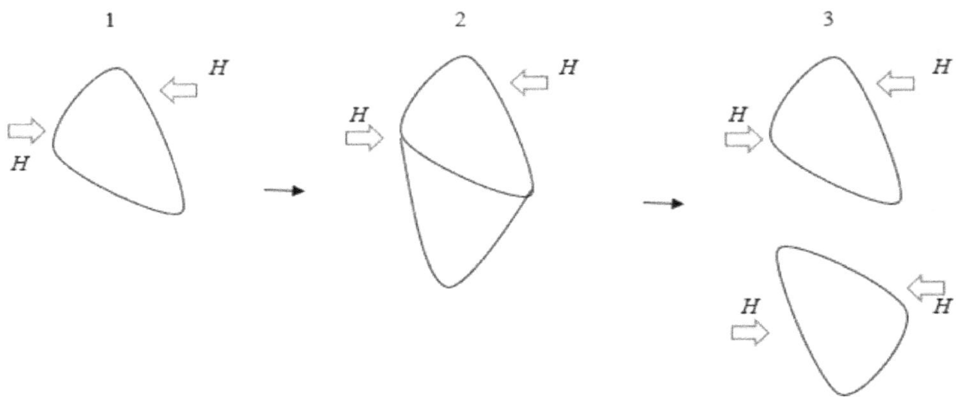

Asexual Reproduction Modeling

Vegetative reproduction and budding. For simplicity, we will show the formation of only one daughter organism (Figure 14). Here, for simplicity, we show only the real part, without the informational component.

Figure 14. Vegetative reproduction and budding: 1 - the beginning of budding (the appearance of a wave induced by external influences, blue), while the initial potential of the soliton decreases; 2 – formation of the daughter organism (its transition to the form of a soliton), the potential of the initial soliton decreases; 3 - the final separation of the daughter organism

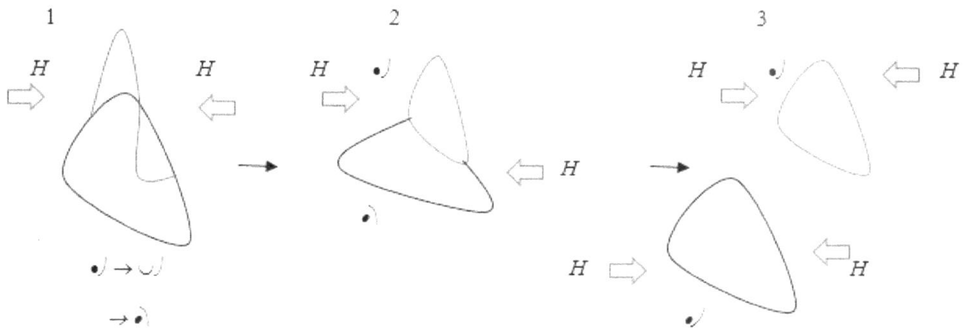

The same figure illustrates ***fragmentation*** - the transition from stage 2 to stage 3 (with an increased component amount).

Reproduction by spores is the same as the model of mitosis (Figure 11), taking into account the note made to it - negative external influences initiate the transition to the state of spores, and favorable external influences initiate the return to the state of a dividing cell.

Modeling of sexual reproduction. Of the features of sexual reproduction described above, the following are especially important for modeling:

- Compliance with the conditions of sexual reproduction (it is necessary to determine which of the interactions ***{DICS}*** can ensure the fusion of solitons and the subsequent development of the resulting soliton - a zygote; the same applies to the information component (chromosomal set).
- The possibility of realizing both ordinary sexual reproduction and its special cases.

Let's assume that the best interpretation of fertilization will be superposition and the occurrence of resonance. Recall that resonance is the frequency-selective response of an oscillatory system to a periodic external influence, which manifests itself in an increase in the amplitude of stationary oscillations and an increase in the responsiveness of the oscillatory system to the action of an external force. The effect is maximum, when the external action frequency coincides with the system's natural frequency, and the attenuation coefficient is zero. For systems of different natures, resonant frequencies are determined by various formulas. Then the conditions for successful sexual reproduction, namely, the presence of the reproductive age of the parents, the periodic readiness of the mother's reproductive apparatus, completed spermatogenesis in the father, and the sexual act itself can be considered as prerequisites for the emergence of resonance (Figure 15).

Figure 15. Sexual reproduction: 1 - from the many spermatozoa, the one who can initiate the resonance receives as a result the additional potential necessary to overcome the three barriers of the egg; 2 - as a result of fusion, a diploid zygote is formed, responsive to external influences that provide it with resources for existence and division

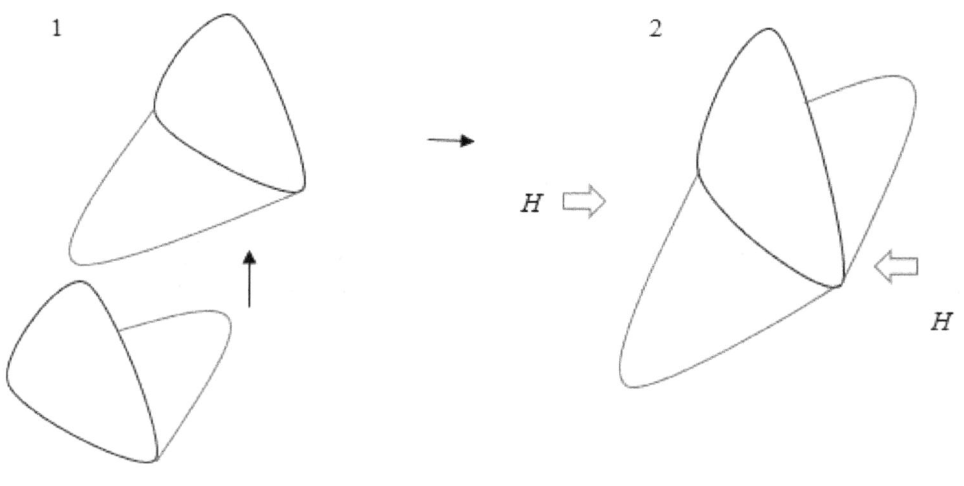

Let's consider the possibility of implementing various options for division and reproduction.

The implementation of meiosis, mitosis, and amitosis follows directly from the comparison of stage 2 with Figures 11-13.

Animal hermaphroditism (plant monoeciousness), animal parthenogenesis (plant apomixis), progenesis, metagenesis, heterogony - it can be assumed that more complex wave models should be used to model them, taking into account not only the effects of resonance but also modulation. As modulating rhythms, it is natural to propose annual and lunar cycles, as well as some rhythms of autoregulation, which are probably also evolutionarily associated with natural rhythms (Zidermane, 1998). Modeling is complicated because there are numerous varieties of reproduction, often significantly different from each other and insufficiently studied.

As an example, let's take a simple definition of gender. (Hughes et al., 2005). Mammals' females have two X chromosomes (XX), while males have different X and Y chromosomes (XY). Although women have two X chromosomes, in somatic cells; one is inactivated. One X chromosome is inherited from the mother, and the second (in women) from the father. However, some male cells contain two (or more) X chromosomes and one Y chromosome (Klinefelter's syndrome); or one X chromosome and two Y chromosomes (XYY syndrome). The cells of some women contain several, more often three or one X chromosome (Shereshevsky-Turner syndrome).

Further, in some mammals, such as the platypus, sex is determined by not one but five pairs of sex chromosomes. Many cold-blooded vertebrates lack sex chromosomes. If they have two sexes, then the sex is modulated by annual rhythms and environmental conditions to a greater extent than genetically. In some reptiles, sex depends on incubation temperature; others are hermaphrodites.

The human X chromosome contains more than 1400 genes, and the Y chromosome contains a little more than 86 genes. The X and Y chromosomes are believed to have evolved from a pair of identical chromosomes (containing over 1,400 genes) when a gene emerged in ancient mammals, one of whose alleles led to the development of a male organism. The chromosomes carrying this allele became Y chromosomes, and the second chromosome became the X chromosome. The X- and Y-chromosomes initially differed by one gene. Subsequently, genes beneficial to males and harmful to females developed on the Y chromosome, or moved to the Y chromosome (translocation). The Y chromosome has lost most of its original genes. In the later stages of Y chromosome degeneration, other chromosomes increasingly use the genes and functions previously associated with it. Several species of rodents have reached the stage where the Y chromosome has disappeared, and they have a new sex determination system. The sex ratio in the population will remain close to 1:1 (Fischer's principle).

Thus, the assumption of combinations of resonance and modulation for modeling sexual reproduction is still a hypothesis, the detailed substantiation of which is beyond the scope of this study.

The meaning of the emergence and preservation of various variants of sexual and asexual reproduction becomes clear when considering the increasing complexity and the inheritance of the structure during evolution.

The increase in complexity and its inheritance is illustrated in Figure 16. During the transition from stages 1 to 2 and 3 and complication, various scenarios (6.2-6.4) or their combinations (6.1) can be implemented with mandatory external influences that replenish the potential of the system (not shown). To ensure the inheritance of a complex structure, more and more complex reproduction methods are needed, including modulation of the rhythms of the external environment.

OPTIMIZATION

Now, having convinced ourselves of the correctness of modeling the processes of evolution, inheritance, and reproduction using various self-organization scenarios, let's look at the evolution phenomenon from a more general position.

A living system, as defined in Chapter 4, has an obvious goal - it is own survival. Moreover, the survival even after the death of an individual organism, by passing its properties to descendants through reproduction and inheritance. Such a system operates in an environment with various resources and restrictions, some of which are involved in its life processes. In contrast, others limit living processes or are neutral.

All this is similar to a classical restricted optimization problem formulation: in a limited region of the parameter space, it is necessary to find the extremal value of the objective function that describes the ratio of optimization criteria.

Figure 16. Complications of the structure of living organisms and inheritance options

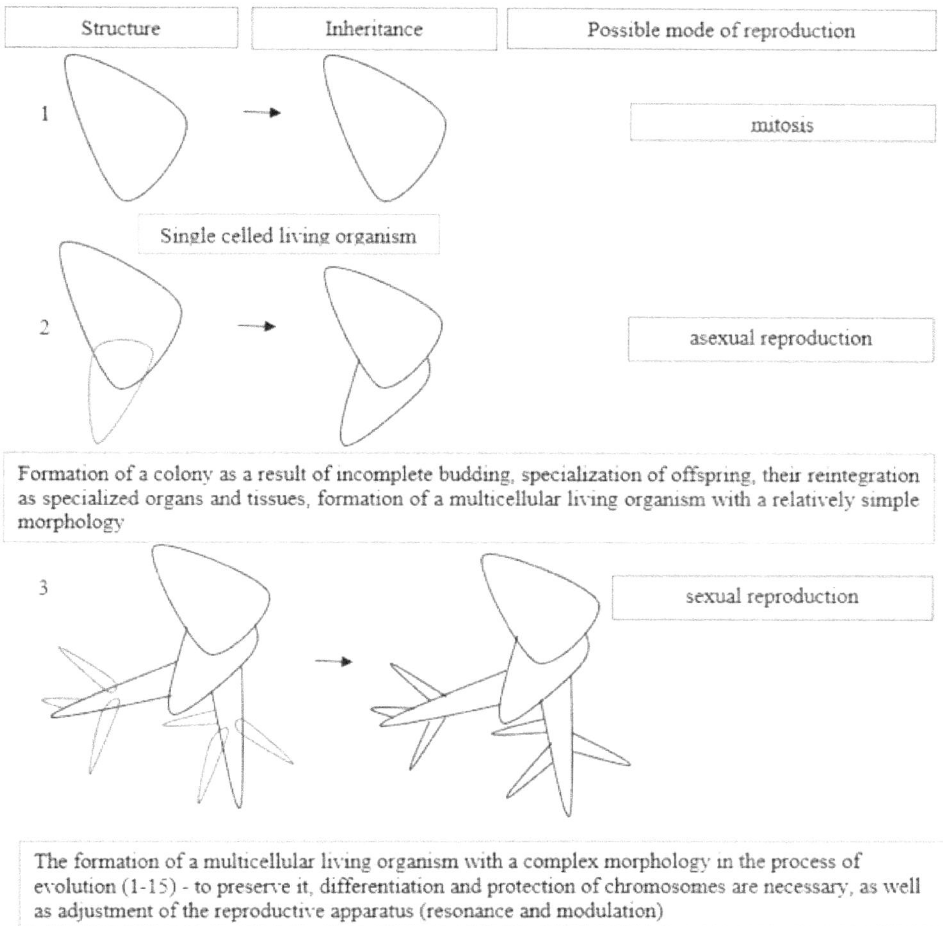

Due to the peculiarities of living systems (Table 8, Chapter 1), indefinite parameters and criteria quantity, and qualitative differences in their units of measurement, classical methods for solving optimization problems are of little use for modeling evolution as an optimization.

Another thing is the optimization methods within the framework of the theory of self-organization. Here, the analog of the objective function is the system itself, the analog of the parameters is potential interactions and resources, the analog of optimization criteria is the resources obtained as a result of actual interactions with the environment, the analog of limits is the potential and actual characteristics of interactions and resources.

The system evolves, that is, it organizes itself under the influence of external factors according to one of the scenarios discussed in Chapter 4. The optimal state of the system defines the maximum of integrity signs, expressed through golden sections, Fibonacci numbers, symmetries, etc. (Chapter 4).

In the course of evolution, the system changes - the appearance of elements at each new level (respectively, new interactions and resources) increases the overall potential of the system, creating conditions for further stratification, reparation, and reproduction, but also increasing asymmetries and chaos within the system.

Thus, evolution aims to preserve the system's existence in interactions with the environment (mechanisms of self-regulation, reparation, and reproduction) and ensure the system has resources.

Following this, we will show the scheme of exchanges and interactions within a conditional ecological system (Figure 17).

Figure 17. A conditional ecological system optimization: resources, self-organization, self-regulation, the direction of evolution. Designations: blue arrows – external impacts; blue curve arrows - corpses and waste products; circled in red – relations of symbiosis, parasitism, etc.; black lines – hierarchic relations in the organism; orange arrows – evolution direction

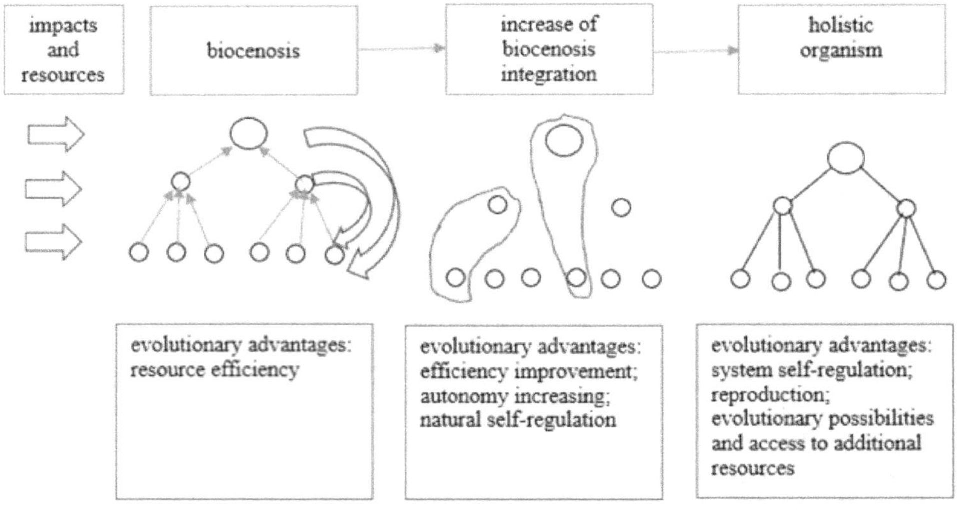

Let's add a few comments.

1. The direction of evolution from the biocenosis to the whole organism is one of the possibilities (along with the differentiation of colony cells) of multicellular organisms; it also occurs in the formation of cells.
2. The unit of evolution is the entire ecosystem of the Earth, including biological and inorganic components; only for the convenience of modeling is it possible to single out individual biocenoses, species, or organisms, which in themselves are not separate structures, but subsystems.
3. At the same time, the Great Oxygen Event in the past or human activity in the present, which assumed a planetary character, stigmatically pushed the Earth's ecosystem development.
4. The immanent cause of the death of an ecosystem or an individual organism is the finiteness of the system, which, according to Corollary 1 of property 4.9, is due to the limited lifetime of the 1st layer.
5. Reproduction and inheritance are the answer of life to the immanent cause of death.
6. Self-regulation and reparation are the reactions of life to external causes of death.
7. It is impossible to estimate the limits of self-regulation for all possible cases due to the variety of self-organization scenarios; We will return to assessing these limits for humans in Chapter 8.

FUTURE RESEARCH DIRECTIONS

The correctness of using the wave model for modeling the human-environment system (Chapters 7, 8) and studying the human life cycle (Chapter 9) follows from the previous. The adequacy of the modeling apparatus creates the basis for its wider application for a detailed study of evolutionary processes - with different calibrations, calculations, and types of models at separate levels. The hypothesis about the role of resonance and modulation seems promising for understanding the process of sexual reproduction, but it needs to be thoroughly tested.

CONCLUSION

The wave model and the theory of self-organization were used to model the evolution of living organisms. At the same time, all 15 identified patterns were explained within the framework of the self-organization scenarios considered in Chapter 4. Considering the fact that inorganic evolution (Chapter 1), synthesis of complex organics (Chapter 2), reproduction, and inheritance also correspond to the wave model abstractions and axioms, it can be stated that the developed modeling apparatus is adequate and can be used in the future to achieve the goals of this work.

The application of the wave model and the theory of self-organization of complex systems allows us to draw the following conclusions regarding the evolutionary process as a whole:

- The complex systems' evolution general regularities exist and explain the purposefulness of evolution. Molecular genetic factors, the struggle for existence, natural selection, and ecological restrictions are particular ways of realizing general regularities.
- By the time of the emergence of life (the Universe had already existed for about 9.8 billion years, and the Earth for about 600 million years, Chapter 1) external factors were formed, and they had evolved over about 4 billion years of life's existence. These factors formed an “advanced matrix”, which guides the evolution of living organisms. Thus, orthogenetic regularities exist but are explained not from creationist positions, but by “aligning” living organisms with the “advanced matrix” of the Universe.
- between the processes occurring at the molecular-genetics level, organismal, population, species (and higher taxa) levels, as well as in ecological systems, there are some similarities (and differences), which manifests another relationship between the levels of the system - self-similarity or fractality.
- the modeling accuracy used in this chapter is sufficient for this study; if necessary, it can be increased, using the dynamic and simulation models, given in Chapter 4.
- the considered models of reproduction and inheritance correspond to their main ways, and show, how in the process of evolution more and more complex structures arise and inherit.
- the assumption was formulated: in the case of sexual reproduction, the best interpretation of fertilization will be superposition, and the condition for success – is the occurrence of resonance.

Consideration of the evolution as not only self-organization, but as optimization also, allows us to explain its purposefulness, and clarify some of its mechanisms. However, that requires further research.

REFERENCES

Alberts, B., Johnson, A., Lewis, J., Raff, M., Roberts, K., & Walter, P. (2002). *Molecular biology of the cell* (4th ed.). Garland Science. DOI: 10.1093/aob/mcg023

Bernstein, H., Hopf, F. A., & Michod, R. E. (1987). The molecular basis of the evolution of sex. In *Advances in Genetics* (pp. 323–370). Elsevier. DOI: 10.1016/S0065-2660(08)60012-7

BrainKart. (2020). Depicts the process of gametogenesis in both males and females, highlighting the importance of meiosis. *Gametogenesis*. BrainKart. Retrieved August 13, 2024, from https://www.brainkart.com/article/Gametogenesis_38015/

Breseghello, F., & Coelho, S. A. G. (2013). Traditional and Modern Plant Breeding Methods with Examples in Rice (*Oryza sativa* L.). *Journal of Agricultural and Food Chemistry*, 61(35), 8277–8286. DOI: 10.1021/jf305531j PMID 23551250

Ecoddington14. (2014). Binary_Fission. *File: Binary fission 2.Svg*https://commons.wikimedia.org/wiki/File:Binary_Fission_2.svg

Fasoula, D. A., Ioannides, I. M., & Omirou, M. (2020). Phenotyping and Plant Breeding: Overcoming the Barriers. *Frontiers in Plant Science*, 10, 1713. Advance online publication. DOI: 10.3389/fpls.2019.01713 PMID: 31998353

Golitsin, Y. N., & Krylov, M. C. (2010). *Cell division: theory, variants, and degradation*. Nova Science Publishers.

Hartung, F., & Schiemann, J. (2014). Precise plant breeding using new genome editing techniques: Opportunities, safety, and regulation in the EU. *The Plant Journal*, 78(5), 742–752. DOI: 10.1111/tpj.12413 PMID: 24330272

Hayflick, L., & Moorhead, P. S. (1961). The serial cultivation of human diploid cell strains. *Experimental Cell Research*, 25(3), 585–621. DOI: 10.1016/0014-4827(61)90192-6 PMID: 13905658

Hughes, J. F., Skaletsky, H., Pyntikova, T., Minx, P. J., Graves, T., Rozen, S., Wilson, R. K., & Page, D. C. (2005). Conservation of Y-linked genes during human evolution revealed by comparative sequencing in chimpanzees. *Nature*, 437(7055), 100–103. DOI: 10.1038/nature04101 PMID: 16136134

Johanson, D. C., & Wong, K. (2009). *Lucy's Legacy: The Quest for Human Origins*. Crown.

Jones, R. E., & Lopez, K. (2013). *Human Reproductive Biology* (4th ed.). Academic Press. [Textbook]

Kovalyov, Y., Mkhitaryan, N., & Nitsyn, A. (2020). *Self-organization of the Human Mind and the Transition from Paleolithic to Behavioral Modernity*. IGI Global International Publisher of Progressive Information Science and Technology Research. DOI: 10.4018/978-1-7998-1706-2

Lewin, B. (2004). *Genes VIII*. Pearson Prentice Hall.

Lodé, T. (2001). *Les stratégies de reproduction des animaux*. Dunod.

Lovett Doust, J., & Lovett Doust, L. (1988). *Plant reproductive ecology: patterns and strategies*. Oxford University Press.

Mayhew, P. J. (2006). *Discovering Evolutionary Ecology: Bringing Together Ecology and Evolution*. Oxford University Press. DOI: 10.1093/oso/9780198570608.001.0001

Mitchison, J. M. (2003). Growth during the cell cycle. In *International Review of Cytology* (pp. 165–258). Elsevier. DOI: 10.1016/S0074-7696(03)01004-0

Molinari, M. (2000). Cell cycle checkpoints and their inactivation in human cancer. *Cell Proliferation*, 33(5), 261–274. DOI: 10.1046/j.1365-2184.2000.00191.x PMID: 11063129

Rosenzweig, M. (1995). *Species diversity in space and time*. Cambridge University Press. DOI: 10.1017/CBO9780511623387

ADDITIONAL READING

Babynin, E. V. (2007). Molekuljarnij mekhanyzm ghomologhychnoj rekombynacyy v mejoze: Proyskhozhdenye y byologhycheskoe znachenye [Molecular mechanism of homologous recombination in meiosis: origin and biological significance]. *Tsitology J*, 49(3), 182–193.

Bogdanov, Y. F. (2003). Varyacyy y evoljucyja mejoza [Variation and Evolution of Meiosis]. *Russian Journal of Genetics*, 39(4), 363–381. DOI: 10.1023/A:1023345311889

Buldakov, L. A., & Kalistratova, V. S. (2003). *Radyoaktyvnoe yzluchenye y zdorovj'e* [Radioactive radiation and health]. Inform-Atom.

Mkhitaryan, N. M., Badeyan, G. V., & Kovalyov, Y. N. (2004). *Erghonomycheskye aspekti slozhnikh system* [Ergonomic Aspects of Complex Systems]. Naukova Dumka.

ENDNOTE

1 Some researchers deny the existence of amitosis.

Chapter 7
The Human Body System, Structure, and Self-Regulation

ABSTRACT

A human is considered a holistic open system, consisting of separate subsystems and components. The body's integrity is ensured by self-regulation and non-additivity of exchanges and interactions of various levels within the organism. This approach differs from the "theoretical set" understanding, presented in Chapter 2, when the human life key structures and processes were reduced to elements and interactions at the cellular and gene-molecular levels. The materials of this chapter are data for modeling interactions in the human-environment system (Chapter 8). They are also necessary for understanding the processes occurring during the life cycle, especially aging (Chapter 9) and the near-death state (Chapter 10).

BACKGROUND

This chapter aims to prepare data for the human body structure modeling as an integral open system (Chapter 8). Here, the emphasis is on the integrity of the system and the mechanisms for ensuring it, while its openness, manifested in various interactions with the environment, is considered in Chapter 8. As in the previous chapters, we will focus on facts and verified sources when considering general information.

Human is a complex system, that includes anatomical, physiological, and psychological components. The anatomical structure of a human is well known; there is a large amount of scientific and educational literature (Gaivoronsky et al., 2011), as well as atlases - printed (Sinelnikov & Sinelnikov, 1996) and electronic (The Human Body, n/d), including those provided in the 3D form (Anatomy 3D Atlas, n/d).

DOI: 10.4018/978-1-6684-8509-5.ch007

Human physiology has also been sufficiently studied (Best et al., 1985; Kositsky, 1985; Guyton & Hall, 2005/1969; Schmidt & Tews, 1983; Filimonov, 2012;). However, some unresolved problems remain here, of which, based on systemic prerequisites, the mechanisms of self-regulation are of the greatest interest (Sherwood, 2012).

Human psychological research is represented by extensive literature (Silverthorn, 2015; Patil et al., 2009). There are even more unresolved problems here than in physiology. Note that psychological functions do not directly depend on either anatomy or physiology (Bednarik, 2013); however, there are undoubted links between them, emphasized by the terms "psychophysiology" or "neurophysiology" (Tsirkin et al., 2023).

Another aspect is human chronobiology and the connection of human biorhythms with rhythms of Nature (Zidermane, 1988; Eisenstein, 2013; Çalıyurt, 2017). They also publish magazines such as Chronobiology International.

The integrity of the human body is expressed as the constancy and consistency of the parameters of the internal environment - homeostasis, which is ensured by various levels and mechanisms of regulation. This is also the subject of extensive literature (McEwen & Wingfield, 2003; Asarian, 2012; Billman, 2020; Libretti & Puckett, 2022). Let's note the problems of determining controlled parameters, their mutual relations, and their modeling (Molenaar, 2001; Mkhitaryan et al., 2004; Milhiet et al., 2011).

THE HUMAN BODY STRUCTURE

Like a cell, the human body is also a complex system (much more than a cell), that exists in a dissipative external environment. As a cell, it includes structures that provide internal and external metabolism, energy, and information, maintain the constancy of the internal environment, move in space, protect, restore, and reproduce. Anatomically and physiologically, the human body is divided into 11 systems sometimes 10 or 14 are indicated, but these are just different classifications). There are also sense organs; however, due to their diversity, they do not form a separate anatomic system (Figure 1).

Figure 1. Human body features and organ systems (According to (Jmarchn, 2009. Based on: Human body features.png. Original: The Wednesday Island Vector and List of systems, n / d)

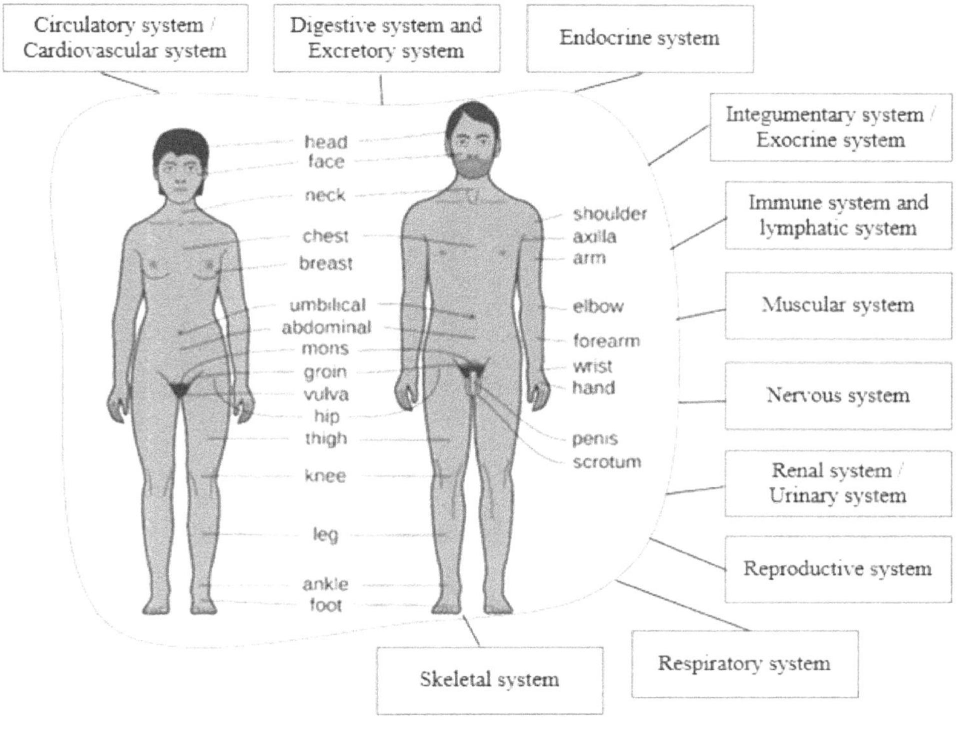

Let us give a brief description of the structure and functions of each of them.

Circulatory System/Cardiovascular System

The circulatory system / Cardiovascular system is shown in Figure 2.

Figure 2. The circulatory system in anterior view (LadyofHats & Mariana Ruiz Villarreal, 2009)

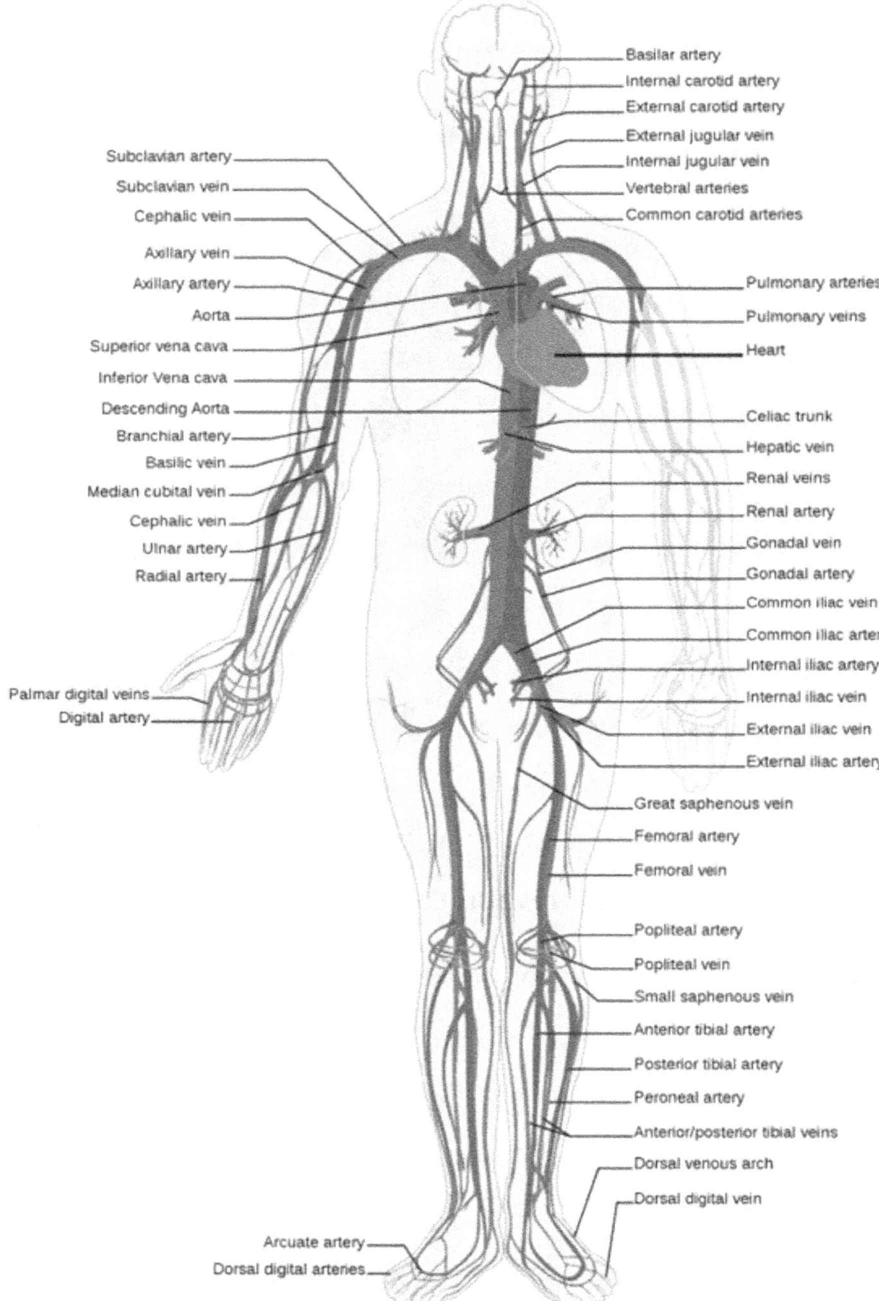

The main functions are: to circulate blood around the body via the heart, arteries, and veins, deliver oxygen and nutrients to organs and cells carry their waste products away, and keep the body's temperature in a safe range. Oxygenated and nutrient-rich blood is distributed to tissues via the arterial system, which branches into smaller and smaller blood vessels from arteries to arterioles to capillaries (where most exchange occurs). Deoxygenated blood and metabolic byproducts are returned from capillaries via venules and veins. The heart functions as a pump to maintain circulation. These substances enter and are removed from the bloodstream through the extracellular fluid.

The components and their functions are the heart, arteries, capillaries, and veins. The *heart* is a discrete organ made up of four separate chambers. The right atrium and right ventricle receive blood returning from the periphery and send it through the pulmonary artery to the lungs for reoxygenation. Once the blood is oxygenated in the lungs, it returns to the left side of the heart through the pulmonary veins. After entering the left atrium, the blood enters the left ventricle and is pumped into the aortic arch for distribution throughout the body. This is how the small and large circles of blood circulation function.

Regulation occurs through stimuli, including changes in blood volume, hormones, electrolytes, osmolarity, drugs, adrenals, kidneys, etc. The parasympathetic and sympathetic nervous systems also play a key role in cardiovascular system regulation.

Digestive System and Excretory System (Figure 3).

Figure 3. Diagram of the human digestive (Mariana Ruiz & Jmarchn, 2006)

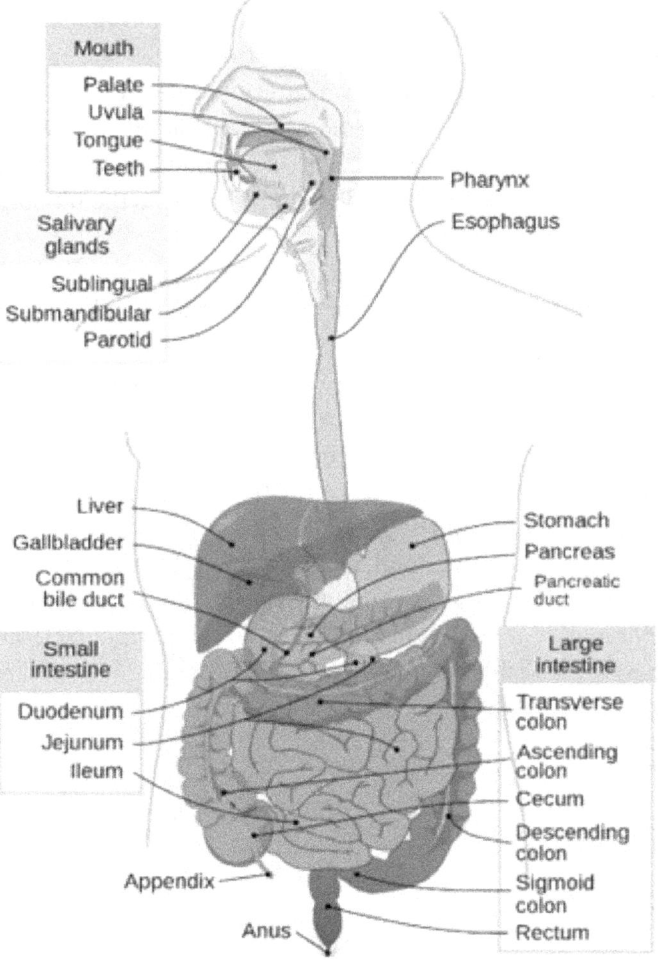

The main functions are: to absorb nutrients, remove waste via the gastrointestinal tract, and eliminate waste from the body (Whitbourne, 1985).

The components and their functions: the mouth, salivary glands, esophagus, stomach, liver, gallbladder, pancreas, and small and large intestines.

Mouth: teeth chew, turning them into small pieces, saliva converts chewed particles into a bolus. The bolus contains electrolytes, salivary amylase (a digestive enzyme), and lysozyme. Salivary amylase converts starch to maltose. The tongue facilitates the food movement during chewing and helps swallow the bolus.

Salivary glands: moisten food and begin chemical digestion of starches

Esophagus: after the mouth, the bolus is pushed down into the esophagus (swallowing and deglutition). The bolus is then advanced, stimulated by waves generated by the contraction of the muscular wall of the esophagus (a process called peristalsis).

Stomach: delays food for four to five hours. In the stomach, food is mixed with gastric juice by the agitation of the stomach walls to form chyme. Hydrochloric acid in the stomach provides an acidic environment that activates the inactive digestive enzyme pepsinogen into the active enzyme pepsin. Pepsin converts proteins into proteases and peptones and food is partially digested inside the stomach.

Small intestine: partially digested food moves into a long tube called the small intestine. The wall of the small intestine produces intestinal juice. Intestinal juice with bile and pancreatic juice acts on partially digested food and converts most of it into liquid. In addition, the small intestine contains various finger-like projections known as villiб which provide maximum surface area for absorption of nutrients into the blood capillaries.

Liver: (the largest gland of the body) secretes bile juice. The hepatic duct in the liver transports bile into the gallbladder.

Gallbladder: stores bile produced by the liver. From the gallbladder, the bile duct empties the bile into the small intestine.

Pancreas: secretes pancreatic juices to help the digestion of proteins and carbohydrates. The accessory ducts of the pancreas are the duct of Santorini, which opens directly into the small intestine and the duct of Wirsung; this duct and the bile duct form the hepatopancreatic duct, which opens into the duodenum.

Food is digested in the small intestine, but the undigested part passes into the large intestine.

Large intestine: absorbs water and minerals from the undigested part of the food. Digestive waste is thrown out through the anus.

Regulation: nervous and endocrine regulatory mechanisms maintain, through mechanical and chemical activity, the optimal conditions necessary for digestion and absorption.

Endocrine System (Figure 4).

The main function is to influence the body's functions using hormones.

The components and their functions (Hiller-Sturmhöfel & Bartke, 1998) are:

Hypothalamus: resides in the brain and controls the endocrine system, using data from the nervous system to determine when other glands, including the pituitary gland, should release hormones. The hypothalamus controls many processes, including mood, hunger and thirst, sleep patterns, and sexual function.

Figure 4. Endocrine system (Anatomy and Physiology for Health Professionals: Endocrine System, n / d)

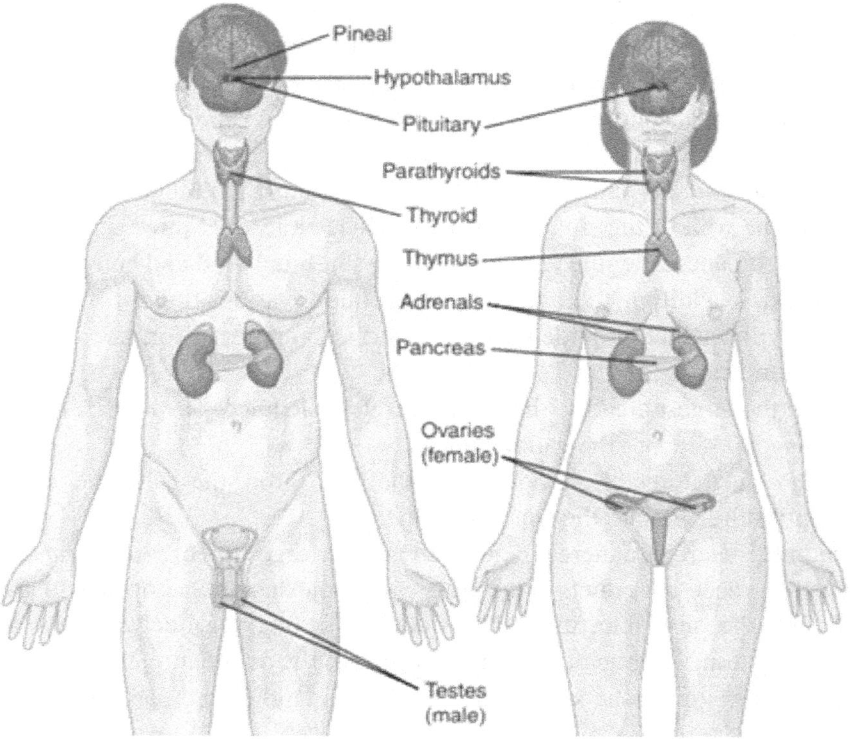

Pituitary: located at the base of your brain; makes hormones that control several other glands such as the thyroid gland, adrenal glands, ovaries, and testicles. The pituitary gland has many different functions, including body growth. It's.

Thyroid: located in the front of the neck. It's responsible for metabolism.

Parathyroid: these four glands control the calcium level, which is important for the heart, kidneys, bones, and nervous system.

Adrenal: two adrenal glands, one on top of each kidney. control metabolism, blood pressure, sexual development, and stress response.

Pineal: manages sleep cycle by releasing melatonin, a hormone that causes one to feel sleepy.

Pancreas: this part of your endocrine system plays a significant role in the digestive system. It makes the hormone insulin that controls the blood sugar level.

Ovaries: two ovaries in the lower abdomen, one on either side, release women's sex hormones called estrogen, progesterone, and testosterone.

Testicles: make sperm and release the hormone testosterone, which affects sperm production, muscle strength, and sex drive.

Regulation: the general scheme is shown in Figure 5.

Figure 5. Schematic representation of negative feedback mechanisms that control endocrine system activity (Hiller-Sturmhöfel, Bartke, 1998) Orange arrow - direct action; blue arrow – feedback

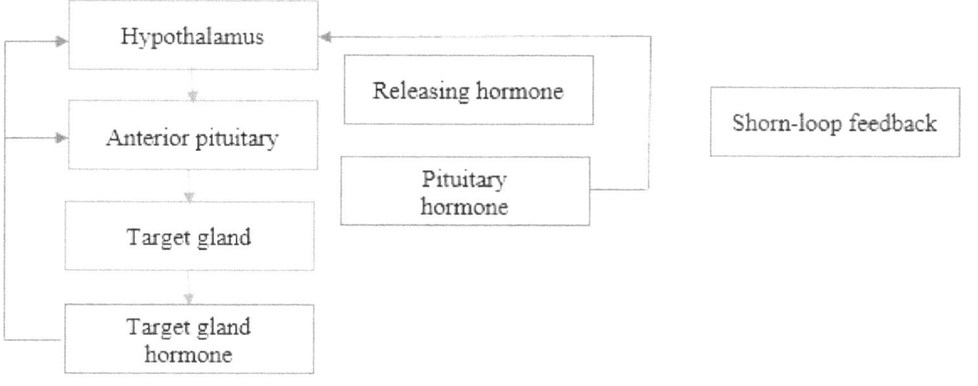

Integumentary System/Exocrine System

The main function is to protect underlying tissues from the skin, nails, and hair.

The components and their functions: skin, hair, nails, sweat, and other exocrine glands.

Skin: guarding underlying muscles and organs, protecting against pathogens, insulation, temperature regulation, sensation, and vitamin D and B synthesis (Figure 6). Skin is composed of the epidermis and the dermis. Below them lies the hypodermis or subcutaneous adipose layer, which is not a layer of skin.

Hair: there are three different types of hair: lanugo, vellus, and terminal.

Nails: protective plate and increased sensitivity of the fingertip.

Glands: sweat glands (the eccrine glands produce sweat; the primary function is body temperature regulation and the apocrine glands, which develop only at puberty and secrete more sweat-containing fats for a month, and then secrete a normal amount of sweat); sebaceous glands; ceruminous glands; mammary glands.

Regulation: by external conditions and limits, and restrictions associated with the conditions of joint work with the Immune system, Digestive system, Cardiovascular system, Nervous system, Respiratory system, and Urinary system.

Figure 6. Anatomy of the skin (National Cancer Institute. Anatomy of the Skin, n / d)

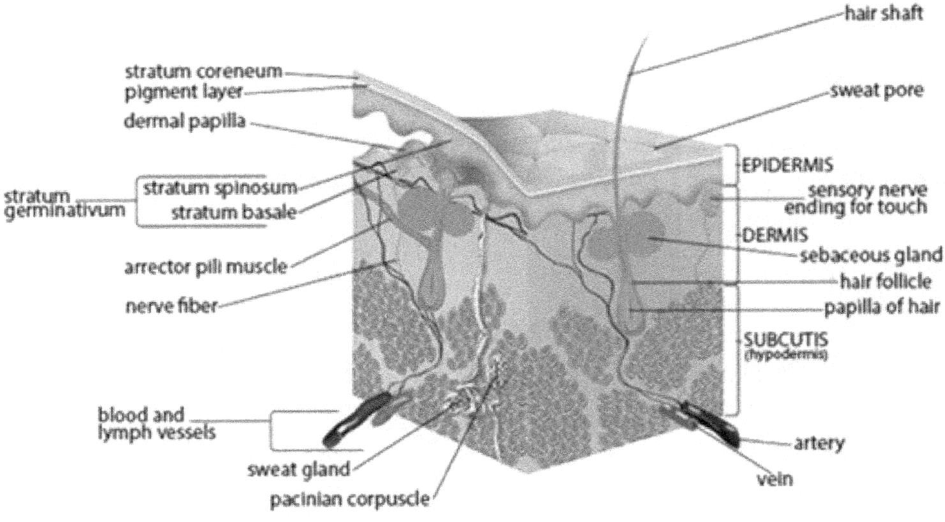

Immune System and Lymphatic System

The main functions are: for the lymphatic system, the transfer of excess fluid into the bloodstream, and the filtration of pathogens from the blood; for the immune system, the destruction or neutralization of pathogens that otherwise cause disease or death.

The components and their functions (Anatomy of the Lymphatic and Immune Systems, n / d)

the lymphatic vessels begin as open-ended capillaries, which feed into larger and larger lymphatic vessels, and eventually empty into the bloodstream by a series of ducts. The lymph travels through about 500–600 lymph nodes near the groin, armpits, neck, chest, and abdomen. A major function of the lymphatic system is to drain body fluids and return them to the bloodstream (Figure 7).

The immune system includes a set of barriers, cells, and soluble proteins that interact with each other in very complex ways. The modern model of the immune function is in charge of three stages, depending on the speed of their action:

- Barrier protection (skin and mucous membranes) instantly prevents pathogenic invasion in body tissues.
- Fast non-specific congenital immune response, consists of specialized cells and soluble factors.

- A slow specific adaptive immune response, includes many types of cells and soluble factors but is primarily controlled by white blood cells (leukocytes: granulocytes, monocytes, and lymphocytes) that help control immune answers.

Blood cells involved in the immune response occur in the bone marrow through differentiation from hematopoietic stem cells. Hematopoietic stem cells are present throughout adulthood. Depending on the function, they distinguish:

- Phagocytic cells that swallow pathogens to destroy them.
- Lymphocytes that coordinate the activity of adaptive immunity/
- Cells containing cytoplasmic granules that help mediate immune answers against parasites and intracellular pathogens, such as viruses.

Figure 7. Lymphatic system (BruceBlaus, 2013. "Medical Gallery of Blausen Medical 2014". WiiJournal of Medicine)

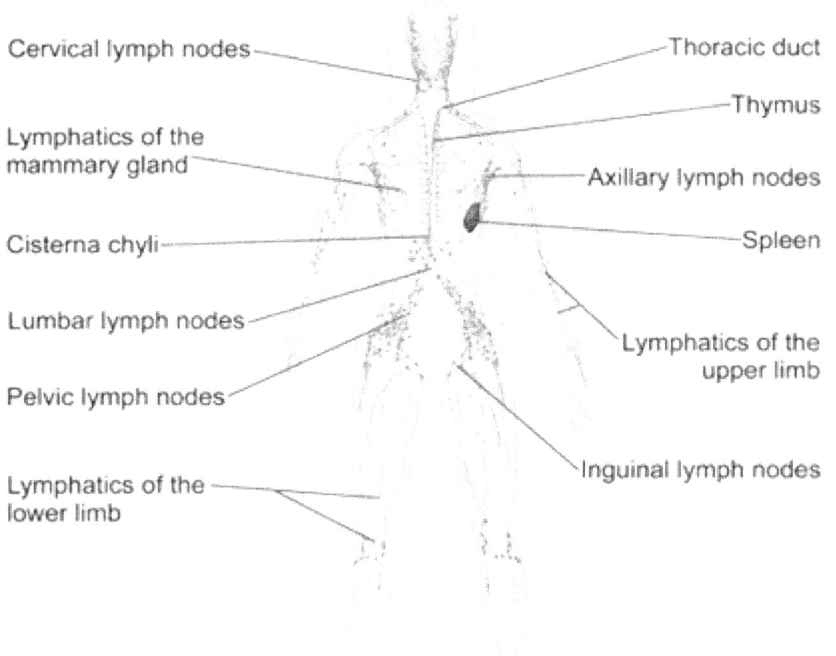

Figure 8 shows the processes involved in the primary immune response.

Regulation: Lymphatic function is variable and modulated by numerous factors including chronic inflammation, tumors, external stimuli such as radiation, age, obesity, and metabolic dysfunction. This variability (Figure 8) can play a key role in immune reactions (Kataru et al., 2019).

The Muscular System (Figure 9 and Figure 10)

The main function is to enable the body to move using muscles.

The components and their functions: are skeletal, smooth, and cardiac muscles.

Skeletal muscle: a striated muscle composed of muscle fibers formed by myofibrils bridged by sarcomeres. When stimulated, skeletal muscles perform a coordinated contraction, shortening each sarcomere. This process consumes large amounts of adenosine triphosphate (ATP). When ATP is used, it is converted to adenosine diphosphate (ADP). Muscle tissue also contains a store of fast-acting creatine phosphate, which, if necessary, can promote the rapid regeneration of ADP into ATP and calcium ions. There are approximately 639 skeletal muscles in the human body.

Figure 8. The processes involved in the primary immune response (Sciencia58 the makers of the single images Domdomegg, Fæ, Petr94 & Manu5, 2020. Using Files - in References)

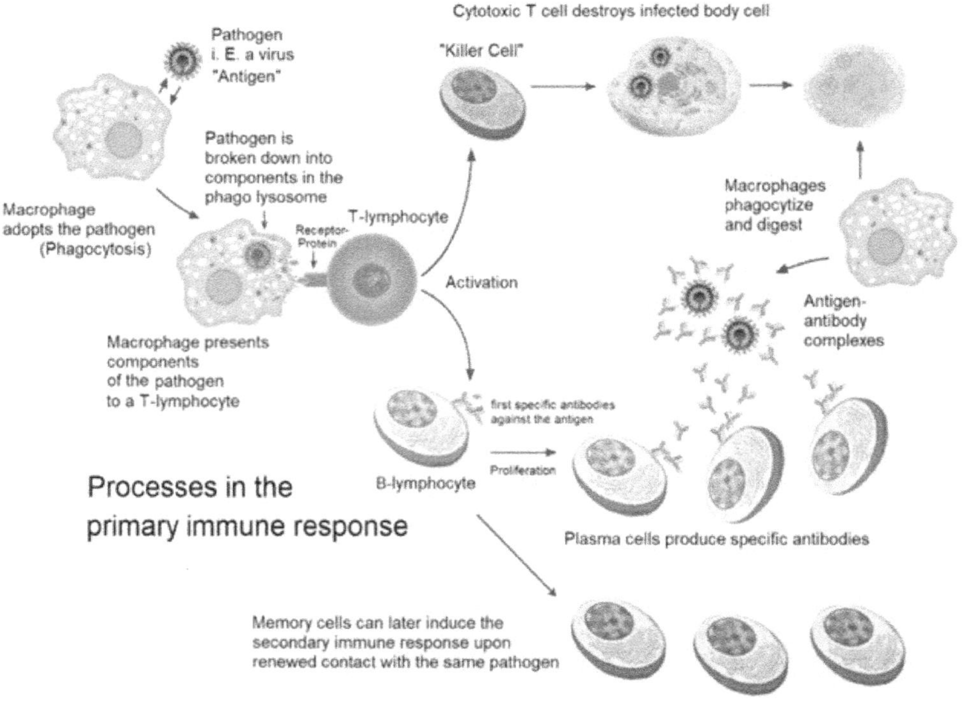

Smooth muscles: are controlled directly by the autonomic nervous system; their movements are involuntary.

Cardiac muscle: is a striated muscle, but unlike the skeletal muscle, its fibers are connected laterally, and their movements are involuntary. The sinus node controls the heart muscle, which the autonomic nervous system influences.

Regulation: the muscular systems are controlled through the nervous system; some muscles, such as the cardiac muscle, may be completely autonomous.

Figure 9. Anterior view of muscles (CFCF, 2013)

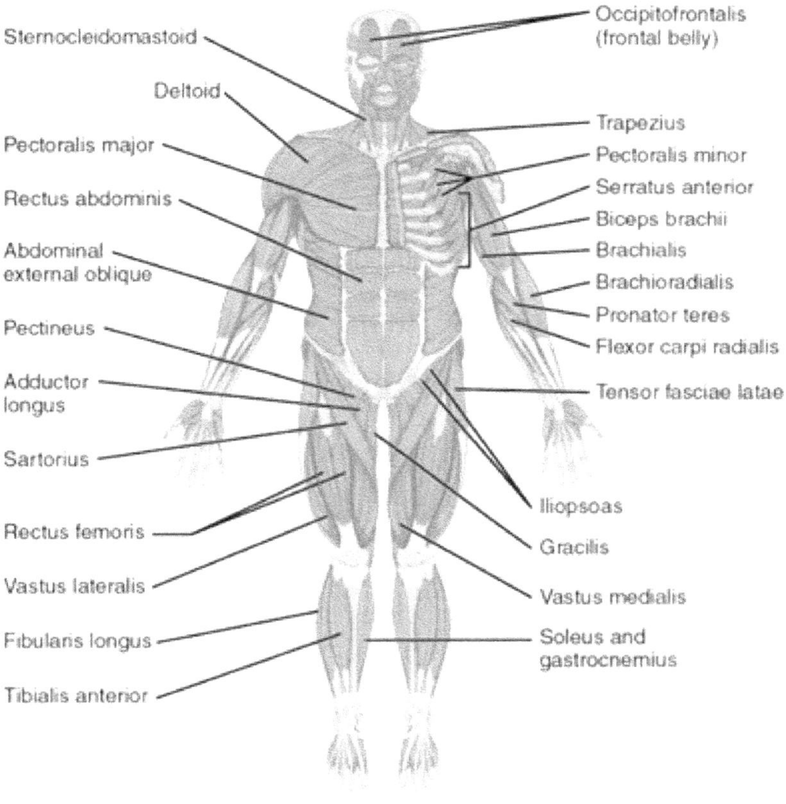

Major muscles of the body.
Right side: superficial; left side: deep (anterior view)

Figure 10. Posterior view of muscles (CFCF, 2013)

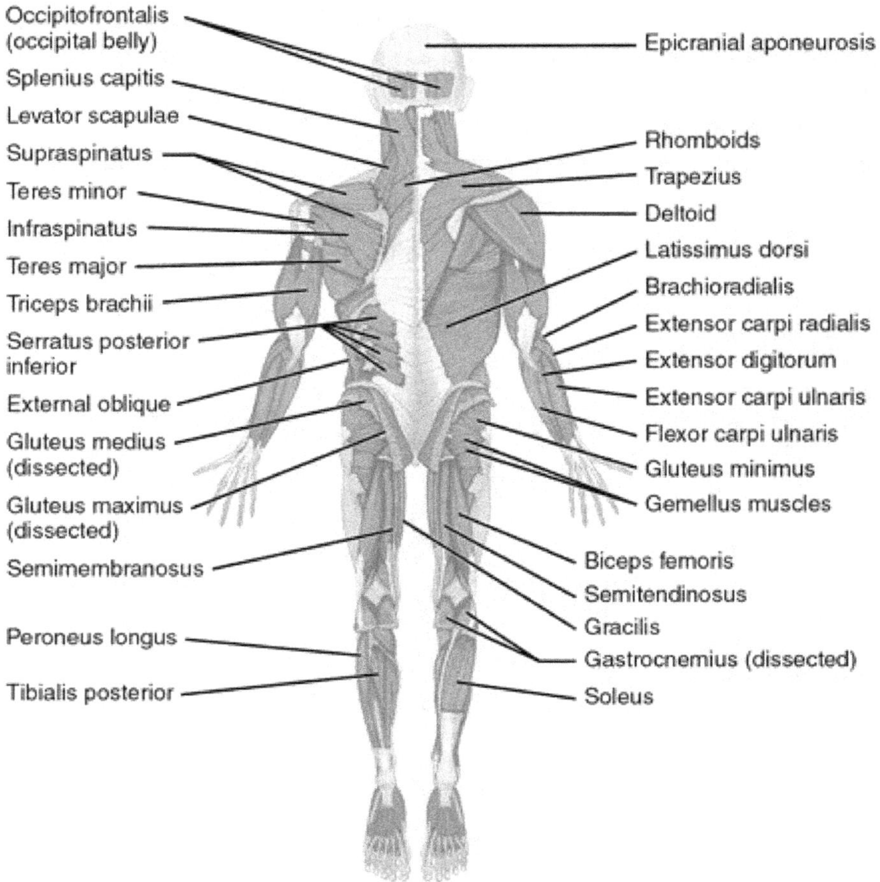

Major muscles of the body. Right side: superficial; left side: deep (posterior view)

Nervous System (Figures 11-13)

The main functions can be grouped as such overlapping functions:

Sensory - contact through receptors with the environment, both external and internal;

Mental - thinking, learning, and memory;

Regulatory (together with the endocrine system) - regulation and maintenance of homeostasis;

Motor - control of various muscle groups;

Integrative - millions of sensory receptors detect changes inside the body (changes in pressure, pH, carbon dioxide concentration, and levels of various electrolytes) and outside the body (temperature, light, sound, etc.). This information is called touch input. Sensory input is converted into electrical signals (nerve impulses), transmitted to the brain. The signals combine to create sensations, produce thoughts, or add to memory. Based on integrated sensory input, the nervous system responds by sending signals to effectors, causing the muscles to contract or causing secretions to be released from the gland. This is the "official" point of view on the integrative activity of the nervous system (National Cancer Institute. Introduction to the nervous system, n / d).

The components and their functions: are *the brain*, *spinal cord*, *nerves*, and *ganglia* (Figure 11). These, in turn, consist of various tissues, including nerve, blood, and connective tissue. Together these carry out the complex activities of the nervous system.

Figure 11. Nervous system diagram (Medium69 & Jmarchn, 2014)

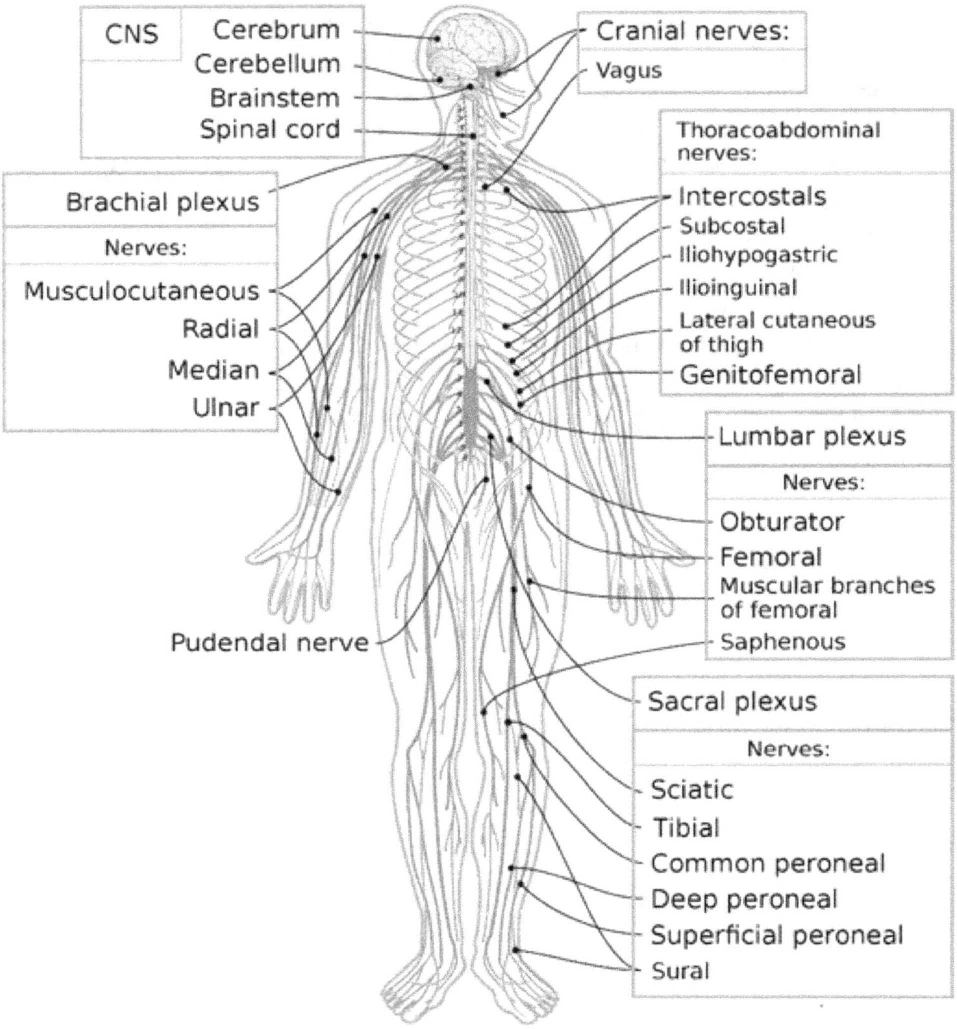

The nervous system consists of two divisions:

CNS - central nervous system (brain – in Figure 12 – and spinal cord) is the integration and command center of the body;

PNS - peripheral nervous system (spinal and cranial nerves) represents the conduit between the CNS and the body; subdivided into the *SNS* - somatic nervous system and the *ANS* - autonomic nervous system (Figure 13).

Figure 12. Neuroanatomy (Sukel, 2019)

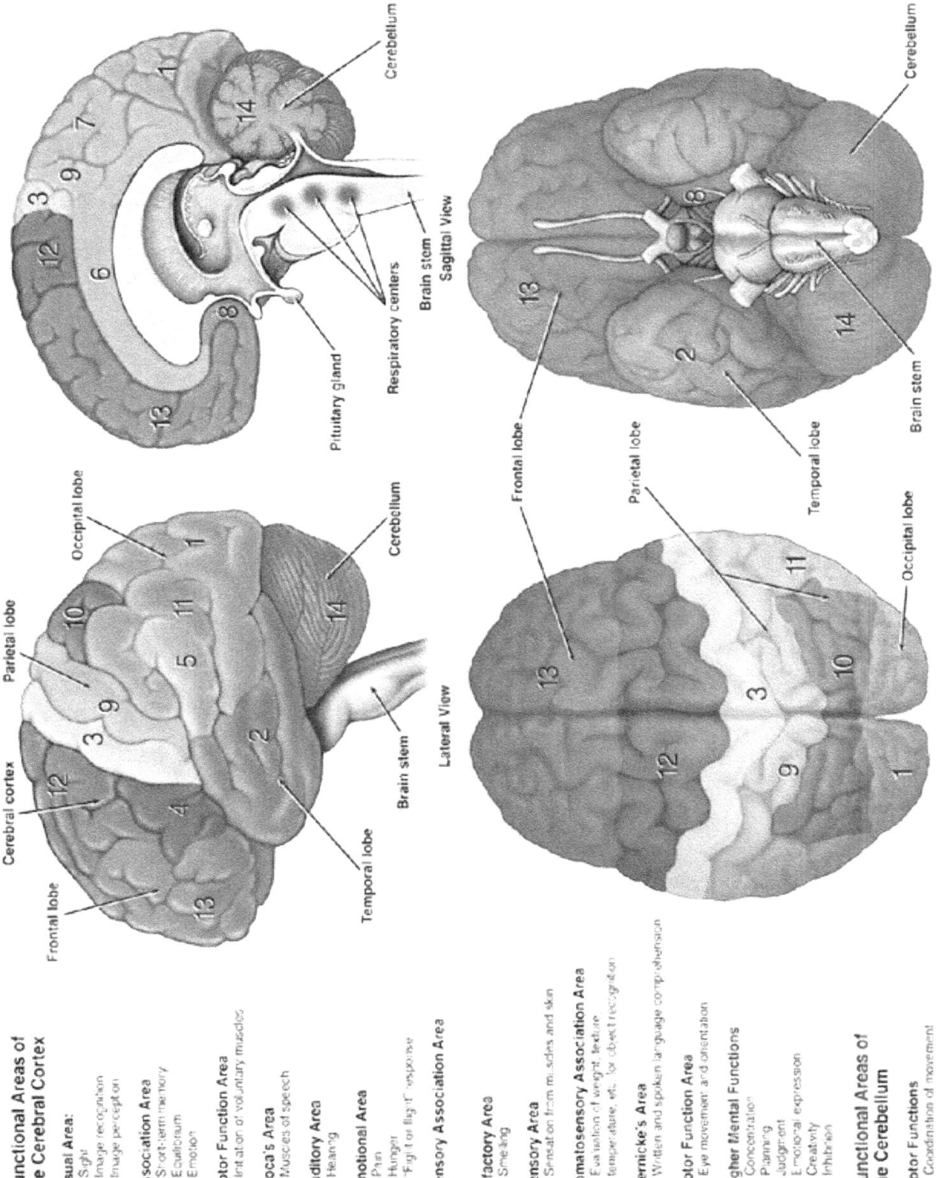

Figure 13. Tre regulation functions of the autonomic nervous system (The New York School of Regional Anesthesia., n/ d)

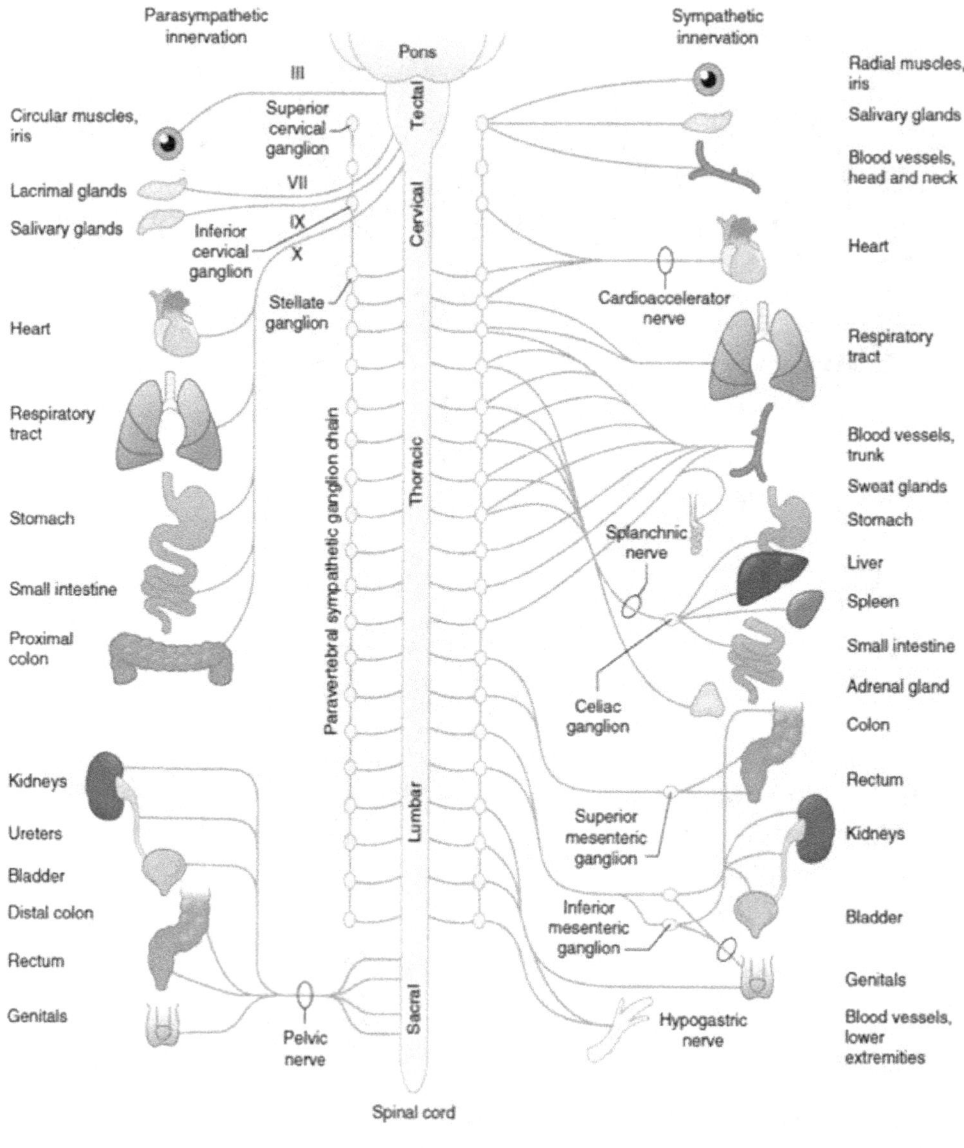

Regulation: the nervous system activity is regulated by the states of the internal and external environments, and also depends on age. It is not always possible to recognize specific regulatory mechanisms.

Renal System/Urinary System (Figure 14)

Figure 14. Renal system / urinary system front view (Johns Hopkins Medicine. Anatomy of the Urinary System, n / d)

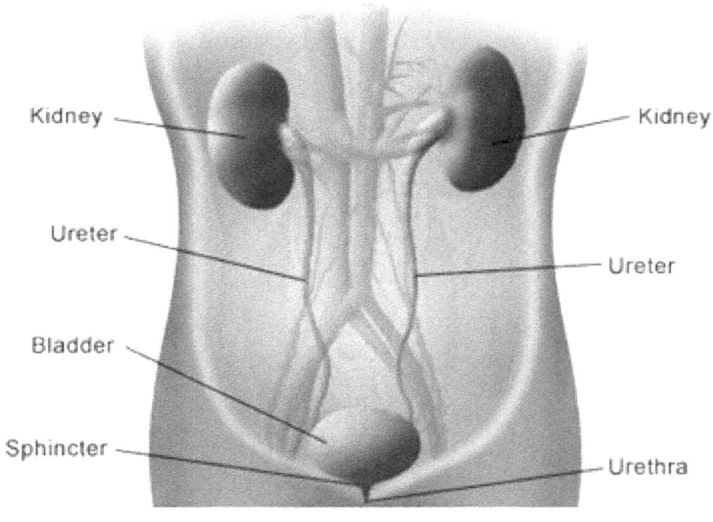

The main functions are: filtering blood to eliminate extra water and salt, toxins, and other waste products.

The components and their functions: the kidneys, ureters, bladder, and urethra.

Kidneys: two kidneys filter blood and make urine.

Ureters: two thin tubes inside the pelvis carry urine from your kidneys to the bladder.

Bladder: holds urine until ready to empty it.

Urethra: carries urine from your bladder out of the body. It ends in an opening to the outside of the penis (in men) or in front of the vagina (in women).

Regulation: by the nervous and endocrine systems.

Reproductive System (Figures 15, 16)

The main functions are: to produce egg and sperm cells; transport and sustain these cells; nurture the developing offspring; and produce hormones.

The components and their functions: the primary reproductive organs (*gonads*, which consist of the ovaries and testes) and secondary, or accessory, reproductive organs (*other organs, ducts, and glands*).

Gonads: producing the egg and sperm cells gametes, and hormones. These hormones function in the maturation of the reproductive system, the development of sexual characteristics, and the regulation of the normal physiology of the reproductive system.

Accessory reproductive organs - transport and sustain the gametes and nurture the developing offspring.

Regulation: by the nervous and endocrine systems.

Figure 15. Genitals of a woman (Tsaitgaist, 2010)

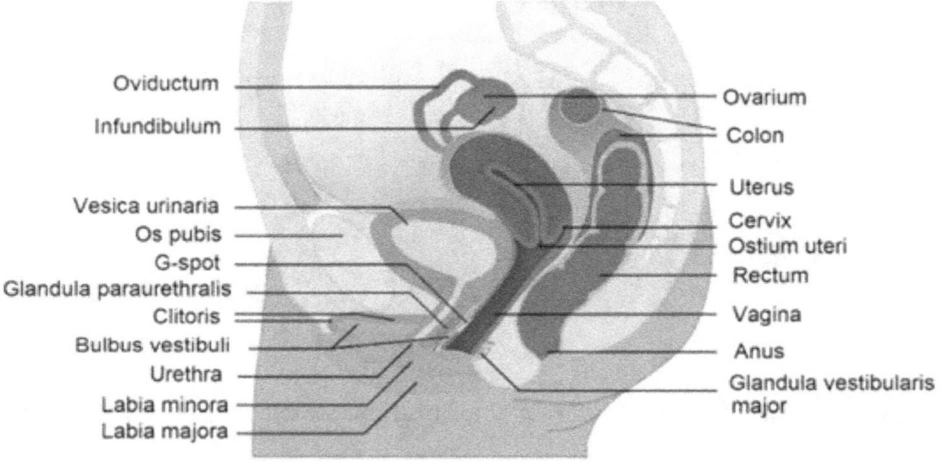

Figure 16. Male genital organs (Tsaitgaist, 2013)

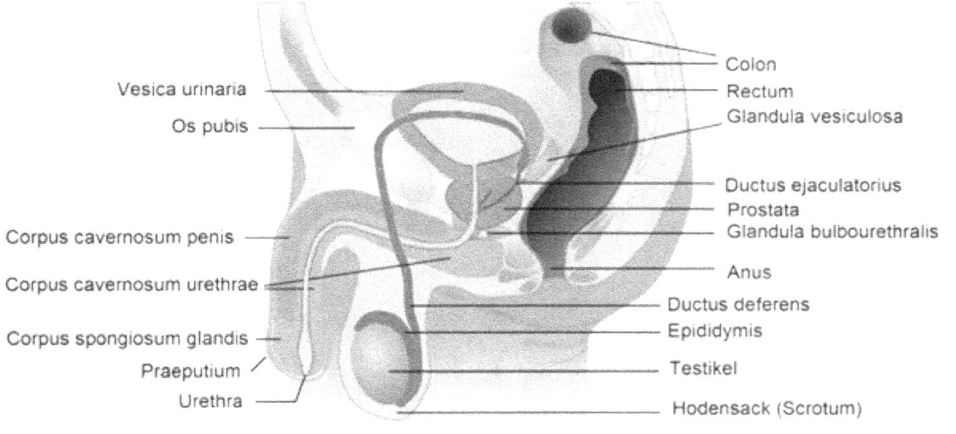

The Respiratory System (Figure 17)

The Respiratory System is the network of organs and tissues that carry out the process of respiration.

The main functions are: to deliver oxygen to the cells; remove waste gases, including carbon dioxide, from the body; protect airways from harmful substances and irritants; to warm air to match body temperature; allow to talk and to smell.

Figure 17. Respiratory system (LadyofHats, Jmarchn, 2007. Using files – in References)

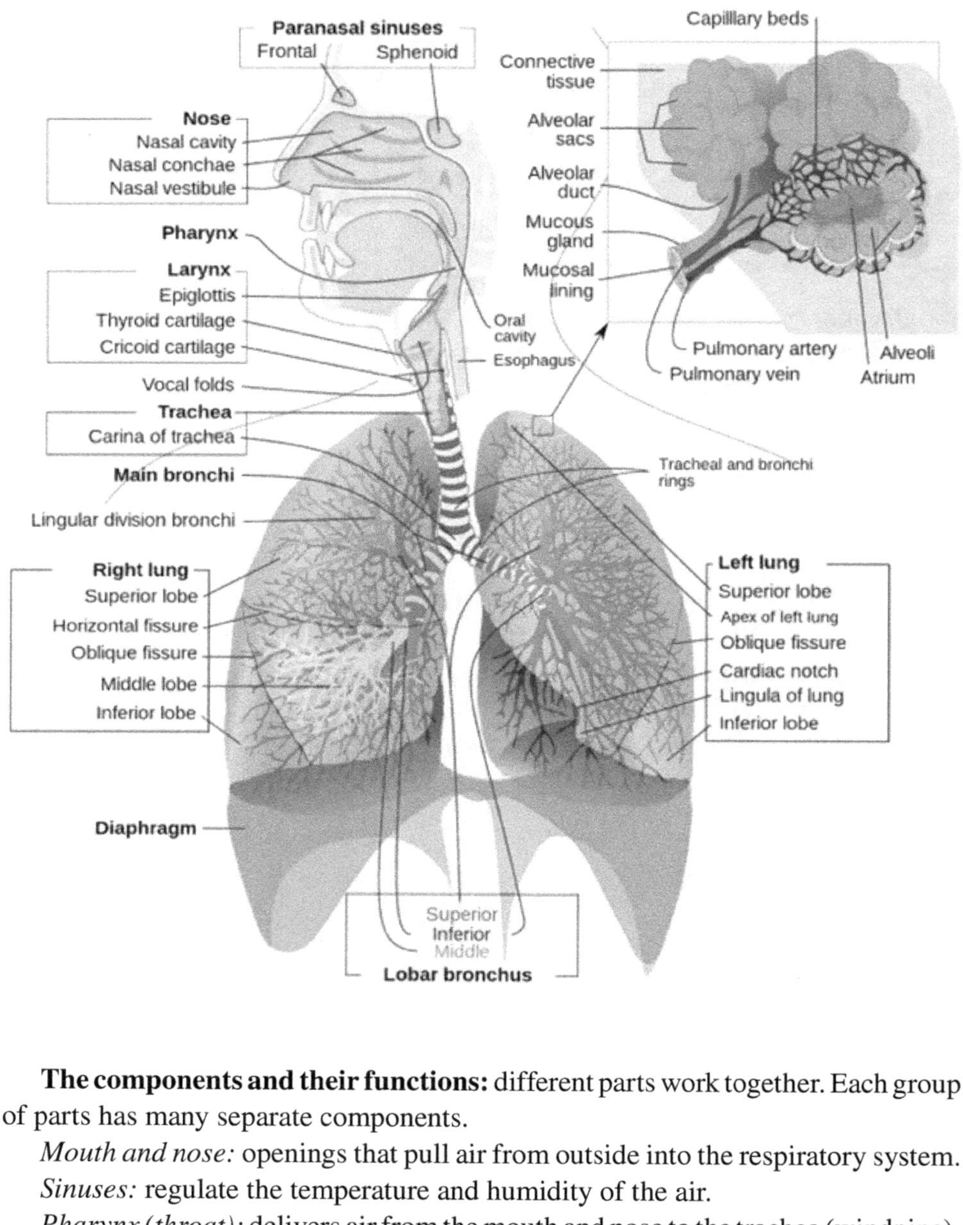

The components and their functions: different parts work together. Each group of parts has many separate components.

Mouth and nose: openings that pull air from outside into the respiratory system.

Sinuses: regulate the temperature and humidity of the air.

Pharynx (throat): delivers air from the mouth and nose to the trachea (windpipe).

Trachea: passage connecting throat and lungs.

Bronchial tubes: Tubes at the bottom of your windpipe that connect to each lung.

Lungs: Two organs that remove oxygen from the air and pass it into the blood.

Muscles (diaphragm) and bones (Ribs) help move the air into and out of the lungs.

Alveoli: tiny air sacs in the lungs, where the oxygen and carbon dioxide exchange take place.

Bronchioles: small branches of the bronchial tubes that lead to the alveoli.

Capillaries: blood vessels in the alveoli walls that move oxygen and carbon dioxide.

Lung lobes: three lobes in the right lung and two in the left lung.

Pleura: surround each lung lobe and separate lungs from the chest wall.

Cilia: tiny hairs that move in a wave-like motion to filter dust out of airways.

Epiglottis: tissue flap at the entrance to the trachea that closes when swallowed to keep food and liquids out of your airway.

Larynx: allows talking.

Regulation: by the nervous and endocrine systems.

Skeletal System (Figure 18)

The main functions are: to support body weight and allow movement: bones, joints, connective tissue, and muscles work together to make body parts mobile; produce blood cells - red and white blood cells are produced in the bone marrow; and protect and support organs; stores minerals - bones hold body's supply of minerals like calcium and vitamin D.

The components and their functions: 206 bones and connective tissue, including cartilage, tendons, and ligaments.

*Bone*s: provide the body shape and support, protect certain organs, serve as a storage site for minerals, and contain bone marrow for the development and storage of blood cells.

Cartilage: covers the tips of bones where they contact, and bones can move without rubbing against each other.

Joints: located where two or more bones in the body come together. There are three different joint types: immovable joints don't let the bones move at all, like the joints between skull bones; partly movable joints, which allow limited movement, like the joints in the rib cage: movable joints, which allow a wide range of motion, like elbow, shoulder, or knee.

Ligaments: hold bones together.

Tendons: connect the ends of a muscle to bone.

Regulation:*by conditions of the external and internal environment and age.*

Figure 18. Human skeletal system (BIO103: Human Biology. Chapter 7- The Skeletal System. Human Skeletal System., n/d)

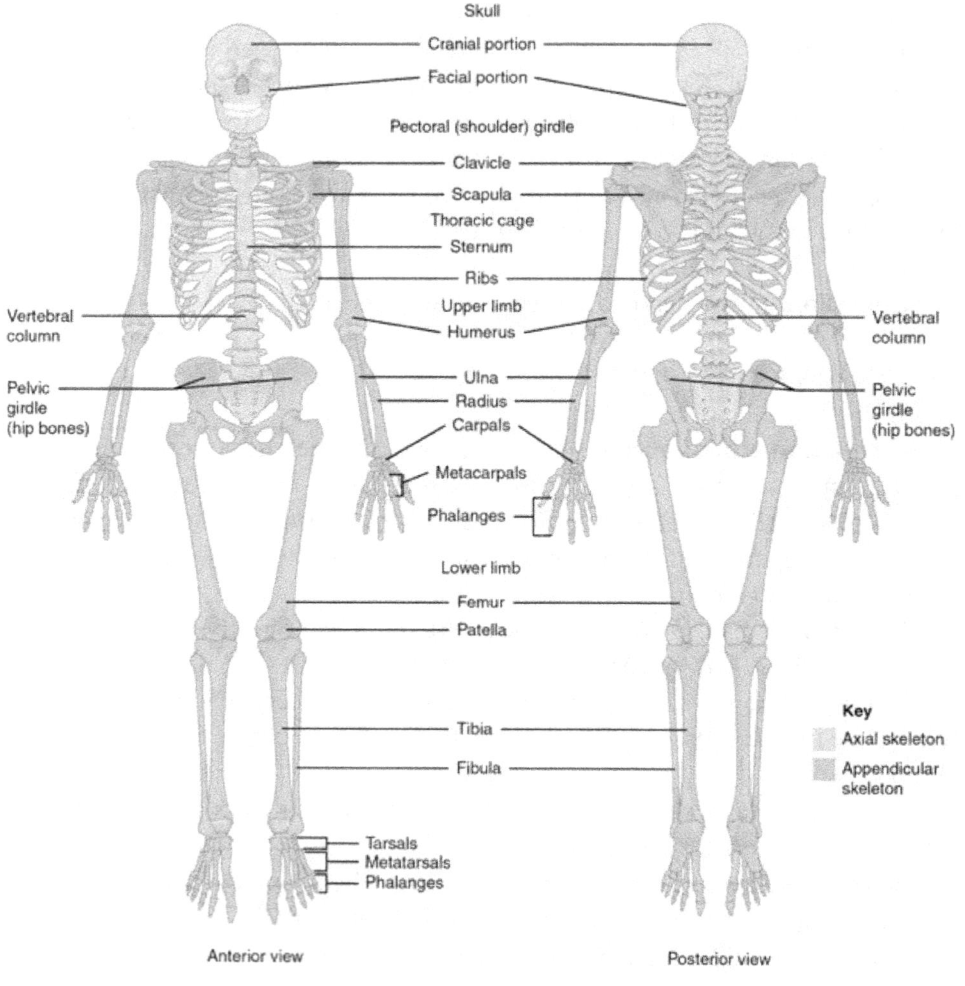

RHYTHMS OF NATURE AND RHYTHMS OF HUMAN

The field of biology that studies periodic (cyclic) phenomena in living organisms, and their synchronization with various natural rhythms, is called chronobiology.

Chronobiology is a complex science and includes anatomy, physiology, genetics, molecular biology, and the behavior of organisms, as well as epigenetics, development, reproduction, medicine, ecology, and evolution.

For this work, three aspects are of greatest interest:

- Natural rhythms that potentially or actually affect a person.
- Human biorhythms that determine our life activity.
- Mechanisms of the rhythms of nature and human biorhythms synchronization.

Rhythms of Nature

Let us give information about periodic phenomena in the solar system and on Earth in descending order of their duration.

Rotation of the Solar System around the center of the Milky Way: one revolution takes about 230 million years. About 20 such revolutions have passed since the solar system's formation. Periodically passing through various regions of outer space, filled with dust and gas to varying degrees, the Solar System is affected by them, which also affects terrestrial life: certain substances become available, the amount of light passing through the atmosphere changes, the climate changes; gravitational effects of large celestial bodies can change orbit or cause meteor bombardment. But this period is too long for a person and even humanity.

Axial precession is a slow and continuous change in the orientation of the Earth's axis of rotation, lasting 25,772 years, caused by gravity. The precession has observable effects - the positions of the south and north celestial poles appear to move in a circle against the background of the stars (in 3200 years, Gamma Cepheus will become the polar star); the position of the Earth in its orbit at the solstices, equinoxes, and so on, is also slowly changing. The precession was known to the ancient civilizations of the Babylonians, Egyptians, and later - the Indians and Maya.

The movement of the planets in orbits has periods (Sidereal periods, years): Mercury - 0.240846; Venus - 0.615; Earth – 1; Mars - 1.881; Jupiter - 11.86; Saturn - 29.46; Uranus - 84.01; Neptune - 164.8; 134340 Pluto - 248.1.

The Solar cycle of magnetic activity is an almost periodic 11-year variation in solar activity as measured by the number of observed sunspots. During the cycle, the levels of solar radiation and solar material ejection, the number and size of sunspots, flares, and coronal loops show synchronized fluctuations from the period of minimum activity to the period of maximum activity and back. The Sun's magnetic field reverses when the cycle is close to its maximum and after two cycles it returns to its original state, completing the so-called Hale cycle. The solar cycle affects many terrestrial phenomena (Chizhevsky, 1976). More long cycles of the Sun are also assumed, which affect, among other things, the climate. The reconstruction of the solar activity over a long time is shown in Figure 19.

The movement of the comets. There are different types of comets - short-period, long-period, non-periodically coming from near (Oort cloud and Kuiper belt) or far space. Give the periods of some short-period comets (years): Halley – 76; Harrington 6.78; Hartley - 36.84; Holmes - 7.09; Kohoutek - 6.67; Tempel 1 - 5.5.

The rotation of the Earth around the Sun (Earth year). A year is the period of revolution of the Earth in its orbit around the Sun. The average length of a calendar year is 365.2425 days. Due to the tilt of the Earth's axis, during the year there is a change of seasons, marked by changes in weather, daylight hours, vegetation, and soil fertility. In temperate and subpolar regions, four seasons are distinguished; in tropical and subtropical regions there are no definite seasons; but yearly rainy and dry seasons are tracked.

The lunar synodic month is 29.53 days long as seen from Earth. Lunar phases (new moon, first quarter, full moon, last quarter) are well observed as the orbital positions of the Moon around the Earth and the Earth around the Sun shift.

Figure 19. 11,000-Year Sunspot Number Reconstruction. The tabulated years correspond to centers of the corresponding 10-year intervals. Negative values are artifacts and are consistent with zero within the error limits (Conscious, 2004)

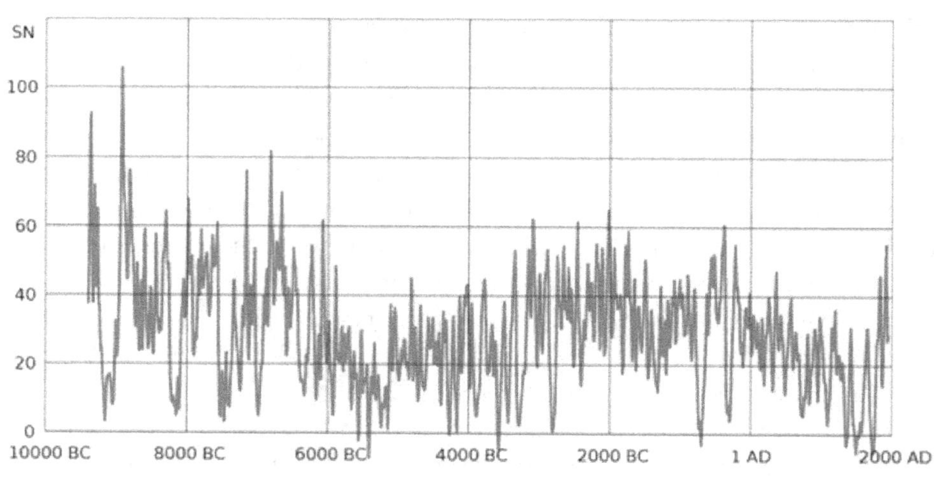

The rotation of the Sun around its axis at a latitude of 26 ° from the equator takes 25.38 Earth days. The speed of rotation of the spots decreases as they approach the poles so that at the poles the period of rotation around the axis can be up to 38 days.

The rotation of the Earth around its axis (Earth Day). It is believed that the Earth slows down its rotation over time, which can be estimated by measuring the sedimentation rings in coral fossils. According to the calculations, the length of the day was: present - 24 hours; Cretaceous period - 23 hours 20 minutes; Trias - 22 hours 40 minutes; Carboniferous - 22 hours; Devon 21 hours 20 minutes; Cambrian (approximately 500 million years ago) - 20 hours 40 minutes.

If we move from astronomical phenomena to the atomic-molecular level, then periodic phenomena also occur, an example of which is photosynthesis discussed in Chapter 2.

Rhythms of a Human

Circadian rhythms (CR) are a set of endogenous autonomous oscillators generated by molecular daily clocks, that synchronize the personal internal time with the external environment during the daily cycle.

Circadian rhythms are mostly light and dark controlled and controlled by a small area in the middle of the brain. The molecular clock of the circadian rhythm system, present in all cells, is composed of oscillating clock-associated proteins that make up the transcriptional/translational feedback loops.

They can affect sleep, body temperature, hormones, appetite, and other bodily functions (Figure 20). Abnormal circadian rhythms may be associated with obesity, diabetes, depression, bipolar disorder, seasonal affective disorder, and sleep disorders such as insomnia (Serin & Acar, 2019; Xie et al., 2019).

Figure 20. Biological clock (YassineMrabet, 2007)

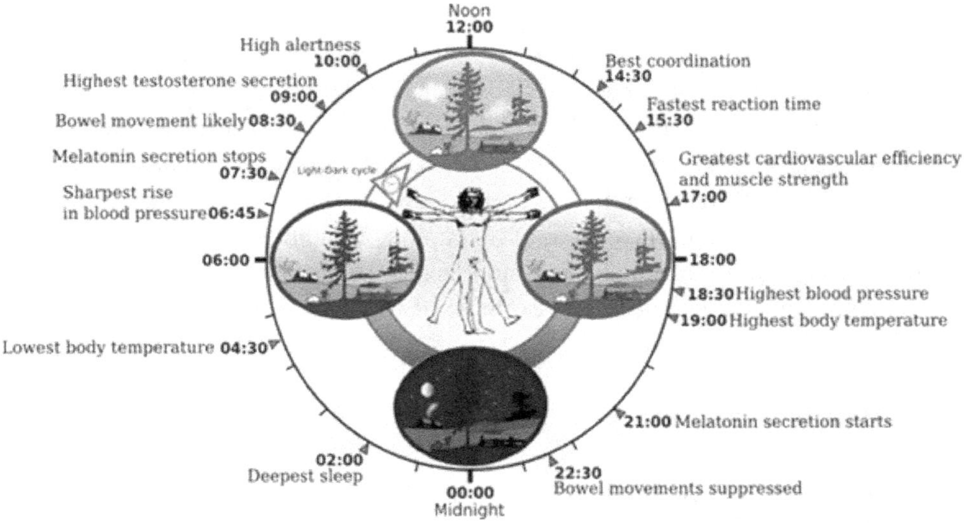

Endogenous processes regulate Circadian rhythms. Exogenous signals may regulate other biological cycles.

Infradian rhythms are cycles lasting more than a day, for example, *circannual* or *annual cycles*, or *lunar rhythms* following the lunar month (29.5 days). These rhythms are present not only in humans, but also in animals, regulating the cycles of migration, reproduction of many plants and animals (in humans - the menstrual cycle), the way of life of marine organisms (how much it depends on the level is determined by the movement of the Moon, Earth, and Sun).

Ultradian rhythms are cycles shorter than 24 hours, such as 90-minute states of light/deep/REM sleep in a 90-minute sleep cycle, a 4-hour nasal cycle, or a 3-hour growth hormone cycle.

Gene oscillations - some genes are expressed more during certain hours of the day than at other times.

Within each cycle, the period during which the process is most active is called the acrophase. When the process is less active, the cycle is in the bathyphase. A particular moment of highest activity is a peak or maximum; the lowest point is the nadir. These phases correlate well with the wave modalities *(æ, ö,)* introduced in Chapter 2.

The Rhythms of Nature and Human Biorhythms Synchronization

The nature of synchronization mechanisms can be evolutionary-biological, cultural-psychological, or physical. Also, they can be classified according to the rhythms that they synchronize.

Evolutionary-biological mechanisms began to form as early as the stage of abiogenic synthesis. The conditions of day and night, or seasons determined the favorable or unfavorable course of chemical reactions (Chapter 1). Their influence on circadian biorhythms is the most studied.

Under the influence of external time signals - "synchronizers" or "entraining agents" - the autonomous cellular circadian clock can place all cells in the same phase of circadian oscillations, called synchronization of circadian rhythms. A description of specific mechanisms is given by (Xie, Tang., Chen. et al., 2019). Synchronizers are:

Light. Most organisms obtain time information through changes in light intensity during the day;

Excitatory stimuli (social interactions, exercise, ingestion of stimulants, and the like) can synchronize CR through "non-light drag". Physiological factors such as glucocorticoids, sympathetic nerves, oxidative stress, hypoxia, cytokines, and temperature are involved here;

Food/feeding. Temporary restriction of food intake during the light phase of the day does not affect the central oscillators. However, fasting and feeding signals reset the circadian clock in peripheral tissues. For example, eating a meal causes insulin to be released into the bloodstream. It causes a clock reset in insulin-sensitive tissues such as the liver, adipose tissue, and muscle. Food restriction can alter tissue anabolic/catabolic cycles, which affect the redox state of cells and further affect in the circadian feedback loop.

Temperature is a weak non-light synchronizer. Many physiological processes depend on ambient temperature, but the CR system has a temperature compensation function and circadian fluctuations remain stable to temperature changes, resulting in approximately 24 hours.

Chemical factors can function as synchronizers: the glucocorticoid secreted by the adrenal cortex serves as a powerful peripheral tissue synchronizer; serum shock has been used to synchronize the circadian oscillations of various tissues, immortalized cells, and fibroblast cells; multiple factors in the blood can stimulate rhythmic fluctuations, etc.

Mechanical stimuli. Discontinuous uniaxial stretching of bone marrow-derived mesenchymal stem cells and adipose tissue-derived mesenchymal stem cells results in synchronization. However, the mechanisms by which mechanical stimuli induce synchronization are still unclear.

Oxidative or hypoxic stimuli can cause the circadian clock to shift in peripheral clocks, such as those in the kidneys and liver, after hydrogen peroxide injection. However, it is not known whether the master clock responds to oxidative stress.

Synchronization with other rhythms has been less studied, but there is evidence of its existence, at least for the solar cycle and lunar rhythms.

Cultural-psychological mechanisms have a powerful synchronizing effect on all biorhythms, except for molecular genetic ones. Thus, the daily cycle of divine services organizes the life of believers, especially monks, during the day. The names of the days of the week, given under the protection of different gods, show how the forms of human activity change in the weekly cycle. Rituals associated with fertility and childbearing are tied to lunar and annual rhythms. Rituals of magic, astrology, and divination involve a person in longer cycles related to the movement of planets and celestial bodies. The appearance of comets has always caused a negative reaction, and the priests used knowledge of the Metonic cycle to predict solar and lunar eclipses, allowing them to demonstrate their connection with the gods.

Let's pay attention to the multilevel and systemic nature of such synchronization, which affects the person and the environment channels of interaction (Kovalyov et al., 2020).

Physical mechanisms. Note two phenomena, that are important when considering synchronization in the framework of the soliton-wave model.

Modulation of oscillations and waves, changes in time of parameters of oscillations and waves. There are several types of modulation: frequency, amplitude, and phase, with slow - there is no correlation with the amplitude, frequency, and phase of the oscillations (Figure 21).

For optical waves, polarization modulation is also used, when the polarization of the radiation is changed.

Figure 21. Modulation of oscillations: 1 - frequency; 2 - amplitude; 3 – phase (Ostrovsky, n/d)

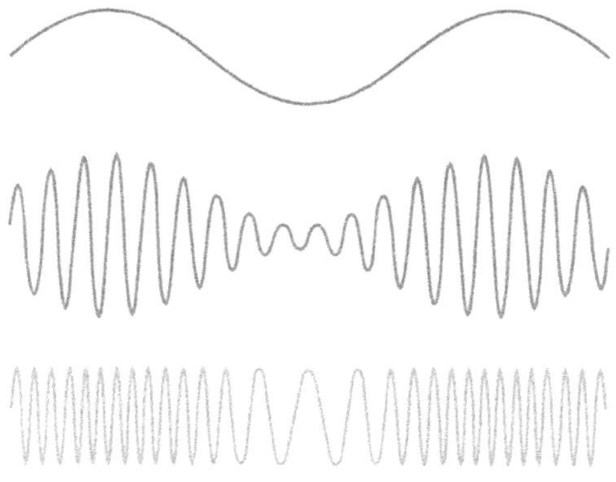

Modulation of the signal *f(t, r)* for the modulation of waves, the amplitude and phase of which they are not only in time but also space, is described as follows (Ostrovsky, n / d):

$$f(t,r)=A(t,r)\cos[\omega 0t-k0r+\varphi(t,r)]$$

where r is the radius vector of a point in space, a wave has both time T and space period $\Lambda=2\pi/|k|$, where k is the wave vector.

The modulation of the wave can be caused by different reasons, for example, variations of initial conditions in time, boundary conditions in space, and slow changes in the parameters of the environment in time and space.

Resonance is the frequency-selective response of an oscillatory system to a periodic external influence. Resonance occurs when the external action frequency ω approaches its frequency oscillation topics $\omega 0$. In this case, the impact frequency is called *resonant frequency* (ω_p). The oscillatory system properties determine its value. Resonance is manifested in a sharp increase in the amplitude of its forced oscillations and an increase in the responsiveness of the oscillatory system to the action of an external force.

The effect is maximum when the frequency of the external action coincides with the system's natural frequency and the attenuation coefficient is zero. For systems of different natures, resonant frequencies are determined by various formulas. In non-linear systems, resonance has essential features: the resonant frequency is non-linear, and the neural system depends on the amplitude of the external influence; with sufficiently strong influences, one-valued correspondence of the amplitude disappears necessary fluctuations in the parameters of the external signal. At the same frequency as the external force, oscillation modes with different amplitudes are possible (Figure 22). The dashed curve in the figure shows a branch corresponding to an unsustainable mode fluctuation. Switching between different modes when rebuilding the frequency occurs in the form of amplitude jumps, has a delay, and depends on the previous states of the system (Klinshov & Nekorkin, n / d).

Figure 22. Resonance curve of a non-linear system (Klinshov & Nekorkin, n/d)

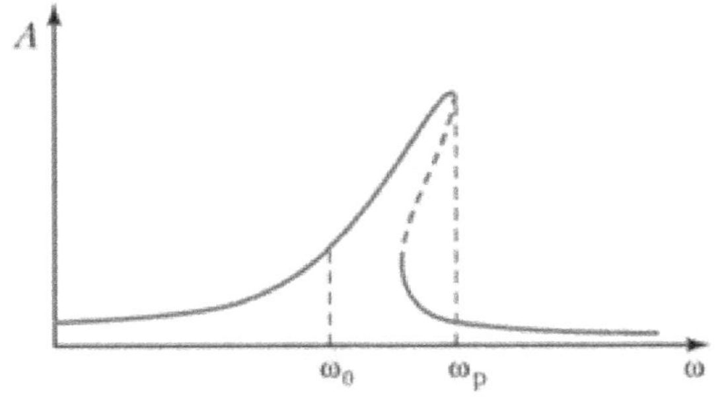

SELF-REGULATION

The cybernetics apparatus is used to understand the human body's self-regulation (Schmidt & Thews, 1983). Most often, feedback control systems are considered (Figure 23).

Figure 23. The feedback control system (Schmidt & Thews, 1983)

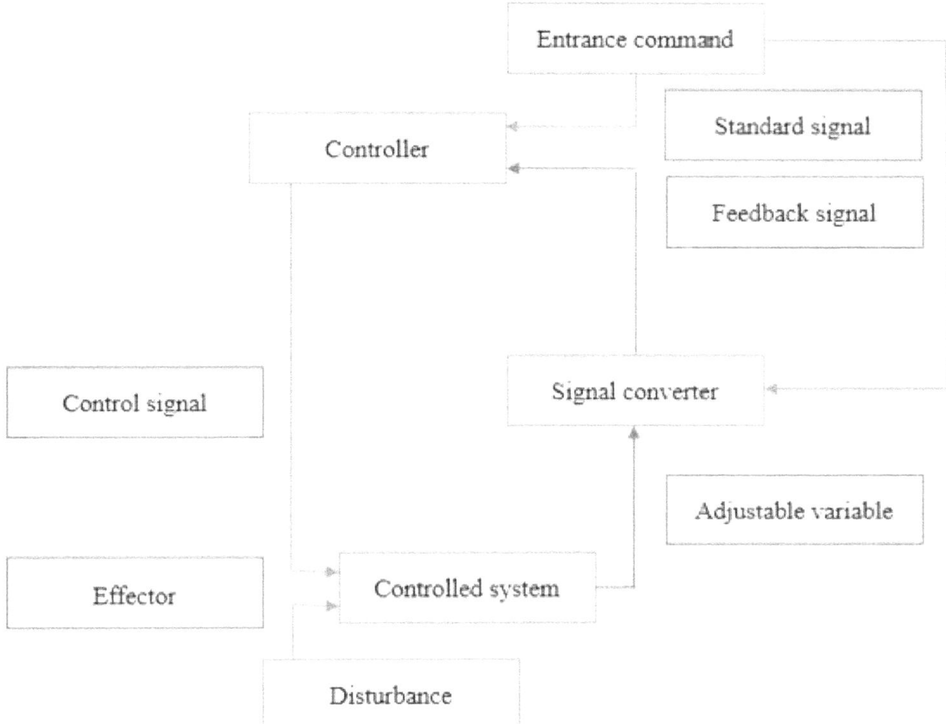

This approach is correct if the modeling object is a simple technical system. However, it is limitedly applicable to understanding the self-regulation of emergent biological systems because of a discrepancy between the set-theoretical and logical premises of cybernetics and the properties of complex open systems (Table 7, Chapter 1).

It was mentioned above about the various systems of self-regulation of the human body, now let's put this information together.

Homeostasis is a state of stability of internal, physical, and chemical conditions maintained by living systems. This state, maintained within certain limits (homeostatic range) despite changes in the environment, diet, or activity level, is determined by the conditions for the body's optimal functioning. Controlled parameters include (McEwen & Wingfield, 2003; Asarian et al., 2012; Billman, 2020; Libretti & Puckett, 2022; Homeostasis, n / d):

- Core temperature.
- Blood glucose.

- Iron levels.
- Copper regulation.
- Levels of blood gases.
- Blood oxygen content.
- Arterial blood pressure.
- Calcium levels.
- Sodium concentration.
- Potassium concentration.
- Fluid balance.
- Blood pH.
- Cerebrospinal fluid.
- Neurotransmission.
- Neuroendocrine system.
- Gene regulation.
- Energy balance;

For a person, as a part of a biological system and a social being, it is also necessary to control:

- Biosphere condition.
- Risk.
- Stress.
- Technology.
- Society's social and cultural environment.

It should be in view that if, within the framework of the set-theoretical understanding, homeostasis is reduced to the need to control such or an even more detailed set of parameters, then within the framework of the theory of self-organization, this is a characteristic of the system as a whole, greater than individual parameters and not reducible to them.

Maintaining homeostasis is considered the main goal of self-regulation, and there are several regulatory mechanisms for this.

Levels of body function regulation. The coordinated work of the body is possible due to the automatic regulation of functions, in which four levels are distinguished (Figure 24).

Figure 24. Levels of regulation of body functions (Nervous system, n/d)

Level 1: The central nervous system is the center for regulating body functions

↓

Level 2: Sympathetic and parasympathetic divisions of the autonomic nervous system: regulation of the functions of internal organs

↓

Level 3: Endocrine system that regulates the functions by secreted hormones

↓

Level 4: Body fluids (blood, lymph, intercellular and intracellular fluids) that carry out nonspecific regulation

The central nervous system (CNS) is the main regulator of processes in the body. Signals coming from the environment (light, sound, smell, and so on), and internal organs (changes in blood supply, oxygen content, etc.). Then they are transmitted along the nerve pathways to the brain and spinal cord centers for analyzing and synthesizing information. The central nervous system works by continuously automatically rearranging the body's work, and creating temporary functional systems (FS) that control the processes occurring in it, carrying out adaptive reactions that lead to a result useful for the body, in response to the influence of changing environmental factors.

All FS have their own "inputs", "outputs", and "input and output variables". Input through input variables affects output variables and the state of the output. Under external influence, the body's main task is to establish the output data of all FS at the optimal level in the current situation.

The peripheral nervous system (PNS) connects the CNS to the body's other parts through cranial and spinal nerves, which divide into many small nerve fibers, forming a network that permeates the entire body.

In addition to dividing the nervous system into central and peripheral, the somatic and autonomic systems are distinguished according to the functions performed.

The actions of *the somatic nervous system* are under the control of consciousness, reacting to information coming from the external environment, and controlling skeletal muscles (body movement). The central part of the somatic nervous system is the cerebral cortex, and the peripheral part is the fibers innervating the striated muscles.

The actions of *the autonomic nervous system* are autonomous, i.e., do not depend on a person's will and control internal organs, smooth muscles, and metabolism. The central division of the autonomic nervous system is located in the hypothalamus, which regulates the autonomic nervous system and endocrine glands through the pituitary gland. The peripheral part of the autonomic nervous system consists of nerve nodes and nerve fibers that innervate the working organ.

The autonomic nervous system is divided into *sympathetic* and *parasympathetic* parts.

The sympathetic system is activated when the body is in a stressful situation: the heart increases its work, blood pressure rises, blood sugar increases, skin vessels narrow, and the digestive organs slow down their activity - a person is ready for action.

The parasympathetic system returns the functions of organs and systems to a state of rest: it reduces blood pressure and sugar content, breathing becomes rarer but deeper, skin vessels expand and the digestive organs are activated.

Now let's illustrate the work of all levels of control for cardiac activity regulation - an example that is not the most difficult (Figure 25).

Figure 25. The heart regulation. orange – gain; blue – weakening; black - ambiguous

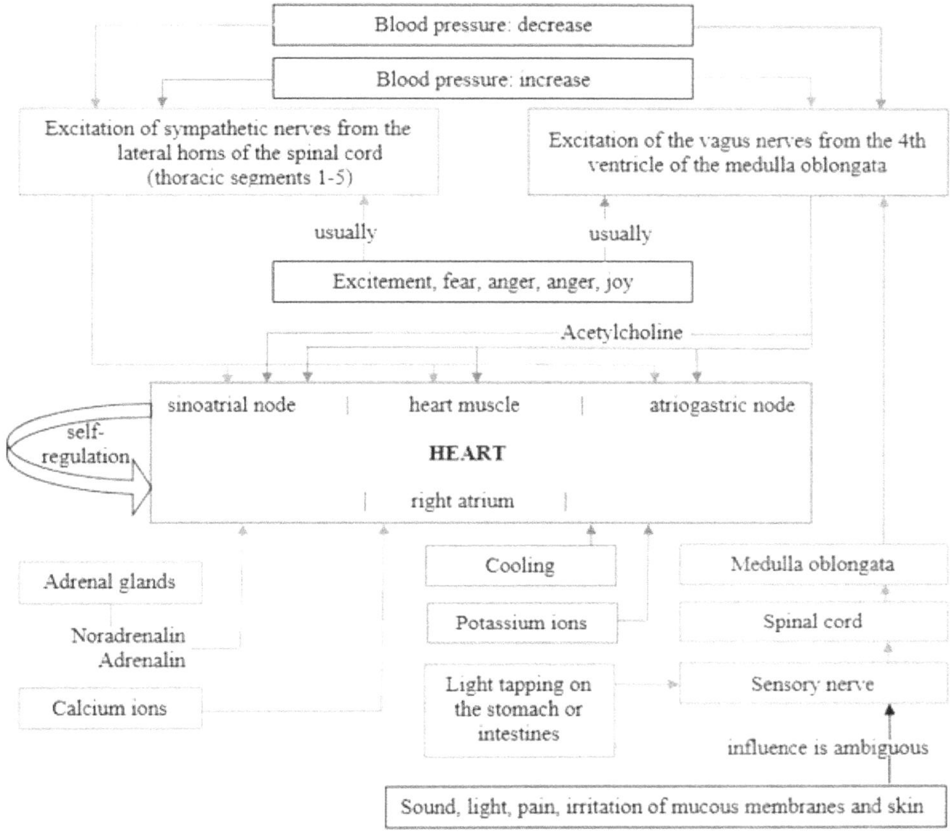

Note that in a situation of real regulation - as in Figure 23 - it is practically impossible to use such models as in Figure 21 - it is impossible to unambiguously determine either the controlled system, or the parameters, or the signals. This is where the emergence of complex biological systems manifests itself.

FUTURE RESEARCH DIRECTIONS

The next task will be to build a human-environment model in Chapter 8. The model should include the general structure of the body, the levels, and channels of the subjective space of a person, the structure of interactions with the environment, and take into account both the events of the evolution of the Universe before the emergence of life, and the results of the evolution of living organisms before the emergence of human. Thus, the model must implement a systematic approach, the wave model paradigm, and the theory of self-organization.

CONCLUSION

The main result of Chapter 7 is the collection of information about the structure of the human body, the coordination of the rhythms of man and nature, and the self-regulation of the human body.

The main conclusion that can be drawn from the above data is that a human is indeed an open, complex, and emergent system, and for his research, it is necessary to apply a holistic systems approach.

Attempts to describe a human as a set of separate systems and elements immediately cause certain difficulties. This is also manifested in the organ systems determination: whether to take the structure as a basis - and then there are separate systems of the skeleton and muscles or functions - and then there is a single musculoskeletal system. This is also manifested in solving the problem of coordinating different rhythms - for example, the modulation mechanism can affect several of them in different ways at once. The same can be said about the mechanisms of self-regulation, which are far from always possible to represent in terms of cybernetics.

Thus, the solution to the problem of constructing a holistic model of a human is necessary both for continuing the study of his life cycle in this work and for solving numerous theoretical and practical problems of a generally scientific nature.

REFERENCES

Anatomy 3D Atlas. (n.d.). https://anatomy3datlas.com/

Anatomy of the Lymphatic and Immune Systems | Anatomy and Physiology II. (n.d.). https://courses.lumenlearning.com/suny-ap2/chapter/anatomy-of-the-lymphatic-and-immune-systems/

Asarian, L., Gloy, V., & Geary, N. (2012). *Homeostasis*. In Homeostasis. In *Encyclopedia of Human Behavior* (pp. 324–333). Elsevier. DOI: 10.1016/B978-0-12-375000-6.00191-9

Bednarik, R. G. (2013). *The Psychology of human behavior*. Gazelle https://www.researchgate.net/publication/291076363_The_psychology_of_human_behavior

Best, C. H., Taylor, N. B., & West, J. B. (1985). *Best & Taylor's Physiological Basis of Medical Practice*. Williams & Wilkins.

Billman, G. E. (2020). Homeostasis: The underappreciated and far too often ignored central organizing principle of physiology. *Frontiers in Physiology*, 11, 200. Advance online publication. DOI: 10.3389/fphys.2020.00200 PMID: 32210840

Çalıyurt, O. (2017). Role of Chronobiology as a Transdisciplinary Field of Research: Its Applications in Treating Mood Disorders. *Balkan Medical Journal*, 34(6), 514–521. DOI: 10.4274/balkanmedj.2017.1280 PMID: 29072179

Conscious. (2004). 11,000-Year Sunspot Number Reconstruction. The tabulated years correspond to centers of the corresponding 10-year intervals. Negative values are artifacts and are consistent with zero within the error limits. Wikipedia contributors. https://en.wikipedia.org/wiki/File:Sunspots_11000_years.svg

Eisenstein, M. (2013). Chronobiology: Stepping out of time. *Nature*, 497(7450), S10–S12. DOI: 10.1038/497S10a PMID: 23698500

Guyton, A. C., & Hall, J. E. (2005). *Textbook of Medical Physiology*. Saunders. (Original work published 1969)

Hiller-Sturmhöfel, S., & Bartke, A. (1998). The Endocrine System. *Alcohol Health and Research World*, 22(3), 153–164. PMID: 15706790

Jmarchn (2009). Organ systems and human sense organs. Own work based on: Human body features.png. https://natworld.info/nauki-o-prirode/osnovnye-sistemy-organov-cheloveka-nazvaniya-funkczii-i-illyustracziyab

Johns Hopkins Medicine. (n.d.). Anatomy of the Urinary System. https://www.hopkinsmedicine.org/health/wellness-and-prevention/anatomy-of-the-urinary-system

Johns Hopkins Medicine. (n.d.). Renal system / Urinary system. Front view. Anatomy of the Urinary System. https://www.hopkinsmedicine.org/health/wellness-and-prevention/anatomy-of-the-urinary-system

Kataru, R. P., Baik, J. E., Park, H. J., Wiser, I., Rehal, S., Shin, J. Y., & Mehrara, B. J. (2019). Regulation of Immune Function by the Lymphatic System in Lymphedema. *Frontiers in Immunology*, 10, 470. Advance online publication. DOI: 10.3389/fimmu.2019.00470 PMID: 30936872

Kovalyov, Y., Mkhitaryan, N., & Nitsyn, A. (2020). *Self-organization of the Human Mind and the Transition from Paleolithic to Behavioral Modernity*. IGI Global International Publisher of Progressive Information Science and Technology Research. DOI: 10.4018/978-1-7998-1706-2

LadyofHats & Jmarchn. Respiratory system. (2007). *File: Respiratory system complete en.svg*. Wikipedia, The Free Encyclopedia. https://en.wikipedia.org/wiki/File:Respiratory_system_complete_en.svg

LadyofHats, & Mariana Ruiz Villarreal. (2009)., The diagram of the human Circulatory system in anterior view. Retrieved August 13, 2024, from https://upload.wikimedia.org/wikipedia/commons/thumb/2/29/Circulatory_System_en.svg/1357px-Circulatory_System_en.svg.png

Libretti, S., & Puckett, Y. (2022). *Physiology, Homeostasis*. StatPearls Publishing. https://pubmed.ncbi.nlm.nih.gov/32644564/

List of systems of the human body. (n.d.). Wikipedia, The Free Encyclopedia https://en.wikipedia.org/wiki/List_of_systems_of_the_human_body

Litwack, G. (2022). *Hormones*. Academic Press., DOI: 10.1016/C2020-0-02802-1

Mariana Ruiz, & Jmarchn. (2006). Diagram of the human digestive (gastrointestinal) tract. https://en.wikipedia.org/wiki/File:Digestive_system_diagram_en.svg

McEwen, B. S., & Wingfield, J. C. (2003). The concept of allostasis in biology and biomedicine. *Hormones and Behavior*, 43(1), 2–15. DOI: 10.1016/S0018-506X(02)00024-7 PMID: 12614627

Medium69 & Jmarchn. (2014). Nervous system diagram. Retrieved August 13, 2024, from https://commons.wikimedia.org/wiki/File:Nervous_system_diagram-en.svg

Milhiet, V., Etain, B., Boudebessem, C., & Bellivier, F. (2011). Circadian biomarkers, circadian genes and bipolar disorders. *Journal of Physiology, Paris*, 105(4–6), 183–189. DOI: 10.1016/j.jphysparis.2011.07.002 PMID: 21767641

Molenaar, P. C. M. (2001). *Systems Modeling*. In *International Encyclopedia of the Social & Behavioral Sciences*. Pergamon National Cancer Institute. https://training.seer.cancer.gov/melanoma/anatomy/

National Cancer Institute. (n.d.). Introduction to the nervous system. https://training.seer.cancer.gov/anatomy/nervous/

Schmidt, R. F., & Thews, G. (Eds.). (1983). *Human Physiology*. Springer-Verlag.

Sciencia58 and the makers of the single images Domdomegg, Fæ, Petr94 & Manu5. (2020). The processes involved in the primary immune response. *File: Primary immune response 1.png*. Retrieved August 13, 2024, from https://commons.wikimedia.org/wiki/File:Primary_immune_response_1.png

Serin, Y., & Acar Tek, N. (2019). Effect of Circadian Rhythm on Metabolic Processes and the Regulation of Energy Balance. *Annals of Nutrition & Metabolism*, 74(4), 322–330. DOI: 10.1159/000500071 PMID: 31013492

Sherwood, L. (2015). *Human Physiology: From Cells to Systems* (9th ed.). Cengage Learning.

Sukel, K. (2019). Neuroanatomy: The Basics. https://www.dana.org/wp-content/uploads/2019/08/anatomy-function-brain-areas-basics-aug-2019-2024.jpg

The Human Body. (n.d.). https://www.healthline.com/human-body-maps

The New York School of Regional Anesthesia. (n.d.). Autonomic Nervous System. https://www.nysora.com/anesthesia/autonomic-nervous-system/

Tsaitgaist. (2009). Drawing of the Male Internal Sexual Anatomy. (n.d.). Wikimedia.org. Retrieved August 13, 2024, from https://commons.wikimedia.org/wiki/File:Male_anatomy_number.svg

Tsaitgaist. (2010). Genitals of a woman. Retrieved August 13, 2024, from https://commons.wikimedia.org/wiki/File:Female_Organs.png

Whitbourne, S. K. (1985). Digestive and Excretory Systems. In *The Aging Body*. Springer. DOI: 10.1007/978-1-4612-5126-2_5

Xie, Y., Tang, Q., Chen, G., Xie, M., Yu, S., Zhao, J., & Chen, L. (2019). New Insights Into the Circadian Rhythm and Its Related Diseases. *Frontiers in Physiology*, 25(10), 682. DOI: 10.3389/fphys.2019.00682 PMID: 31293431

YassineMrabet. (2007). Biological clock. Wikimedia.org. Retrieved August 13, 2024, from https://commons.wikimedia.org/wiki/File:Biological_clock_human-af.svg

ADDITIONAL READING

Chizhevsky, A. L. (1976). *Zemnoe эkho solnechnыkh burj* [Earth echo of solar storms] (2nd ed.). Thought.

Kositsky, G. I. (Ed.). (1985). *Fyzyologhyja cheloveka* [Human Physiology]. (3rd ed.). Medicine.

Sinelnikov, R. D., & Sinelnikov, Ya. R. (1996). *Atlas anatomyy cheloveka v 4-kh tomakh*. Atlas of Human Anatomy in 4 Volumes. Medicine.

Zidermane, A. A. (1998). *Nekotorye voprosy khronobyologhyy y khronomedycyny: obzor lyteratury* [Some issues of chronobiology and chronomedicine: a literature review]. Zinatne.

Chapter 8
Human–Environment Model, Self–Regulation, and Rhythms Coordination

ABSTRACT

A soliton-wave model of "man-environment" is constructed, describing the structure of a human and the levels of interaction with the environment. To check the adequacy and practical usefulness of the model, calculations were performed, and examples of self-regulation and coordination of various rhythms were given. The results of the modeling correlate with the known data. The accuracy of the "man-environment" model corresponds to the tasks of the human life cycle studying, in particular understanding the processes of aging and dying as the destruction of integrity, including due to desynchronization and loss of coordination of processes and rhythms (Chapters 9 and 10). In addition, the practical application of the model is possible for human-machine systems optimization, in particular for increasing operator productivity and developing information display systems.

BACKGROUND

Sources on anatomy, physiology, psychology, and chronobiology are discussed in the previous Chapter 7. Let's list the main ones:

- Human anatomical structure (Gaivoronsky et al., 2011), atlases (The Human Body, n/d), including those provided in the 3D form (Anatomy 3D Atlas, n/d).

DOI: 10.4018/978-1-6684-8509-5.ch008

- Human physiology (Guyton & Hall, 2005; Ravi Kumar Patil et al., 2009; Schmidt & Tews, 1983);
- Human psychology and neurophysiology (Tsirkin et al., 2023).
- Chronobiology (Frank, 2008; Zidermane, 1988).

Since a human is considered an integral complex system, it is essential to understand the self-regulation and synchronization processes. Therefore, used publications on homeostasis and regulatory systems (McEwen & Wingfield, 2003; Asarian, 2012; Billman, 2020; Libretti & Puckett, 2022). Also, there are various examples and calculations (Lüttge & Souza, 2019; Thurner & Klimek, 2018; Artyukhov, 2009; Bodnar, 2005; Vasyutinskiy, 1990; Ovchinnikov, 1966; Soroko, 1984; Stakhov, 1986; Urmantsev, 1988; 2017).

The next problem is the reconstruction of the external environment and its influences. In addition to the presentation of empirical and theoretical material, it is necessary to assess how objective the relevant theories are, and how much they are speculative. Therefore, as examples of ancient speculative theories, based, however, on a long period of observation, the ideas of Samkhya and Taoism (Radhakrishnan, 2009/1923-1927; Maspero, 2007; Hansen, 2000), modern materialism (Burns, 2002; Beamish, 2007; Oizerman, 1979) and, finally, ideas of modern physics (Peebles, 1998); Novikov, 1988; Weinberg, 2000; Deutsch, 2001; Mukhi, 1999; Green, 2011/1999; Randall, 2011; Gukov, 1998; Aleksandrov, 2004; Einasto, 2012; Ignatiev, 2012). Their comparison has a certain epistemological significance, and the factual material is interpreted within the framework of the wave model. The combined ideas about a person, the evolutionary process, and the organization of the environment make it possible to build a model of a person-environment, which is the main result.

Various ergonomic examples were used to test the model (Selye, 1982; Korobko, 1997; Human Factor, 1991-92; Kovalyov, 1997).

HUMAN-ENVIRONMENT MODEL

Self-Organization Scenario

When modeling relations in the “man-environment” system, we will proceed from the following provisions:

1. The system “human–environment” is considered from the standpoint of the general theory of systems as an integral, complex, open system, which is in a state of dynamic equilibrium. The environment refers to the “close”, cognizable world that directly affects the processes of human life.

2. The system is modeled as S-space.
3. The self-organization is modeled as a set of stratification-convolution scenarios.
4. The components, connections, and interactions are modeled S-elements, S-sets, relations, and operations displayed in the OM1 form.
5. The set of stratification-convolution scenarios is determined by the composition of {S} and {O} - "remote" parts of the universe.
6. The problem-solved features determine calibrations.

The "human-environment" model (HEM) can be considered realistic provided that it can describe and make predictions regarding such facts and phenomena:

- Heterogeneity, autonomy, and, at the same time, the integrity of the components of the system (homeostasis of the organism, self-sufficiency of nature).
- The presence of qualitatively different interdependent levels of organization (physiological and mental processes, substance and flight).
- Prevalence of symmetry, constant *ф*, and derivative relations as factors of system integrity.
- Qualitative difference in the interactions of components at different levels - both their coordination and the global nature of the action (systems of regulation, feedback).
- Dismemberment of time scales of processes for different levels - and their mutual consistency.

It follows from the given data on the human body that the most general case of the organization takes place, that is, Sp is formed as a result of the interaction of {S} and {O}, and scenarios are described as (6.1-6.4), that is, they are similar to evolution scenarios (Chapter 6).

The general plan of the structure of the human body is simple and clear: the body consists of a symmetrical head and torso; each half of the body has a pair of limbs; each of the limbs consists of three parts (shoulder, forearm, hand and thigh, lower leg, foot, respectively), connected by movable joints and ends with five hoops, consisting of three phalanges, also connected by joints.

This structure is represented by a PC-graph corresponding to scenario (6.2), that is, (1S, 10), (Figure 1). It is logical to relate the soliton part of the SWM to the body.

Figure 1. The general plan of the structure of the human body

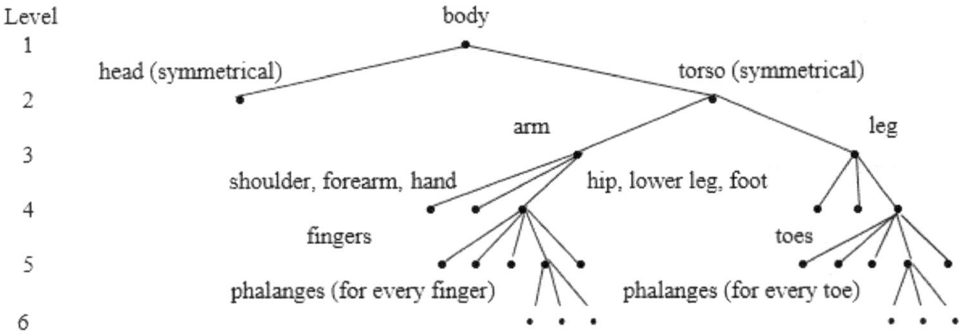

This graph implements the (1S, 10) scenario discussed in Chapter 4.

A Human Subjective Space

Now let's see if it is possible to group human sensations similarly. Let us present the facts relating to the physics, physiology, and psychology of perception. The main human senses are vision and hearing.

Let's start with ***color perception***. There are seven primary colors. Is there a physical reason why there are seven, or maybe it is due to the structure of the eye or the physiology of perception?

1. *The physical nature of color*. Color depends on the wavelength of light, which is determined by the temperature of the emitting body. It is also necessary to remember the corpuscular nature of light, which determines the discreteness of radiation, depending on which energy level the photons belong to. Since there are many such levels, in reality, we get a certain average length corresponding to a certain body temperature, and we can approximately assume that the series of wavelengths in the part of the spectrum perceived by the eye from violet to red is continuous. Thus, in a series of waves, no peaks or dips separate the ranges corresponding to the primary colors from each other.
2. *Sensitivity of the eye*. In the human eye, about 110-130 million receptors - "rods"- are responsible for the perception of light in general, and 5-7 million "cones" are responsible for the perception of color. The eye cannot perceive a still image; even when we look at one point, the eyes still move, making from 20 to 70 movements per second. With a perfectly transparent atmosphere, retinal rods could react to the light of a candle more than 30 kilometers away. The eye can distinguish 130-250 pure color tones and 5-10 million mixed shades. There

is only one perception maximum in the yellow-green color region, not seven, as expected.

Consequently, no physical or physiological prerequisites exist for isolating exactly seven colors. The yellow-green color corresponding to the best perception, is added by ergonomists to the seven basic ones, and thus there are eight.

Interestingly, in various historical eras, the main colors were considered:

- two (dark and light or warm and cold);
- three (preserved in many national flags, in the RGB color rendering system, etc.);
- five (corresponding to the elements or primary elements).

Sound perception. Humans perceive sound non-linearly; there are areas of relatively more and less high sensitivity than the sound pressure level measured using instruments (the first is measured in phons, the second in decibels). The human ear can pick up sounds with a frequency of 10-20 Hz to 15-20 kHz and is most sensitive to the range of 2000 - 2300 hertz, and a person can distinguish 3-4 thousand sounds of different heights. Speech range 1-3 kHz. The best ear for music (the ability to distinguish height) falls in the 80-600 hertz region. Here, the ear can distinguish between sounds with a frequency of 100 and 100.1 hertz.

As with color perception, the recognition of tones as high and low (2 gradations) or high, medium, and low (3 gradations) is in no way connected with either the objective level of continuously changing sound pressure or with the maxima and minima of the sensitivity of the hearing organ. In music, eight fundamental tones (an octave) are distinguished, with a continuous change in pitch.

Other sensory perceptions have been less explored. It is known, for example, that there are about 900 taste buds on the tongue, and their sensitivity depends on temperature.

The area of the olfactory zone of the nose is 5 square centimeters. There are about a million olfactory nerve endings here. For an impulse to arise in the olfactory nerve fiber, about 8 molecules of an odorous substance must get to its end, and for the sensation of smell, at least 40 nerve fibers must be excited.

A human also classifies these sensations regardless of the physical impact and sensitivity of the receptors. Different sources give a different number of tastes, aromas, and other things, which indicates the unformedness of clear sensations.

Sense organs. Receptors perceive hunger and thirst, responsible for balance, coordination, and orientation. They are not grouped into organs but send impulses to the human brain, which classifies them when a certain threshold of sensitivity is reached. The number of receptor sources is determined differently (from 21 to 33).

But there are only five external organs. And this is also inexplicable from the point of view of physics or physiology. Why, for example, sight, hearing, and touch are different senses, if they all perceive external vibrations?

Perception of space. The human eye is a sensitive instrument; there is evidence that it can perceive even single photons. Then, taking into account the difference in the sizes of objects such as:

- Human body – 1.
- Body cells - 10^{-5}.
- Simple molecule - 10^{-9}.
- Atom - 10^{-10}.
- Atomic nucleus - 10^{-14},

one would expect that we see each other not as solid three-dimensional bodies but as separate points of light. But what we see are solid bodies, and this needs an explanation.

On the other hand, physicists do not explain the phenomenon of the three-dimensionality of space. Moreover, if I. Newton considered space as three-dimensional, then A. Einstein's special theory of relativity is defined in a four-dimensional space-time continuum (G. Minkowski's space), and T. Kaluza operated on a five-dimensional space (the addition of the fifth dimension to the four-dimensional continuum made it possible to derive Maxwell's equations of their equations of general relativity) (Kaluza-Klein Theory, n /d). According to E. Noether's theorem, there is a connection between symmetries of various types and conservation laws (Table 1) (Noether's Theorem., n / d). Modern physicists have calculated that ten-dimensional space (string theory) is needed to implement all known conservation laws (Mukhi, 1999), and some theories even consider twenty-six dimensions (bosonic string theory). Therefore, the perception of three-dimensional bodies as integral and three-dimensional in three-dimensional space must be explained.

Table 1. Correspondence of symmetries, invariances, and conservation laws according to the Neter theorem

Transformation	Invariance	Conservation law for
Time broadcast	Uniformity of time	Energy
Space broadcast	Homogeneity of space	Impulse
Rotation of space	Isotropy of space	Moment of impulse
C, P, CP, and T symmetry	Isotropy of time	Parity
Lorenz group	Lorentz covariance	Mass center motion

continued on following page

Table 1. Continued

Transformation	Invariance	Conservation law for
Calibration conversion	Gauge invariance	Charge

(Teorema Neter. [the Neter theorem], n / d)

The same applies to the perception of time - we divide it into past, present, and future. The dimensions of space are not completely equal (the third is less accessible due to gravity), so the present is less accessible to us compared to the past and future[1].

The binarity of judgments (good-bad, yes-no, white-black, etc.) underlies mathematical logic. However, its artificiality, relativity, and inconsistency in many cases with reality were clarified in ancient India (darshans of Nyaya and Vaisheshika), in ancient China (Taoism), and in ancient times (several aporias of Zeno). Also, modern specialists in pattern recognition within the framework of mathematical logic cannot solve, for example, the problem of "turning a tadpole into a frog", but are forced, as in many other cases, to resort to fuzzy logic, introduce categories of feature classes, etc. However, the binarity of assessments and conclusions demonstrates an unusual persistence and needs to be explained.

Unity of truth. A person is inclined to complete his judgments with a single true statement - God is one, the Universe is one, any decision is the only true one, and so on. In most cases, such a reduction is not based on any objective facts and provable judgments but corresponds to our intuitive feeling.

Based on the above, it is possible to model the structure of sensations with a graph (Figure 2). It is logical to correlate the wave part of SWM with the subjective space.

Figure 2. Grouping of human perceptions

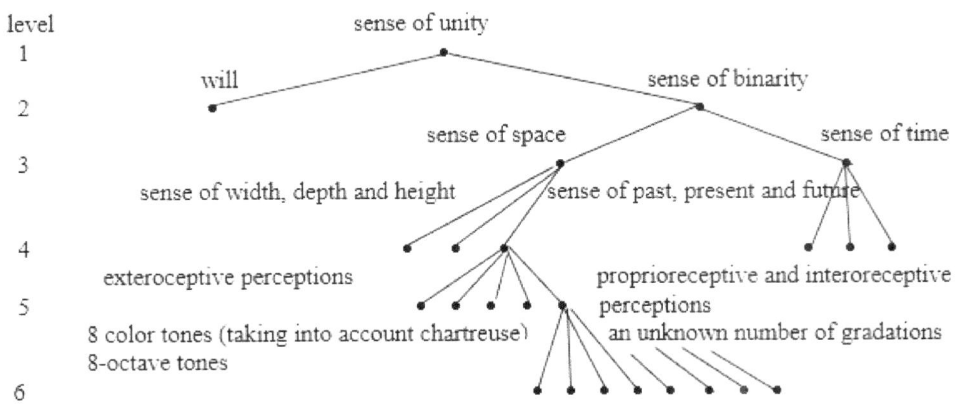

This graph is an implementation of scenario (1S, 1O).

Let's use the graph in Figure 2 to correlate levels and channels with the structures and "abilities" of the human psyche, responsible for the perception of the environment. We will call their totality *the subjective space of a human* - to avoid misunderstandings associated with the concepts of "psyche", "intellect", "mentality", "soul", "and spirit". So:

1. Subjective space is a gestalt conditioned by the general laws of self-organization of complex systems, structuring interactions with the environment.
2. This gestalt includes a set of levels, structures, and channels that are not reducible to each other, each of which has its object of perception.
3. Namely:
   a. The relationship of unity with the environment is intuitively perceived (1channel).
   b. Through the feeling of one's "ego" - the internal unity of the subject (1 channel).
   c. Will and reason affect the object, and the response (2 channels) is evaluated.
   d. The nous operates with spatial and temporal categories (2*3 channels – three dimensions of space, past, present, and future time, grouping sensations as intro-, proprio-, and exteroceptive ones).
   e. Organs of perception determine specific reactions to direct influences (1*5+1*(?)+1*(7) - there are five external sense organs; it is not known exactly how many internal (hunger, thirst, etc.) Sensations exist and how many sensations refers to borderline (for example, a sense of balance). We note once again, that internal and borderline sensations do not have specific organs.
   f. Further there is a change in the principle of organization, perceptions become homogeneous, and separate organs of perception are no longer formed (although their rudiments can be seen, for example, in different responses to the light of rods and cones of the human eye). Nevertheless, it is possible further, with some approximation, to trace the correspondence of sensations to the graph in Figure. 2. Thus, the structuring ends here.
4. The graph or PC-diagram of the self-organization scenario *(1S, 1O)* describes the ratio of the channels. The number of channels is defined by the Fibonacci number series - 1-1-2-3-5-8. For example, the mind operates with binary categories according to the "good/bad" principle, the spatiotemporal relations of the mind are ternary, the number of external sense organs is five, etc. Note that the scope of a particular channel covers all subsequent levels, and channels related to subsequent levels inherit one or another of its features since they are formed on its basis. This explains the above quantitative characteristics of the

perceptions of external influences by the senses, the three-dimensionality of space and time, and so on. Thus, we state:

a. consciousness ("ego") manifests at the second and subsequent levels.
b. reason manifests itself from the third and subsequent levels.
c. the mind (nous) manifests itself from the fourth level and on subsequent levels.
d. the fifth level corresponds to the sense organs, and the number of external organs of perception (exteroceptors) is 5.

5. Data about the ability to distinguish shades of sensory sensations are insufficient for unambiguous conclusions. It can be assumed that the quantities of homogeneous sensations are related to Miller's numbers. Independent organs are not formed here (Table 2).

Table 2. Levels and channels of human subjective space

Levels of perception	Objects of perception	Characteristic of perception	Channels, structure, and their number
1	Unity	Humans and the environment are not separate, but the possibility of separation potentially exists.	Intuition 1
2	Separation	Awareness of a person as a self-sufficient whole	+ Ego (mind, consciousness) 1
3	Impact and Reaction	Impact on "not yourself", the response of the environment in response	+ Will (action-reaction) and reason (good-bad) 2
4	Space and time	Organizing impacts on the environment and its reactions by space and time categories	+ Nous and emotions (tracing forms and changes) 2*3
5	Sensory impacts	Organizing the effects of the environment, internal and borderline sensations by color, sound, etc.	+ Feelings and emotions 2*3*5(?)
6-7	Tones of color, sound, etc.	Distinguishing between shades, tones, etc.	+ Feelings and emotions 2*3*5*8(?)

6. All these channels are present and active constantly, but only one is dominant at some period, which determines the corresponding "state of consciousness". In turn, interaction with a specific object determines the content of the state of consciousness - in the form of an idea, action, word, or emotion, including those related to the past (memory) or future (prediction) time.
7. In psychophysiology, there is no exact understanding of what is meant, for example, by "intuition" - whether it is a sensation, feeling, or ability, whether it corresponds to any physical structure and where such a structure is located, how it relates to "understanding" and "knowledge." Therefore, fundamentally

different definitions are offered - by indicating the object of perception, the level of organization, and the characteristics of perception. Thus, each level or channel (S-set or S-element) corresponds to one or another stage of stratification of the S-space and is determined according to Table. 2.

8. We emphasize a certain autonomy of the subjective space both to the physical body of a person (corresponds, except level 6, to the general plan and does not correspond to the structure of the nervous system described by the network graph (6.4)), and the structure of the external environment (more on this below). It is this autonomy, expressed by the sense of one's ego, that explains the emergence of several cognitive and other abilities, for example, the imagination can present images that do not correspond to the environment.
9. An external influence is necessary to start updating the structure (first group axioms), i.e., some metabolic process. The preservation of integrity in this case requires the ratio of the potentials of the levels, which is determined based on the proportion of the golden section. The corresponding weighting coefficients of each level are proportional to the series 1, 0.618, 0.382, 0.236, 0.146, and 0.09. They also determine the contribution of each of the perception channels to holistic image creation (Figure 3).

Figure 3. Weighing coefficients of perception channels

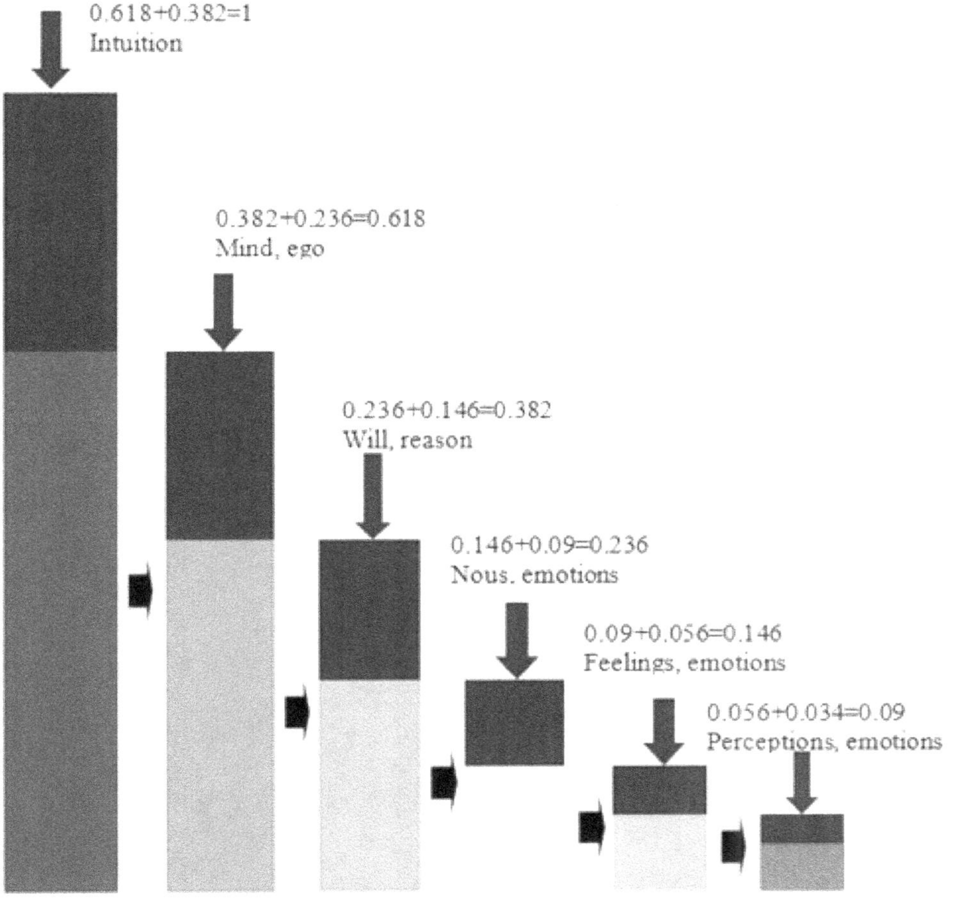

At the same time, since the potential of the previous level is greater than all the elements of the next one, there is a qualitative difference in the characteristics of both levels (non-additivity), which makes it possible to perceive heterogeneous and different-quality objects. So, the inheritance of the previous level characteristics is due to the transfer of its "qualitatively colored" potential, and the appearance of new features is due to external compensations.

10. We also note the great similarity between a person's mental component and the general structure of his body compared to the animals (examples are considered in (Kovalyov et al., 2020), which allows us to draw the following conclusions:

a. the balance of the mind and body (integrity) allows a more complex impact on the environment at the level of individuals;
b. this balance is an evolutionary advantage of man, providing an increase in the likelihood of survival;
c. striving for the most holistic ratio of body and mentality, expressed by the scenario (1C, 1O), as corresponding to the general laws of self-organization, is the most important factor in evolution, determining, along with environmental conditions, the direction of natural selection (Chapter 4).

Human models differ depending on which invariants need to be preserved:
The model in the form of a graph (i.e., OM1, Chapter 4) allows you to show the structural correspondence of the general plan of the body structure and subjective space (Figure 4).

Figure 4. Correspondence of the general plan of the body structure and subjective space

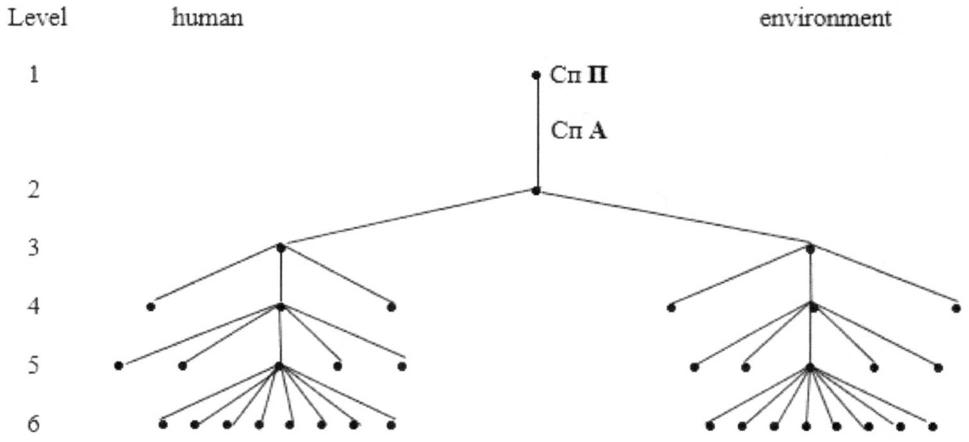

The human body is relatively stable and reacts slowly to environmental influences, while the subjective space is permanently changeable and its reaction is much faster. In this case, the structure of both bodies and channels does not change.

These characteristics can be shown using the images of solitons and waves (Figure 5). The explanations for why waves are obtained in one case, and solitons in the other, should be sought in the properties and influences of the external environment.

Dynamic models (OM2) would emphasize the static nature of the body and even better the dynamism of the subjective space.

It is possible to use simulation models (OM3), which allow you to display many invariants and improve accuracy. However, their creation and verification are much more complicated than in the previous cases.

The Structure of The Environment

There is a problem of finding adequate information: different sciences (for example, philosophy and physics) define such fundamental categories as matter, space, time, and so on in different ways. Even in the natural sciences (chemistry, physics, biology), there is a significant difference in interpretations. In addition, we should not forget that modern science, based on experiments and the correct processing of their data, is very young. In other historical epochs, other paradigms took place due to different activities of channels of subjective space (Kovalyov et al., 2020).

Therefore, the authors adopted the following approach: first, natural philosophical doctrines will be presented as an example of a "view from another era", then one of the modern philosophical approaches, and finally the data of modern physics. A comparison of these representations will be useful both for reconstructing the structure of the environment and for interpreting the epistemological process.

Figure 5. Soliton-wave model of the correlation of the body structure plan and the channels of the subjective space: 1 - the different nature of these components is transferred as a wave and a soliton; 2 - the different nature of the components is conveyed as a coincidence of waves and solitons (Figure 4, Chapter 4). Solitons are shown in black and waves in blue

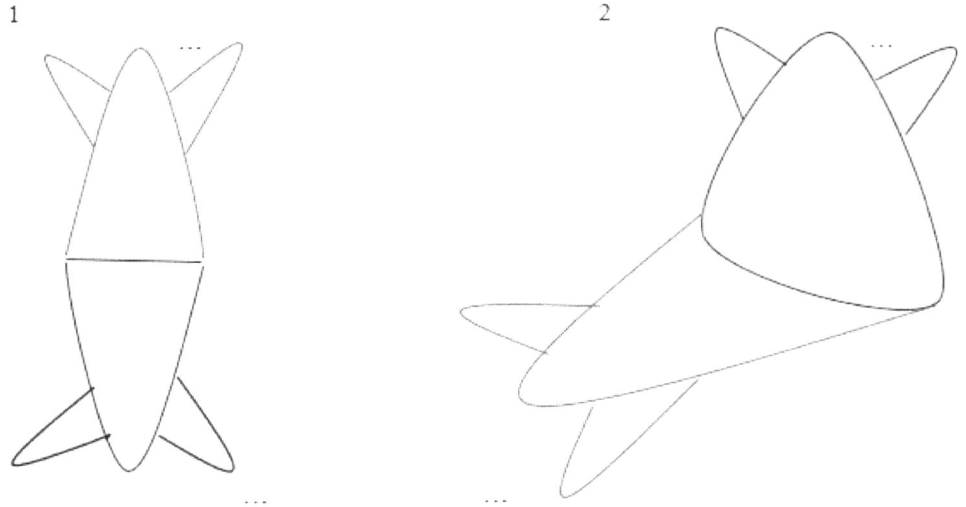

Sankhya view. Sankhya - one of the six classical schools of Indian philosophy (Radhakrishnan, 2009/2023-2027) - is dualistic, recognizing the original existence of the spiritual (Purusha) and material (Prakriti) components of the Universe. The presence of Purusha initiates the emergence of various structural units of Prakriti related to the organization of the environment, the structure of the human body, and the development of consciousness[2] (Figure 6).

Without going into the interpretation of the structures in Figure 5 (we refer to (Radhakrishnan, 2009/2023-2027) for comments, we note the presence of some systemic patterns of the scenario (1S, 1O): Fibonacci numbers 2, 3, 5, symmetry, feedback.

Taoist perspective. The key concepts of Taoism are defined as follows (Maspero, 2007; Hansen, 2000).

Tao (way) is a kind of coordination of changes in the Universe, which makes it possible to perceive the world as a whole. Regarding the self-organization theory, Tao is an intuitively perceived modality of potential existence.

It is believed that Tao gives rise to a unit, a unit - a deuce, the components of which, having set in motion, become three and give rise to the entire manifested world.

Figure 6. The system of interaction between man and the environment according to Sankhya. The presence of Purusha (indicated by the dotted line) predominantly affects the conscious and subtle structures. Prakriti is structured (indicated by solid lines) at first as 3 gunas[3], and their combinations lead to the appearance of material elements of the environment and body organs

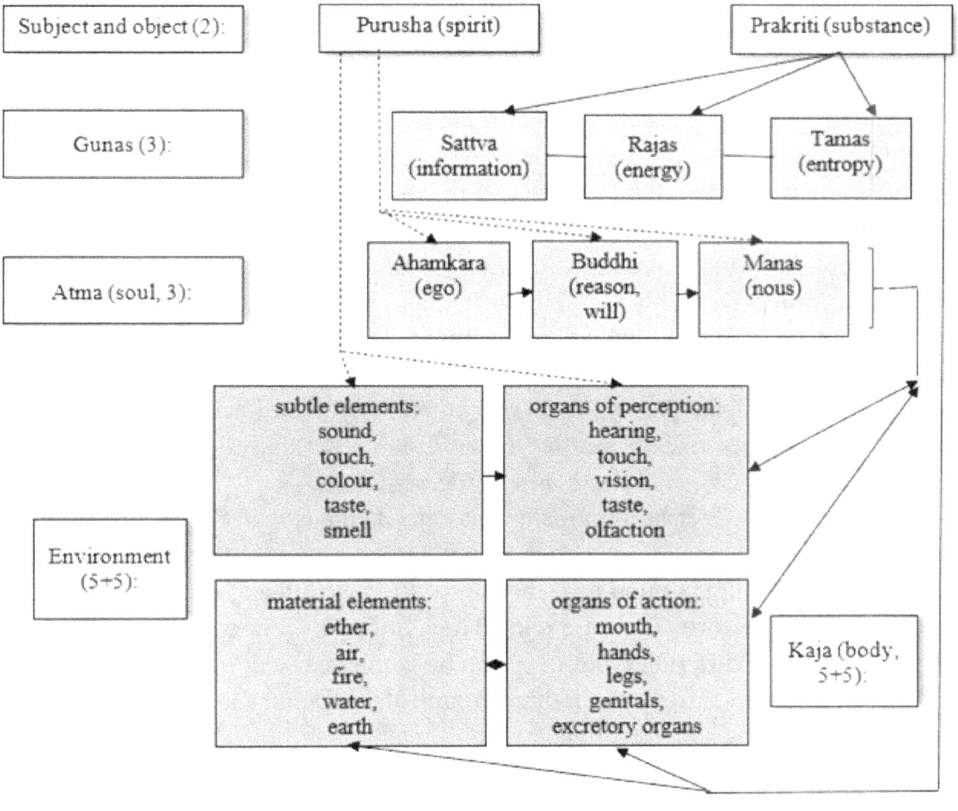

The unit generated by Tao is considered a modality of actual existence.

Further, the deuce - *tai chi* - a pair of opposites of *yang* and *yin*. Yang corresponds to the sky, dragon, sun, heat, light, spirit, life, active and masculine, left side, odd numbers. Yang symbolizes everything light, dry, and high: mountain, heaven, solar animals, and birds. Yin - initial waters, tiger, feminine, moon, soul, depth, negative, soft and flexible, north, darkness, death, paired numbers. In the field of human thinking, yin is the irrational female intuition, yang is the rational common sense of a man; yin is the passivity of the sage immersed in himself, yang is the activity of the active ruler.

The perception of tai chi as a harmony of opposites, constantly turning into each other, correlates with the channels of the third level - will and reason.

The next category is *San Cai*. The world is a trinity (San Tsai), expressed in the images of Heaven, Human, and Earth. Sky-Father is the vault of heaven and everything subtle, unknowable, lying in the spheres of other worlds, the worlds of spirits, the worlds of deities, and the worlds of ancestors. Mother Earth gives life to everything that is born on it. The Earth is a living organism; in the image of the Earth, all the material objects surrounding people are merged. Man in this Trinity is the point of contact between Heaven and Earth, where the laws of both appear (a complex system in a dissipative environment). The unity of everything existing considers that all is created and is powered by *qi*. These are the origin of the world, the creator and the material, the idea and the realization. These form the spiritual and material worlds and man. These exist always in the form in which it is necessary according to the conditions of the system where it is manifested. Qi exists in a single trinity: Heavenly Qi, Humans Qi, and Earth Qi. Heavenly qi manifests itself in our world as time, which, filling any idea, allows it to arise and develop. Human qi manifests as the body's life force, flowing along the meridians connecting acupuncture points and passing from one organ to another. Lack or excess of qi in the body leads to its disease. Earth Qi manifests itself as physical interaction and ensures the existence of the Earth as a living organism.

If we "translate" these arguments into the language of the theory of self-organization, then here we are dealing with three-dimensional space and time, perceived by the human mind as a channel of the fourth level. Qi is related to potential, and its transformations are related to calibrations. For example, at the level of common sense, different types of qi can be compared with the Hindu Huns, the information-energy-entropy triad, three groups of receptors, and ternary categories inherent in this level of perception.

Graphically, the idea of ternarity is expressed as trigrams surrounding Tai Chi (Figure 7).

Figure 7. Trigrams and Tai Chi (Bagua)

The next concept is *Wu-Xing*, five elements. The extreme *yang* and *yin* correspond to the elements of fire and water. The cycle of their mutual transformations includes two intermediate stages, symbolized by the metal and wood elements. A circle of *yin* and *yang* transformations, which have a center, is formed. The emblem of the center is the element of the earth. Thus, Tai Chi unfolds into a five-year structure that combines the duality of yin-yang and the triad of creation. According to the scenario (1C, 1O), this diagram correlates with the level and channels of the external sense organs.

I-Ching. Everything in nature happens in cycles. For example, a wave on the water evolves cyclically: rising phases (*yang*) are replaced by falling phases (*yin*), symmetrical projections, and depressions (two more combinations of *yang-yin*), the process develops in three-dimensional space and time, which increases the number of combinations (trigrams) to eight. Denoting yang with a long one and yin with two short lines, we get the classic image of 8 trigrams I-Ching (Figure 8).

Figure 8. Formation and image of 8 trigrams (8 Trigrams Meaning and Definition in Qi Men Dun Jia, 2022, n /a)

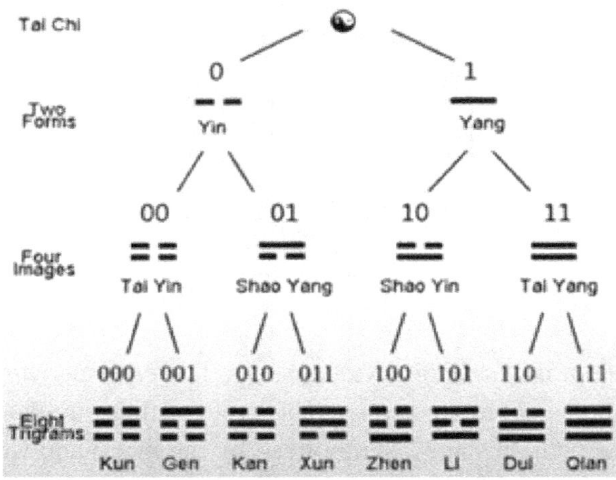

At the same time, it can be assumed which phase will be next - all yang will be replaced by yin and vice versa. Compiling hexagrams from two trigrams in this way, we get 64 hexagrams that describe both the current state of the wave and its development. So, a cyclic process, regardless of its nature, has an invariant – the 64th hexagram.

As you can see, the correspondence to the scenario (1S, 1O) is even closer: modalities appear, the Fibonacci number series is presented more fully than in Sankhya (1, 1, 2, 3, 5, 8), symmetries are more diverse, in general, the description is closer to the abstraction of a wave, as well as an open system (especially San Tsai, where a person acts as an open system, and S and O - Heaven and Earth.

View of modern materialism (Matter (philosophy), n/d; Burns, 2002; Beamish, 2007; Oizerman, 1979). Matter denotes the original, real substance to subjectively real consciousness. Matter exists objectively independently of our sensations and is reflected by them.

The attributes of matter, the universal forms of its existence are movement, space, and time, which do not exist outside of matter, just as there can be no material objects that would not have these attributes.

Forms of matter movement are the main types of movement and interaction of material objects, expressing their integral changes. Each body has several forms of movement. There are 3 groups containing many specific forms of movement:

- In inorganic nature (7): spatial movement; motion of elementary particles and fields - electromagnetic, gravitational, strong and weak interactions, processes of transformation of elementary particles, etc.; movement and transformation of atoms and molecules, including chemical reactions; changes in the structure of macroscopic bodies - thermal processes, changes in aggregate states, sound vibrations, and more; geological processes; change in space systems of various sizes: planets, stars, galaxies, and their clusters.

- In wildlife (5): metabolism; self-regulation, management, and reproduction in biocenoses and other ecological systems; interaction of the entire biosphere with the natural systems of the earth; intraorganismal biological processes aimed at ensuring the preservation of organisms, maintaining the stability of the internal environment in changing conditions of existence; supraorganismal processes express the relationship between representatives of various species in ecosystems and determine their abundance, distribution zone (range), and evolution.
- In society (2): diverse manifestations of people's conscious activity; all higher forms of reflection and purposeful transformation of reality.

It is believed that the higher forms of movement arise based on the lower ones and include them in a transformed form, that is, there is unity and mutual influence. The higher forms are qualitatively different from the lower ones and cannot be reduced to them. These theses are contradictory from the point of view of set theory. Still, they can be described within the framework of the theory of self-organization (stratification of S-space and the transition from S-sets to S-elements not "from bottom to top", but "from top to bottom").

Further, the universal properties of matter are (7):

- Indestructibility and indestructibility.
- Eternity of existence in time and infinity in space.
- Immanence of motion and change of matter, self-development, a transformation of some states into others.
- Determinism of all phenomena.
- Causality - the dependence of phenomena and objects on structural relationships in material systems and external influences, on the causes and conditions that give rise to them.
- Reflection - manifests itself in all processes, but depends on the structure, interacting systems, and the nature of external influences. The historical development of reflection leads to the emergence of its highest form - abstract thinking.
- Qualitative inexhaustibility of the properties of matter – matter cannot be defined by listing some finite set of its properties.

Thus, the list of properties is open.

Three laws of the existence and development of matter are also postulated:

- Unity and struggle of opposites.
- The transition of quantitative changes into qualitative.
- Denial.

View of modern physics. (Matter, n/d; Peebles, 1998); Novikov, 1988; Weinberg, 2000; Deutsch, 2001; Mukhi, 1999; Green 2011/1999; Randall, 2011; Gukov, 1998; Aleksandrov, 2004; Einasto, 2012; Ignatiev, 2012). Physics also operates with the concept of matter, defining it as a set of all the contents of space-time that affects the properties of space-time. The definition of matter expanded with the development of various fields of science, from objects described by the properties of mass, charge, temperature, etc. in absolute space and time, to the understanding of their mutual connection in the evolution of the Universe.

This evolution, in a very brief summary, includes: the existence of some initial point of singularity, where known physical laws do not apply and the categories of space, time, matter, and so on do not apply; in the course of an event called the Big Bang, the unfolding of the Universe begins, gradually bringing it into the form familiar to us.

Hypothetically, at the earliest stages of development, there were pre-quark superdense material formations, the components of which are string-like objects, to which all elementary particles, that is, all known forms of matter, can be reduced, moreover, the difference between the substance and field forms of matter at this level is erased.

At later stages, the following forms of matter are distinguished:

- matter: hadronic matter (consists of many constituent particles – hadrons); baryonic matter (consists of baryons); matter in the classical sense (consists mainly of fermions; dominates in the solar system and the nearest star systems); antimatter (consists of antiparticles); neutron matter (consists mainly of neutrons and is devoid of atomic structure; the main component of neutron stars); other types of substances having an atom-like structure (for example, a substance formed by mesoatoms with muons); Quark-gluon plasma is a superdense form of matter that existed at an early stage in the evolution of the Universe.
- field, unlike matter, has no internal voids, and has absolute density: electromagnetic field; gravitational field; quantum fields of different nature.

We note the corpuscular-wave dualism of particles (can be compared with the soliton-wave nature of the bodily and subjective, mental, components of a person considered above), as well as the fact of the predominance of matter in the Universe (can be compared with the chiral purity of amino acids and other organic compounds, Chapter 1).

The quantum field can be considered as a universal form of matter, to which both substances and gravitational and electromagnetic fields can be reduced. At the same time, there is a fuzzy division into real fields (lepton and quark fields of fermionic nature) and fields of interactions (strong gluon, intermediate bosonic weak, and photon electromagnetic fields of bosonic nature, this also includes the hypothetical field of gravitons). However, the Higgs field is difficult to assign to any of these categories.

In addition, material objects of an obscure physical nature are assumed: dark matter and dark energy are introduced to explain some astrophysical and cosmological phenomena (moreover, according to the WMAP data used in the framework of the Lambda-CDM model, the Composition of the Universe is estimated as follows: 74% of dark energy, 22% dark matter, 3.6% of interstellar gas and only 0.4% of the matter of stars, planets, etc.) (Dark matter[REMOVED HYPERLINK FIELD], n/d).

Further, a classical substance can be in one of the states of aggregation: gaseous, liquid, solid crystalline, solid amorphous, or in the form of a liquid crystal. In addition, a highly ionized state of matter called plasma is distinguished, as well as Bose-Einstein condensate and quark-gluon plasma. If we consider them, the number of states of matter will equal 8.

Each fundamental fermion (each type of quark and lepton) has its spinor field within the framework.

Fundamental bosonic fields are carriers of fundamental interactions and are called gauge fields. The following types are known:

- Electroweak.
- Electromagnetic field.
- Electric field.
- A magnetic field;
- Field - carrier of weak interaction.
- Gluon field.
- Field of nuclear forces.
- Gravitational field.

A hypothetical fundamental interaction is also considered, which may be associated with dark matter and/or dark energy.

Fundamental interactions are described by two modern generally accepted theories: the general theory of relativity and the Standard Model. Their unification has not yet been achieved due to the difficulties of creating a quantum theory of gravity. Various approaches are used to combine fundamental interactions: string theory, loop quantum gravity, and M-theory; some manage to avoid the initial singularity. Accounting for conservation laws and their relationship to symmetries (Noether's theorem, Table 1) is an argument favoring string theories and emphasizes the systemic factor's role.

We also note the anthropic principle of cosmology - the ratio of physical constants is such that only under such conditions is it possible for a person to appear as a result of the evolution of the Universe - the modern reading of the natural philosophical thesis about the unity of the macro- and microcosm speaks of the unity of the Universe.

Let's summarize. First, we note the greatest systemic nature of ancient natural-philosophical theories compared to modern materialism and physics data. This can be explained by the state of consciousness of that era, characterized by great potential and relatively more developed intuition, ego, will, and reason compared to the modern state of consciousness (Kovalyov et al., 2020). Since both Samkhya and Taoism, although based on long-term observation, did not have modern verification methods, and can be characterized as speculative, the influence of the state of consciousness during their creation is much greater than on modern theories.

This confirms the previously stated judgment that the structuring and activity of the subjective space are relatively autonomous to environmental influences and are largely determined by systemic factors within the scenario (1S, 1O) framework.

Secondly, modern materialism arose at the peak of the development of a practical mind (nous), prone to analysis, too subject to formal logic, but with a weak sense of the unity of the system, therefore this view itself is internally contradictory and of little use when compared with the real world.

Thirdly, modern physics, and more broadly, modern science, is based on a paradigm formed in the time of Newton, Leibniz, and Descartes; even if we add Galileo and ancient mathematicians to this series, we get a period that is too short compared to the time of human existence, life on Earth and the Universe. Therefore, it is not surprising that ideas about the world around us are in the process of formation. The presence of numerous competing and incompatible theories points directly to this; the need to introduce corrections - 96% of dark energy and dark matter in the composition of the Universe - to remain within the framework of modern theories, shows their imperfection and the need to replace them with more realistic ones. All theories are based on intuitive categories - Purusha, Tao, matter, energy, mass, or charge.

Modeling

Therefore, when reconstructing environmental effects, many uncertainties arise. If we proceed from the fact, that the theories have in common the recognition of the long evolution of the Universe by the time of the emergence of life and the appearance of man (Chapter 1), and interpret the events within the framework of the theory of self-organization, then we can propose the following hypothetical scenario (Figure 9).

Figure 9. Hypothetical reconstruction of external factors before the emergence of life

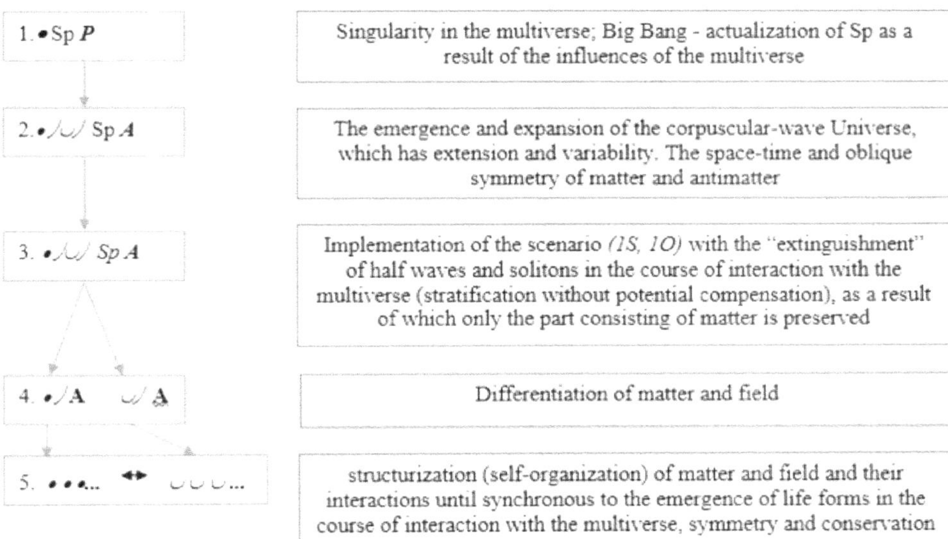

Further, the evolution of life took place already in the realities determined by events 1-5; its events correlated with it in the likeness (fractally), plus there were general laws of self-organization—specific life forms adapted to the characteristics of their habitats (Chapter 5). Let us show the correspondence between the events of the evolution of the Universe and life (Figure 10). The black arrows show the necessary conditions, and the blue arrows show the similarity relations.

Finally, let us build a human-environment model, bearing in mind that evolutionarily a person relates to his ancestors in likeness, which takes the form of inheritance (Chapter 5), and with the beginning of psychosocial evolution - cultural commonality and direct learning (Figure 11, structures are shown, their substance and field character).

SELF-REGULATION

In general, self-regulation is the ability to maintain homeostasis through coordinated responses. Examples of such reactions are osmoregulation in the kidneys, removal of metabolic waste products by the kidneys, lungs, sweat glands, and gastrointestinal tract, regulation of body temperature, for example, through sweating, regulation of blood glucose levels by the liver as well as the pancreas, and so on.

As special cases of self-regulation, stress resistance (a set of qualities that allow the body to endure the action of adverse factors without outbursts of emotions that affect well-being up to mental disorders and physical pathologies), immunity (the body's ability to maintain biological individuality by recognizing and removing foreign substances, bacteria, and viruses) can be considered, as well as their tumor cells) and regeneration (the ability to repair damaged tissues).

All of them are characterized by certain threshold values - the minimum (self-regulation does not begin), and the maximum (self-regulation becomes impossible.)

Figure 10. The similarity of the evolution of life and the evolution of the universe

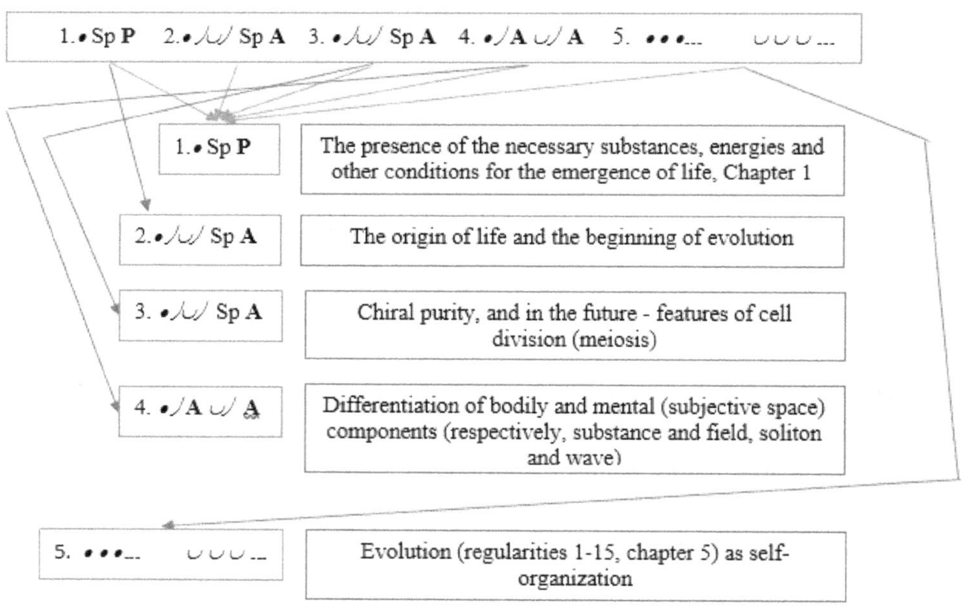

Figure 11. Human-environment model

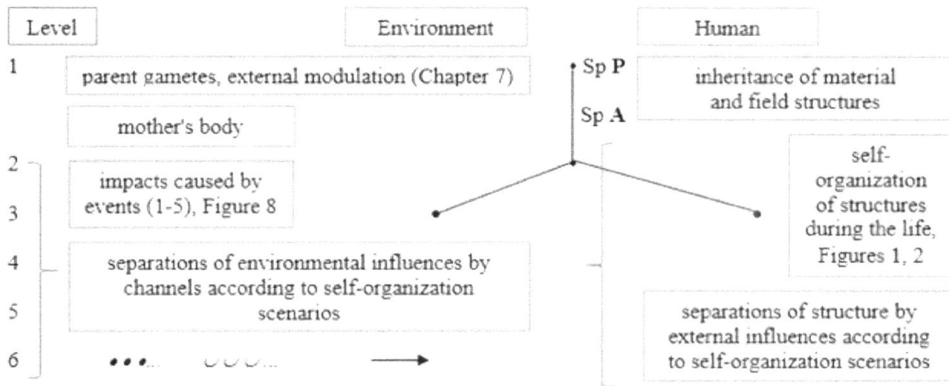

In humans, the epidermis and its derivatives regenerate well - hair and nails, bone, and liver (with the loss of up to 85% of the organ, the remaining fragments begin to increase in size due to an increase in the size of the cells themselves; this is how the liver completely restores its original mass), fingertips and, contrary to popular belief, nervous system. In lower animals, much more comprehensive cases of regeneration are known - up to the restoration of the entire organism from a small fragment. Naturally, it would be highly desirable to achieve the same efficiency of regeneration in humans - and such hopes are associated with the progress of research in the field of cellular regeneration, for example, (Spalding et al., 2005; Sender & Milo, 2021), and, especially, the use of stem cells for medical purposes (Sagaradze et al., 2019; Karagyaur et al., 2019).

Let us consider the mechanisms and limits of self-regulation from the point of view of the theory of self-organization.

The Emergence of Connections and Exchanges

Let, because of external influence, one of the elements, for definiteness, of the i-th layer Sp^+ be damaged. Then:

1. Since the integrity is violated, the system will either cease to exist ($i=1$) or restore integrity due to the redistribution of potentials between the Sp^+ layers. This redistribution is possible within certain limits and corresponds to the phenomena of regeneration, self-regulation, and fatigue inherent to living organisms. For layers with numbers less than i, the influence of the external stimulus will be

less and less significant. For Sp^-, similar changes will occur due to symmetry. The overall potential *Sp* will decrease.

2. In the process of redistribution, the modalities of existence and state will change, transformations • and • will occur, as well as interactions of S-elements. This will lead to the emission of waves and the displacement of solitons, i.e., to the exchange processes between the environment and the system. Since the qualitative characteristics of the layers are different, the processes of exchange and change of layers with numbers *i* and *i+1* for Sp^+ and Sp^- will also be qualitatively different.
3. Various exchanges and changes (maintenance of dynamic balance, restoration or destruction of systems and organs) can be described by operations ***{DICS}*** adequate to the characteristics of these processes. For example, the normal, paradoxical, and ultraparadoxical reactions of the nervous system to a change in the magnitude of the stimulus are interpreted as special cases of *S*. Since the implementation of operations is subject to restrictions (conditions 4.1-4.4, Chapter 4) associated with the potentials of the elements, this leads to a stratification of interactions corresponding to the stratification of *Sp*. According to Noether's theorem, the conservation laws will be fulfilled within each layer boundary, determined by its inherent symmetry.
4. The most sensitive will be interactions at resonant frequencies. Their compensation requires redistributing the largest part of the potential, and the impact will be felt for the next few layers. This effect explains the mechanism of action of adverse effects of one type on various body systems. Thus, we have the following provisions:

Regulation 8.1.*Reason for existence: homeostasis and regeneration; forward and reverse links; qualitative differences in the processes of exchange between the components of the system "man-environment" in conditions of unbalanced external influences is the need to preserve its integrity.*

Regulation 8.2.*Favorable should be considered such as influences of the external environment that contribute to the restoration of the integrity of* Sp^+*; unfavorable - influences that violate the integrity of* Sp^+.

The Limits of Self-Regulation

The limits of self-regulation will be determined based on the dynamic balance of the system maintain conditions (formalized as preventing further stratification (or convolution). This corresponds to a change in the potential for each element of each layer within the boundaries of $\pi \div \pi\, \Phi$. Thus, for indicators derived from the potential, we have the following situation:

Regulation 8.3.*The limits of change in indicators related to one level should not exceed the ratios 1:Φ for scenario (6.2) or derivatives from it for scenarios (6.3), and (6.4).*

In a more general case, when the adaptation process captures structures of several levels, the need to maintain integrity leads to:

Regulation 8.4.*If for the last of the levels involved in the adaptation process, the allowable limits for changing indicators satisfy the conditions of provision 8.3, then for the previous levels (1, 2, ..., k, counting from the last one), they are calculated:*

$$\frac{\pi_{i,n} \cdot i_n}{\Phi^k \cdot i_{n-k}} \div \frac{\pi_{i,n} \cdot i_n}{\Phi^{k-1} \cdot i_{n-k}}, \tag{8.1}$$

where i is the number of elements of the last layer n.

From here it follows directly

Regulation 8.5.*The conditions that make possible pathological processes are the excess of the limits specified in regulations 8.3 and 8.4.*

Regulation 8.6.*The conditions that make regeneration possible are external influences that make it possible to return the potential of vital signs to the limits indicated in (8.1).*

As seen from (8.1), even small changes in the potential for elements of the first levels can destroy elements of subsequent levels. Therefore, one should be especially careful when organizing specific interactions for the first levels.

Stratifications of time. According to properties 4.6-4.8 and their consequences, the units of time for each layer naturally decrease as the stratification (4.12); there are scenarios for which the lifetime of each layer, measured in its units, is the same (4.13). Moreover, the lifetime of any element and any layer does not exceed the S-space as a whole lifetime (Property 4.9 and corollaries).

It is known that the time scales for different levels of the organism are not the same. The organism's chronobiological rhythms, physiological systems, organs, cells, and biochemical reactions are consistent with astronomical phenomena (Zidermane, 1988). When moving to the next members of this series, the time scale decreases. This testifies to the natural dismemberment of a single stream of time into separate rhythms while maintaining the integrity of the " man-environment" system. Therefore, we have:

Regulation 8.7.*The division of time into flows corresponding to the levels of the system organization is due to the need to preserve its integrity.*

Thus, HEM corresponds to the known data on the structures and processes of the "man-environment" system and predicts qualitative, and sometimes quantitative patterns. Let's illustrate this with examples of self-regulation.

Adaptations and Pathologies

During evolution, under the influence of external influences, there are exchanges with the environment, changes in relatively unstable (waves) and relatively stable (solitons) states, and changes in the absolute values of the *Sp* potential and components. Let us show the correspondence between these changes and the processes of adaptation and pathology.

According to the ideas of G. Selye, refined in the course of subsequent studies (Selye, 1982), human adaptation to long-term adverse effects includes (Figure 12), the change in the level of resistance is shown in the form of a qualitative graph):

- Individually optimal values of physiological parameters characterize the initial rate.
- The anxiety stage immediately following the stimulus's appearance, during which the indicators change, approaching acceptable limits, and resistance increases.
- The stage of resistance, replacing, after the transitional stage, the stage of alarm, characterized by the relative stability of indicators at the level of their limit values.
- The exhaustion stage following the transitional stage, is characterized by a drop in resistance, expressed in the suppression of vital functions and the output of physiological parameters beyond acceptable limits and is already accompanied by pathological changes.

Figure 12. Development of stress

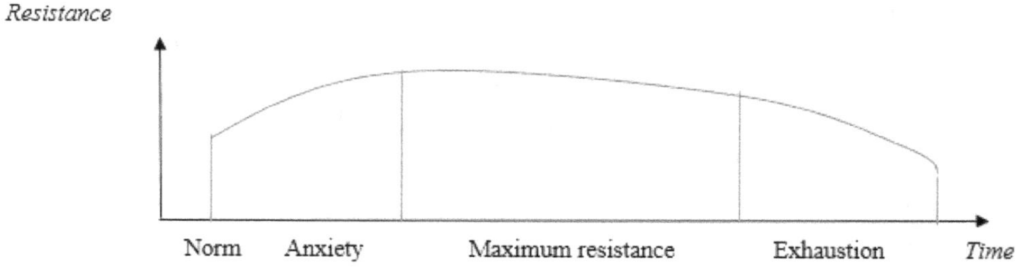

It can be seen that these stages correspond to the evolution of Sp:

1) ***H°P(AP(•A))***, i.e., optimal distribution of characteristics, actually existing C-elements for "established" external influences.
2) ***H°P(AP(•A Aæ•A))***, i.e. a change in external influences causes the excitation of waves (activation of a complex of adaptive reactions), which corresponds to the transition to the stage of anxiety.
3) ***H°P(AP(•A))***, i.e., the waves have turned into solitons, for which a certain potential has been expended, which must be restored from the outside.
4) ***H°A(A(•A ö P))***, i.e. solitons pass into waves, the modality of the state of which changes from extinction to disappearance, and the modality of existence - from ***A*** to ***P***. The transition from extinction to disappearance marks the beginning of the development of pathology.

The schemes and characteristics of the "critical" stages of the *Sp* evolution are adequate to the stages of stress, from which the following hypothesis, which needs experimental verification, directly follows:

The manifestations and sequence of stages of stress are based on the patterns of evolution of an integral open system with changes in external influences, limited by the requirements of maintaining the organization. Stress characteristics (duration of stages, consistency and allowable limits of change in indicators, level of resistance) can be obtained as static and dynamic characteristics of the calibrated Sp wave model and approximately calculated as invariants of the calibrated OM Sp.

This hypothesis adoption as a working one allows one to develop measures to prevent the implementation of the depletion stage based on the conditions for maintaining a constant (on the verge of stratification-convolution) state of Sp. Consider again the problem of maintaining optimal performance.

Calculation of physiological and ergonomic parameters. The "golden section" characteristics and derivative ratios have been used to calculate some ergonomic parameters, optimal values, and allowable limits. The general rule is formulated as follows: "*Parameters of various functions of his body that are favorable for a person can be obtained by dividing the range between the maximum possible or permissible limits of the corresponding functional characteristics by the golden ratio*" (Korobko, 1997, p. 348). There are many correlations, realizing one of the self-organization factors (golden section), for example, calculations of permissible values of sound volume or temperature, or limit values of systolic and diastolic pressure and heart rate (Korobko, 1997, pp. 247-256, 263, 301). Let's try to formulate the general rules for such calculations.

The Vital Activity Indicators

Generally accepted methodology to select, measure, and process indicators does not exist: a set of methods from 8-12 basic ones is used, and several dozens of variations), which complicates the study. There are no formulated principles for determining the working capacity according to the indications of the functional state of a person. The number of indicators themselves reaches 100 or more. Their influence on the organism as a whole and each other and their correlation with different levels of regulation remains unclear (Figure, Chapter 7).

Interpretation of such problems within the framework of HEM gives the following:

1. The choice of a representative indicators group is determined by: the organization levels, the number of elements at each of, them, and the interaction operations that determine the course of self-regulation.
2. Measurement conditions (measurement in laboratory conditions; in production; selection of measurement intervals; correlation, indicators measured at certain intervals), as well as measurement accuracy are a consequence of the openness of *Sp*; its evolution under the influence of external factors; variability of the coordinate system and S-sets and elements in the process of evolution and the very act of measurement, conservation of invariants during the transitions *Sp* OM, metrization of the OM.
3. The ranking of indicators is reduced to the definition of weighting factors following the levels of the organization.
4. Reducing the number of experiments and measurements is achieved by calibrating HEM, and their detailing, using different types of OM.
5. The correlation of different quality indicators among themselves and their use for building models of physiological processes is determined by the symmetry and harmony relations, the dynamics of changes in potential and other characteristics, and the consistency of their changes based on integrity conditions.

Now let's look at this sequence in more detail.

The choice of representative indicators is carried out as follows:

1. Preliminary studies are carried out to determine the nature of the main human activity and the composition of the involved levels of the organization. The corresponding PC diagram is constructed.
2. The number of measured indicators for each of the levels is determined and specified by the predicted parameters,
3. Selection of adequate methods is carried out.

4. The problem of specialization and universalization is solved based on the following considerations:
   a. A universal (group) indicator for specialized (single) indicators, interpreted as a set of characteristics of S-elements. The membership of S-elements is determined according to the PC-diagram. We emphasize that the group indicator will not be reduced to a mechanical sum of single ones. Accordingly, the methods differ.
   b. Detailing of measurements, i.e., using a group indicator or a set of individual indicators is determined by the study objectives. At the same time, we would remember the large lability of single indicators.

Measurement conditions include:

1. Choice of the coordinate system (Cs) by axioms 4.1 and 4.2 (Chapter 2) instead of metric axioms E^n. Here a dilemma arises: the need for accuracy requires the choice of sensitive, i.e., variable Kc and the corresponding means of measurement, while the need to compare the results obtained at different times and under different conditions - on the contrary, constant Kc. this implies

Regulation 8.8.*It is necessary to choose Cs so that during repeated measurements it would be possible to bring it to a state close to the initial one, i.e., axiom 3.9 was observed.*

Further, it is necessary to balance external conditions and measurement methods. From this follows not only the usual requirement to carry out experiments under the same conditions and based on the same technique but also the following rule:

Regulation 8.9.*If it is impossible to provide the same experimental conditions, it is necessary to adjust the applied methods set, based on integrity conditions.*

2. The variability of the characteristic due to the act of measurement leads to the inadequacy of the measured data to the real characteristics of the processes. This is expressed in the loss of sensitivity of instruments because of a change in the scale of processes, in a qualitative change like the phenomena under study, and leads to their distortion and misunderstanding. Therefore, it is necessary:

Regulation 8.10.*To ensure the correctness of the measurement procedure, it is necessary to minimize its influence on the measured characteristic by:*

- *The use of measuring instruments that do not violate normal production conditions.*

- *Using a system of external compensation for changes occurring due to measurement.*
- *Changes in the set of measuring instruments as the characteristic under study changes, based on integrity conditions.*

3. The development of the process can lead to a change in the number of its independent characteristics. Therefore:

Regulation 8.11.*It is necessary to provide the possibility of changing the number of methods used, depending on the emergence of new characteristics as the process under study develops by the conditions of integrity.*

4. It should be in mind, that even if recommendations 8.8-8.11 are followed, the measurement accuracy is limited by theoretical conditions arising from the soliton-wave nature of *Sp* and the axioms of the wave model.

The ranking of the measured indicators is provided by:

1. Bringing indicators of different quality to homogeneous units of measurement that characterize any common consequence, which can be associated with potential, and assigning weight coefficients derived from potential to indicators.
2. Determining the potential of each specific indicator, and then determining the appropriate weighting factor.

The replacement of experimental data with calculated ones is possible when, after statistical processing of the results, the connection between the basic hypothesis describing the development of the process under study and one of the scenarios for organizing *Sp* is clear. Necessary:

1. Rank characteristics by layers.
2. Determine the quantity and composition of unmeasured characteristics.
3. Calculate their relative values according to the formulas given in Chapter 3.
4. Determine the relationship between the measured values and the potentials.
5. Calibrate and determine the absolute values of non-measured parameters.

Examples of such a calculation are given in (Mkhitaryan et al., 2004). It is possible to use the obtained data to test basic hypotheses, determine individual deviations of indicators from calculated values, etc.

Coordination of indicator changes. Elucidation of the psychophysiological processes indicators' coordination patterns is a difficult task. The reasons lie in the large number and diversity of indicators, which reflects the complexity of a person as a psychophysiological system. There is no hope of the process's essence achieving an understanding but even the identification of all coordination in the foreseeable future. The previous applies also to the particular task of studying the self-regulation of a human.

However, the regularities of the system organization take place, therefore, it is legitimate to use the HEM to determine the mechanisms of coordination, as well as to predict the development of adaptation processes, it is legitimate Regulations 8.1-8.7, (8.1).

Coordination is based on the symmetry of organization and indicators. Components of the same level that belong to the same S-set or different S-sets of symmetric branches of Sp^+ are symmetric. In the latter case, the symmetrical elements correspond in pairs. Let symmetric elements $p_1, p_2, \dots, p_n$ (pairs of elements $(p_1, ps_1), \dots, (p_n, p^s_n)$) be characterized by sets of indicators $\{x_1\}, \dots, \{x_n\}, \{x^s_1\}, \dots, \{x^s_n\}$. Then, with the development of physiological processes, one should expect a simultaneous increase, decrease, or preservation of constant indicators of symmetrical elements or pairs:

$$\{sign\Delta x_1\} = \{sign\Delta x_2\} = \dots = \{sign\Delta x_n\} \tag{8.2}$$

$$\begin{aligned} \{sign\Delta x_1\} &= \{sign\Delta x_1^s\} \\ \{sign\Delta x_2\} &= \{sign\Delta x_2^s\} \\ &\dots \\ \{sign\Delta x_n\} &= \{sign\Delta x_n^s\} \end{aligned} \tag{8.3}$$

Examples of this consistency kind are an increase in the depth of breathing and heart rate with an increase in physical activity, a decrease in the frequency of brain and heart rhythms in deep sleep, etc.

Coordination based on skew-symmetric relations. For components belonging to the same level, but belonging to skew-symmetric branches, this process takes place. In this case, the indicators of pairwise skew-symmetric elements either remain constant or change with different signs:

$$\begin{aligned} \{sign\Delta x_1\} &= -\{sign\Delta x_1^s\} \\ \{sign\Delta x_2\} &= -\{sign\Delta x_2^s\} \\ &\dots \\ \{sign\Delta x_n\} &= -\{sign\Delta x_n^s\} \end{aligned} \tag{8.4}$$

Examples are the processes of formation of a dominant focus of excitation, characterized by the development of inhibition on the periphery, as well as ultra-paradoxical reactions to small stimuli.

Coordination is *based on ownership relationships*. Such coordination occurs between the S-set and its S-elements. It consists of a unidirectional change in indicators corresponding to their general characteristics $\hat{x}$, i.e., only to the characteristics of the S-set:

$$\left\{sign\Delta\hat{x}_{1,i}\right\} = \ldots = \left\{sign\Delta\hat{x}_{n,i}\right\} = \left\{sign\Delta\hat{x}_{i-1}\right\}, \tag{8.5}$$

where i is the number of elements layer, and i-*1* is the number of the S-set layer.

In this case, the limits of indicator change are determined by the conditions (8.1).

Perhaps this scheme underlies the formation of "sensitivity thresholds" - the accumulation of changes in the receptors of the first level (S-set) when the limits (8.1) are exceeded leads to a qualitative change in the states of the receptors of the next level (S-elements).

Coordination is *based on the axiom of state*. Such coordination is a generalization of the previous three. It occurs most often, since the processes of interaction with the environment affect several levels of the organization at once, even if only one external factor is acting.

Therefore, determining the correspondence of indicators to organizational levels is a complex experimental task, as well as identifying symmetrical characteristics, the planning of which can be facilitated by bringing indicators to a single measure, analyzing their sensitivity, etc.

Now consider the case of going beyond the limits (8.1).

Coordination based on stratification-convolution scenarios. Here, external influences are needed that are strong enough to transform the potential possibility of changing the modality of the state ↓ into an actual one, or, conversely, restoring destroyed Sp^+ structures, or forming new ones. This is expressed formally in the parameters i and n in (8.2-8.5) change and the inclusion in the lists of agreed indicators of the characteristics Sp^-, i.e., implementation of integrity conditions for the entire *Sp*.

Examples of such coordination are the formation of neural circuits of conditioned reflex connections in response to a repetitive stimulus, or, more broadly, the adaptation process as a whole.

The task of representative indicators choosing and determining their correlations was considered by the authors using the methodology as an example (Integral Performance Assessment, 1990), and the results were published in (Mkhitaryan et al., 2004).

Presenting research results in a visual form is essential for their correct interpretation. In this regard, two diagrams are proposed based on a simplified (without emergence and heterogeneity) HEM - Figure 13 shows a simple but more universal version, and Figure 14 is a more complex one, designed to represent psychological testing data (subjective space breakdown into area corresponds to the data of the well-known projective test "Tree").

Figure 13. Diagram for presentation and comparison of indicators (universal). Here it is convenient to compare the measured values of the indicators with the theoretical ones for each level. Blue, orange, and green colors show the measured values related to different levels

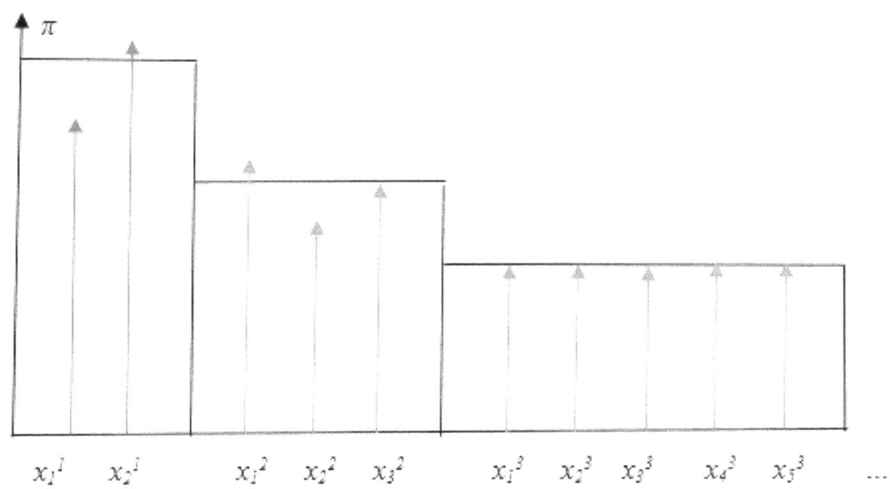

Figure 14. Diagram for the presentation and comparison of subjective space indicators

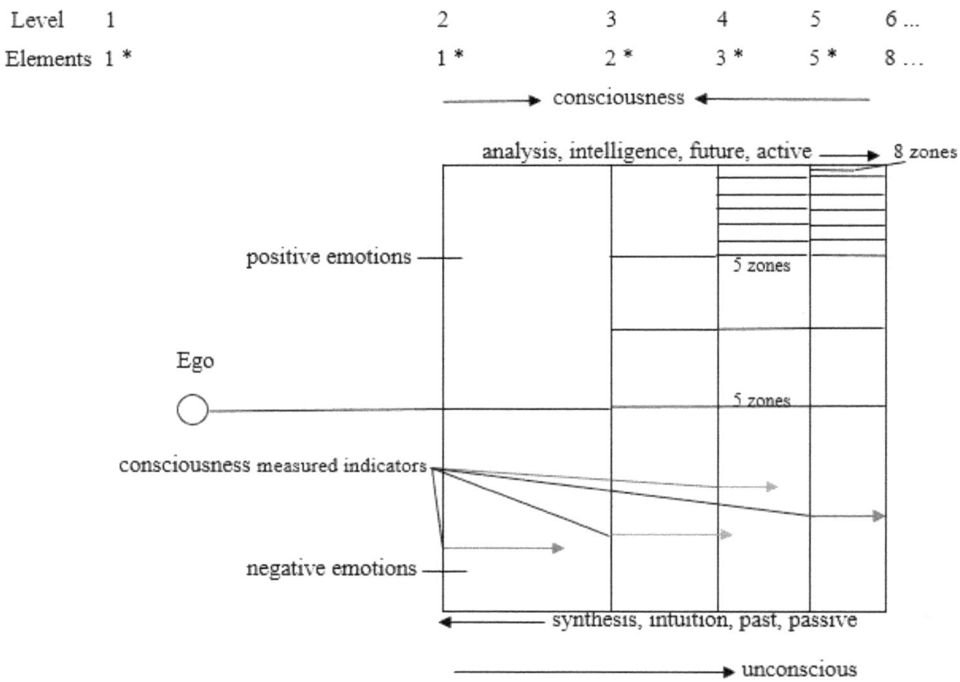

Let's comment on the diagram in Figure 13.

1. The top lines show the number of levels and characteristics required to determine the psychological portrait of a person.
2. The upper half shows, what a diagram looks like, which depicts certain psychological characteristics (this structure is not shown below).
3. The boundaries between the levels (length calibration) show the calculated, according to the distribution of HEM, or the average, according to experimental data, the values of the characteristics for the contingent, group, or individual.
4. The set of vectors below (not all characteristics are shown) reflects the degree of development of the corresponding mental qualities (the decoding may look something like this: "an optimist with above-average intelligence, poor eyesight; the ability to distinguish primary colors is normal).

After displaying the research data, the model will represent the psychological portrait of the person (group).

RHYTHM COORDINATION

Rhythm coordination includes the following tasks:

- Determination of the characteristics of rhythms as cyclic and wave processes.
- Determination of correlations of rhythms with states of consciousness, variations of homeostasis, and specific systems and organs.
- Determination of internal and external factors influencing rhythms.
- Determination of means of rhythm coordination to achieve a purposeful and predictable effect.

Because of the diversity and extensiveness of research conducted worldwide, we will limit ourselves to two examples: coordination of brain rhythms and purposeful influence on rhythms.

Brain Rhythms Coordination

Brain research is of great importance. It includes global international projects (for example, Human Brain Project, n / d) national programs, university research groups programs, and the like. Naturally, published works are very numerous (Krause, 1992; Buzsáki, 2007; Buzsáki & Watson, 2012; Simpkins & Simpkins, 2011; Srimaharaj et al., 2018; Gaillard et al., 2021). The discussion about linking different types of activity and rhythms to certain areas of the brain continues; at the same time, many researchers are inclined to the idea that the connections of brain regions are so numerous and diverse that it is necessary to consider the brain as a whole (Tononi & Edelman, 1998; Li et al., 2022).

The brain rhythms characteristics and their correlation with states of consciousness are known (Buzsáki, 2007), (Figure 15).

Data on the frequency ranges of brain rhythms differ from different authors. For example, (Frank, 2008) the following values are given: α-rhythm: 8–13 Hz, β-rhythm: 13–35 Hz, and γ-rhythm: 35 Hz and higher, and (Mahmoud et al., 2014): Δ-rhythm: 0.5-4 Hz, Θ-rhythm: 4-8 Hz, α-rhythm: 8-13 Hz, β-rhythm: 13-30 Hz. The existence of rhythms with both higher and lower frequencies is discussed (Buzsáki & Watson, 2012).

Figure 15. Brain rhythms and states of consciousness (Srimaharaj et al., 2018)

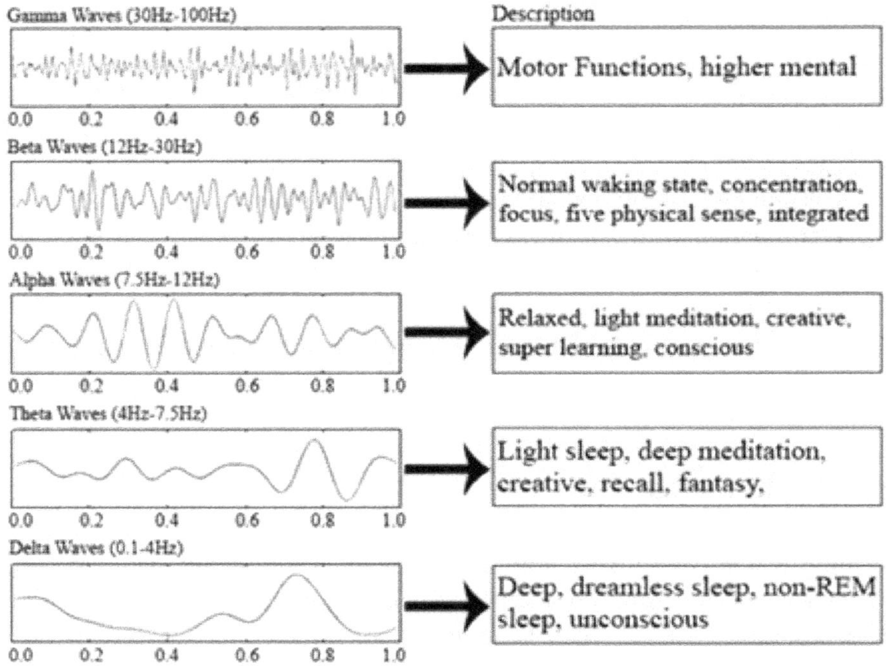

Now let's evaluate this picture from the standpoint of the theory of self-organization and the HEM model (Figure 10). It is obvious that all the listed rhythms correspond to 4-6 levels of organization of the subjective space of a person and, to put it simply, "serve his everyday consciousness within the framework of his daily cycle. "Easy" meditative practice does not violate the overall picture, and there is no reason to believe its correlation with "expanded" states of consciousness. The boundaries of each rhythm's ranges are close to the Fibonacci numbers, i.e., self-organization occurs.

Accordingly, we can assume the existence of "very long" waves and solitons corresponding to levels 1-3, the length of which, at least, corresponds to the life span of a particular individual.

Note that such rhythms have never been measured; existing methods for measuring brain activity do not allow this (Krause, 1992).

Internal and external factors influencing rhythms. It is believed that brain rhythms are internal oscillators; it is also known that they change their characteristics under the influence of external factors (Foster & Roenneberg, 2008; Chakraborty, 2014), as well as the activity of the person himself (Buzsáki & Watson, 2012; Gaillard et al., 2021). Thus, they "adjust" to the frequency characteristics of the interactions.

What are the mechanisms for such adjustment? Let's consider the rhythms change in their change (Figure 16).

Comparing Figure 16 with Figure 19, Chapter 7, we note signs of 3 types of modulation, as well as resonance; hence it follows that it is modulations (including the rhythms of levels 1-3) and resonance that are the mechanisms of "adjustment".

Rhythm coordination to achieve a purposeful effect. Such targeted influences include, for example, meditation practices (Simpkins & Simpkins, 2011; Srimaharaj et al., 2018). We will return to their detailed discussion in Chapter 11.

Figure 16. Ten seconds of simulated EEG data in the five differently named frequency bands of neural oscillations, or brainwaves: delta, theta, alpha, beta, and gamma (Krol, 2020)

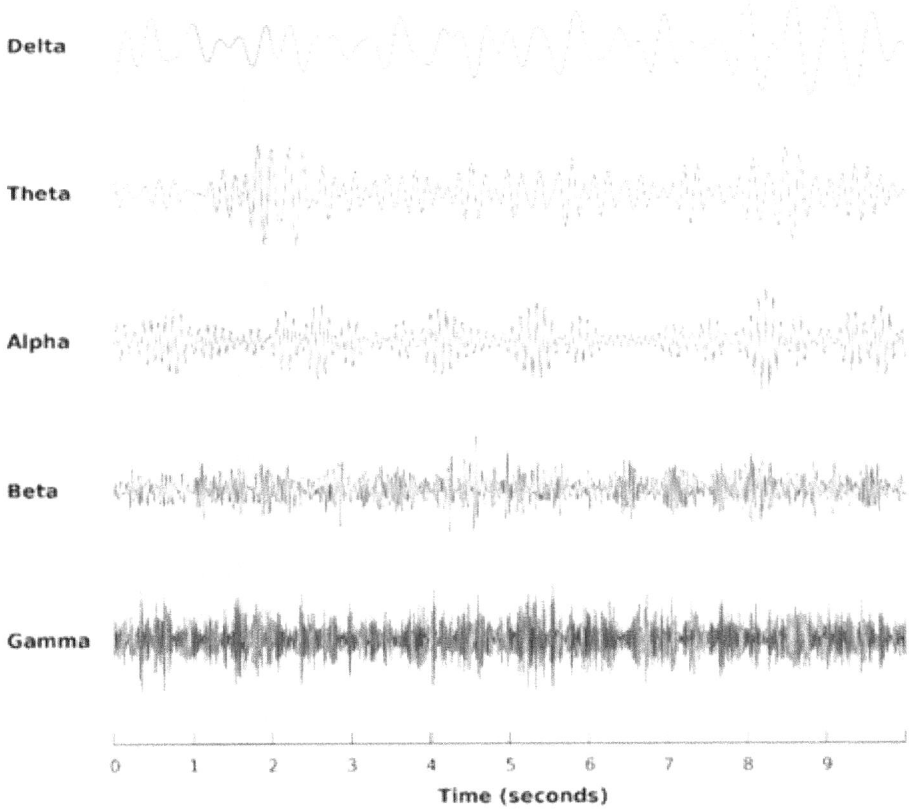

FUTURE RESEARCH DIRECTIONS

It seems possible to use the man-environment model in the following areas:

- Study of the human life cycle (Chapter 9).
- Practical application to optimize man-machine systems to improve the performance of the operator and the development of information display systems.

CONCLUSION

The main result of the chapter is the man-environment model. The model includes the general structure of the body, levels, and channels of the human subjective space, and the structure of interactions with the environment, and takes into account both the events of the evolution of the Universe before the emergence of life, and the results of the evolution of living organisms before the emergence of man. Thus, the model implements a systematic approach, the wave model paradigm, and the theory of self-organization.

We note several features of the model.

The first is its original dualistic soliton-wave character. Soliton-wave abstraction can be calibrated in different ways, depending on the conditions of the problem being solved. The most important interpretations for this study are corpuscular-wave dualism, the material-field nature of a person, bodily and mental duality, and information-physical pair; others are possible.

The second is the relative autonomy of the subjective space (mental, field component), the organization of which, after being singled out as a structure, is determined to a greater extent by the laws of self-organization than by environmental influences. The interaction mechanism with the environment is due to the presence of symmetries between the structures of the environment and the person, belonging to the same level. Self-organized subjective space separates environmental influences by the characteristics of wave interactions (Chapter 4), determining the gestalt of perception of the surrounding world. This is the epistemological significance of the man-environment model.

Third, the systemic relations of a person's bodily and mental components determine the limits of self-regulation, the conditions for the occurrence of pathologies, resistance and the development of stress, labor productivity, brain activity, and states of consciousness. Examples and calculations performed using self-organization scenarios confirm the reliability of the model.

Thirdly, the systemic relations of a person's bodily and mental components determine the limits of self-regulation, the conditions for the occurrence of pathologies, resistance and the development of stress resistance, labor productivity, brain activity, and states of consciousness. The influence of these relationships on aging and dying will be discussed in Chapters 9 and 10.

It is necessary to understand exactly what regulations 8.1-8.7 mean and the calculation and examples given before. They are derived from the properties that characterize the system as a whole (Table 8, Chapter 1), so they do not give fixed values, but only determine the boundaries of the change in values, and only for those cases when the systems are emergent and exist in fairly stable conditions.

In extreme situations, when time factors and accumulated reserves matter, the values may go beyond these limits. If the exit is short-term, there is compensation due to the reserve, and the body will need rest and recovery, but pathology does not occur. If such a situation is prolonged, compensation will become insufficient, and pathology will become inevitable.

Fourth, the wave characteristics of the human-environment model make it possible to understand better the mechanisms of coordination and changes in human biorhythms. It also has practical value, making it possible to purposefully develop means of influencing the human condition, in particular, increasing the comfort of the living environment and housing, optimizing the activities of operators of man-machine systems, and the like.

REFERENCES

Aleksandrov, S. Yu. (2004). Lorentz-Covariant Loop Quantum Gravity. *Theoretical and Mathematical Physics*, 139(3), 751–765. DOI: 10.1023/B:TAMP.0000029699.54716.0e

Anatomy 3D Atlas. (n.d.). https://anatomy3datlas.com

Beamish, R. (2007). *Dialectical Materialism.* In *The Blackwell Encyclopedia of Sociology*. Blackwell Pub. DOI: 10.1002/9781405165518.wbeosd062

Burns, T., & Dietzgen, J. (2002). The History of Marxism. *Science and Society*, 66(2), 202–227. DOI: 10.1521/siso.66.2.202.20997

Buzsáki, G. (2007). The structure of consciousness. *Nature*, 446(7133), 267. DOI: 10.1038/446267a PMID: 17361165

Buzsáki, G., & Watson, B. O. (2012). Brain rhythms and neural syntax: Implications for efficient coding of cognitive content and neuropsychiatric disease. *Dialogues in Clinical Neuroscience*, 14(4), 345–367. DOI: 10.31887/DCNS.2012.14.4/gbuzsaki PMID: 23393413

Chakraborty, U. (2014). Effects of different phases of the lunar month on humans. *Biological Rhythm Research*, 45(3), 383–396. DOI: 10.1080/09291016.2013.830508

Einasto, J. (2012). *Dark matter.* In *Astronomy and Astrophysics*. https://www.eolss.net/sample-chapters/c01/E6-119-21-00.pdf

Foster, R. G., & Roenneberg, T. (2008). Human Responses to the Geophysical Daily, Annual and Lunar Cycles. *Current Biology*, 18(17), R784–R794. DOI: 10.1016/j.cub.2008.07.003 PMID: 18786384

Frank, M. G. (2008). *Brain Rhythms. Springer eBooks* (pp. 482–483). DOI: 10.1007/978-3-540-29678-2_727

Gaillard, C., De Sousa, C., Amengual, J., Loriette, C., Ziane, C., Hassen, S. B. H., Di Bello, F., & Hamed, S. B. (2021). Attentional brain rhythms during prolonged cognitive activity. bioRxiv *(Cold Spring Harbor Laboratory)*. DOI: 10.1101/2021.05.26.445730

Green, B. (2011). *The Elegant Universe: Superstrings, Hidden Dimensions, and the Quest for the Ultimate Theory* (Reprint edition). Norton & Company.

Gukov, S. (1998). Introduction to string dualities. *Physics Uspekhi*, 41(7), 627–638. DOI: 10.1070/PU1998v041n07ABEH000416

Guyton, A. C., & Hall, J. E. (2005). *Textbook of Medical Physiology*. Saunders. (Original work published 1969)

Hansen, C. D. A. (2000). *Daoist Theory of Chinese Thought: A Philosophical Interpretation*. Oxford University Press. DOI: 10.1093/oso/9780195134193.001.0001

Human Brain Project. (n.d.). https://www.humanbrainproject.eu/en/

Karagyaur, M., Rostovtseva, A., Semina, E., Klimovich, P., Balabanyan, V., Makarevich, P., Popov, V., Stambolsky, D., & Tkachuk, V. (2020). A Bicistronic Plasmid Encoding Brain-Derived Neurotrophic Factor and Urokinase Plasminogen Activator Stimulates Peripheral Nerve Regeneration After Injury. *The Journal of Pharmacology and Experimental Therapeutics*, 372(3), 248–255. DOI: 10.1124/jpet.119.261594 PMID: 31888957

Kovalyov, Y., Mkhitaryan, N., & Nitsyn, A. (2020). *Self-organization of the human mind and the transition from paleolithic to behavioral modernity*. IGI Global International Publisher of Progressive Information Science and Technology Research. DOI: 10.4018/978-1-7998-1706-2

Krause, W. (1992). Measuring mental performance: An old idea and a new approach. *Zeitschrift für Experimentelle und Angewandte Psychologie*, 39(1), 114–128. https://pubmed.ncbi.nlm.nih.gov/1598763/ PMID: 1598763

Krol, L. R. (2020). Ten seconds of simulated EEG data in the five different frequency bands of neural oscillations, or brainwaves: delta, theta, alpha, beta, and gamma. *File: EEG Brainwaves.svg - Wikimedia Commons*. https://commons.wikimedia.org/wiki/File:EEG_Brainwaves.svg

Lim, Q., Ver Steeg, G., Yu, S., & Malo, J. (2022). Functional Connectome of the Human Brain with Total Correlation. *Entropy (Basel, Switzerland)*, 24(12), 1725. DOI: 10.3390/e24121725 PMID: 36554129

Lin, Y. C., & Tung, C. (2013). Acupuncture Qi Flow and Points Measurement. In *Springer eBooks* (pp. 153–158). DOI: 10.1007/978-1-4614-5275-1_9

Lüttge, U., & Souza, G. M. (2019). The Golden Section and Beauty in Nature: The perfection of symmetry and the charm of asymmetry. *Progress in Biophysics and Molecular Biology*, 146, 98–103. DOI: 10.1016/j.pbiomolbio.2018.12.008 PMID: 30557534

Mahmoud, S. A., Balla Elamien, M., Zaatar, A., & El Tinai, M. (2014). Low noise low power instrumentation amplifier for biomedical application. *University of Sharjah. Senior Design Project Report*. https://www.researchgate.net/publication/303487700_Low_noise_low_power_instrumentation_amplifier_for_biomedical_application

Mukhi, S. (1999). The theory of strings: An introduction on JSTOR. https://www.jstor.org/stable/24104471

Radhakrishnan, S. (2009). Indian Philosophy. Oxford University Press. (Original work published 1923)

Randall, L. (2005). *Warped Passages. Unraveling the Mysteries of the Universe's Hidden Dimensions*. Ecco.

Ravi Kumar Patil, H. S., Makari, H. K., Gurumurthy, H., & Sowmya, S. V. (2009). *A Textbook of Human Physiology*. I K International Publishing House.

Sagaradze, G., Basalova, N., Kirpatovsky, V., Ohobotov, D., Nimiritsky, P., Grigorieva, O., Popov, V., Kamalov, A., Tkachuk, V., & Efimenko, A. (2019). A magic kick for regeneration: Role of mesenchymal stromal cell secretome in spermatogonial stem cell niche recovery. *Stem Cell Research & Therapy*, 10(1), 342. Advance online publication. DOI: 10.1186/s13287-019-1479-3 PMID: 31753023

Schmidt, R. F., & Thews, G. (Eds.). (1983). *Human Physiology*. Springer-Verlag.

Selye, H. (1976). Stress without distress. In *Psychopathology of Human Adaptation* (pp. 137–146). Springer US. DOI: 10.1007/978-1-4684-2238-2_9

Sender, R., & Milo, R. (2021). The distribution of cellular turnover in the human body. *Nature Medicine*, 27(1), 45–48. DOI: 10.1038/s41591-020-01182-9 PMID: 33432173

Simpkins, C. A., & Simpkins, A. M. (2011). *Neuroscience Findings: How Meditation Can Change the Brain*. DOI: 10.1002/9781118155684.ch02

Spalding K. L., Bhardwaj R. D., Buchholz B. A., Druid, H., & Frisén, J. (2005). Retrospective birth dating of cells in humans. *Cell, 15*(122), 133-43. DOI: 10.1016/j.cell.2005.04.028

Srimaharaj, W., Chaisricharoen, R., Chaising, S., & Sittiprapaporn, P. (2018). *Classification of human brain attention focused on meditation, effected by L-theanine acid in Oolong tea*. DOI: 10.1109/ICDAMT.2018.8376536

The Human Body. (n.d.). https://www.healthline.com/human-body-maps

Thurner, S., Hanel, R., & Klimek, P. (2018). *Introduction to the Theory of Complex Systems*. Oxford University Press. DOI: 10.1093/oso/9780198821939.001.0001

Tononi, G., & Edelman, G. M. (1998). Consciousness and complexity. *Science, 4*(282), 1846-51. DOI: 10.1126/science.282.5395.1846

Trigrams Meaning and Definition in Qi Men Dun Jia. (2022). https://dougleschan.com/8-trigrams/8-trigrams-meaning-and-definition-in-qi-men-dun-jia/

Weinberg, S. (1993). *The First Three Minutes: A Modern View of the Origin of the Universe*. Basic Books.

ADDITIONAL READING

Artyukhov, V. V. (2009). *Obshhaja teoryja system: samoorghanyzacyja, stabyljnostj, raznoobrazye, kryzysy* [General Systems Theory: Self-Organization, Stability, Diversity, Crises]. Librokom.

Bodnar, O. Ya. (2005). *Zolotyj pereriz i neevklidova gheometrija v nauci ta mystectvi*. [Golden section and non-Euclidean geometry in science and art]. Ukr. technologies.

Chelovecheskyj factor. (1991-1992). [The human factor]. V. 1-6. Mir.

Deutsch, D. (2001). *Struktura realjnosty* [The structure of reality]. RHD.

Ignatiev, Yu. G. (2016). *Klassycheskaja kosmologhyja y temnaja energhyja* [Classical cosmology and dark energy]. Publishing House of Kazan University.

Korobko, V. I. (1997). *Zolotaja proporcyja y problemy gharmonyy system*. [The Golden Proportion and the Problems of Harmony of Systems]. Izd. Association of buildings universities of the CIS countries.

Maspero, A. (2007). *Daosyzm* [Taoism]. Nauka.

Mkhitaryan, N. M., Badeyan, G. V., & Kovalyov, Y. N. (2004). *Erghonomycheskye aspekti slozhnikh system* [Ergonomic Aspects of Complex Systems]. Naukova Dumka.

Novikov, I. V. (1988). *Kak vzorvalasj Vselennaja*. Nauka. [How the Universe Exploded]

Ovchinnikov, N. F. (1966). *Pryncypy sokhranenyja* [Preservation principles]. Nauka.

Soroko, E. M. (1984). *Strukturnaja gharmonyja system* [Structural harmony of systems]. Nauka.

Urmantsev, Y. A. (1988). *Evoljucyonyzm yly obshhaja teoryja razvytyja system pryrodы, obshhestva y myshlenyja* [Evolutionism or the general theory of developing systems of nature, society, and thinking]. Institute of Plant Physiology of the Academy of Sciences of the USSR.

Urmantsev, Y. A. (2017). *Symmetryja pryrody y pryroda symmetryy: Fylosofskye y nauchnye aspektы OTSU* [Symmetry of nature and the nature of symmetry: Philosophical and scientific aspects of GTSU].

Vasyutinskiy, N. A. (1990). *Zolotoe sechenye* [The golden ratio]. Young Guard.

Zidermane, A. A. (1998). *Nekotorye voprosy khronobyologhyy y khronomedycyny: obzor lyteratury* [Some issues of chronobiology and chronomedicine: a literature review]. Zinatne.

ENDNOTES

1 Memory serves to move in the past, foresight in the future, and there is no movement in the present - it moves along with our consciousness from the past to the future.

2 Reminds a popular (but incorrect) formulation of the principle of relativity.

3 Other interpretations are also possible, for example, consciousness, activity, inertness.

Chapter 9
Life Cycle From Conception to Senility

ABSTRACT

The events of the human life cycle from birth to old age are considered. Particular attention is paid to the changes occurring at birth and aging. Four iterations of the soliton-wave model of the human life cycle are constructed. Calculations of the duration of age periods are given. Variants of its theoretical and practical use are considered, in particular for assessing and selecting various strategies for prolonging life. Models of the most important life cycle events and periods are constructed. The modeling results correlate with known data. The immediate goal will be to work with near-death states using the soliton-wave model (Chapter 10).

BACKGROUND

The human life cycle is not universal for living organisms: there are 4 types of cycles according to ploidy (haplophase - bacteria that do not have sexual reproduction; haplophase with zygotic reduction -green and other algae, and mushrooms; diplophase with gametic reduction - multicellular animals; haplo-diplophase (diploid and haploid generations) - higher plants. In addition, life cycles are distinguished by the number of generations: simple (one generation, for example, in humans), complex - two or more generations (alternating generations of polyps and jellyfish in cnidarians or marites, sporocysts and redia in trematodes). Thus, the life cycle depends on the methods of reproduction (Polishchuk, 2004; Granovich et al., 2010; Life cycle (biology), n / d); Bell & Koufopanou, 1991; Magi, 1976; Valero et al., 1992).

It is necessary to distinguish the life cycle as a characteristic of the species from ontogenesis - the development of an individual from its appearance to death or division.

DOI: 10.4018/978-1-6684-8509-5.ch009

The beginning of the life cycle is, of course, the emergence of a new living organism. However, this moment itself is defined ambiguously: if for most people the countdown is from the moment of birth, then for some it is from the moment of conception. In addition, there is a legislator's point of view - this is the period when abortion is still allowed (Mishatkina et al., 2008). According to modern medical standards, abortion is performed at up to 20 weeks of gestation or, if the gestational age is unknown, with a fetus weighing up to 400 g. fetuses 1 week apart are listed in (Your-baby/fetal-development, n / d).

Events throughout the life cycle include coordinated anatomical, physiological, and psychological changes. Numerous studies have been devoted to each of these areas. As in the previous sections, we will state the proven facts outlined in the textbooks, however, we will also provide links to relevant scientific publications. Sources on anatomy and physiology (Nikityuk & Chtetsov 1990; Gamova, 2010; Lysova & Aizman, 2018; Tyurikova & Tyurikova, 2018; Sergeev et al., 2020; *Human Physiology/Development: Birth Through Death*, 2022). Sources on developmental psychology (Erickson, 1996; Sapogova, 2001; Khukhlaeva, 2006; Rean et al., 2015). These features are also of great practical importance, for example, when designing a comfortable urban environment in terms of age-related psychological characteristics (Mkhitaryan et al., 2021), and physiological characteristics (the concept of inclusive design (Joyce, 2022; Khlistun, 2015).

Particular attention is paid to the aging process. It should be noted that among the scientific community, there is no unity of views on the causes of aging and the methods of life extension studied by gerontology (Theories of aging, n / d); Vina et al., n / d). Two concepts have been proposed:

- Evolutionary - aging is not a necessary property of living organisms but developed as a result of evolution since it gives certain advantages to the population as a whole (Brian, 1952; Drenos & Kirkwood, 2005; Gavrilov & Gavrilova, 2002; Flatt & Schmidt, 2009; Frankel & Rogina, 2021; Le Bourg, 2001).
- Accidental damage to cells - aging is the result of a natural process of damage accumulation, which the body fights using various mechanisms, and the differences in aging are explained by the different effects of this fight (free radical theory of D. Harman (Harman, 1972; 1981), telomere theory by A.M. Olovnikov (Olovnikov, 1973), hormonal-genetic theory (Dilman, 1972) and others).

At present, the second approach is the main one; some theories are difficult to attribute to any of these concepts; the total number of theories of aging is more than 300.

FROM CONCEPTION TO BIRTH

The intrauterine period of human development (prenatal ontogenesis) begins with the formation of a zygote and ends with the birth of a child. Here are facts related to the period of intrauterine development (Lubimova & Nikitina, 2019; O'Rahilly & Müller 2010); illustrations of embryonic and fetal development throughout pregnancy at 1-week intervals are provided in (Marple, 2023).

Periodization and Characteristics

The embryonic (the first 2 lunar months - 8 weeks) and fetal (from the 9th week to birth) periods are distinguished in the prenatal period.

The embryonic period includes the following events:

- Fertilization (formation of a zygote).
- Crushing (formation of blastula).
- Gastrulation (leading to the formation of two- and three-layer embryos and laying the axial complex).
- Formation of the main complex of rudiments of organs and tissues.
- Organogenesis and histogenesis.

After fertilization and the formation of a zygote, cleavage occurs in the oviduct; as a result, cells called blastomeres are formed. Blastula formation is very fast. A cavity filled with liquid is formed in the center of the embryo. The thin surface layer of cells surrounding the blastocoel is called the trophoblast; cells inside the trophoblast do not communicate with the fluid in the oviduct and uterus and remain attached to one of the sections of the blastula wall. From these cells, the body of the embryo will subsequently develop. On the 4-5th day after fertilization, the embryo enters the uterine cavity and swims freely in it for 2-3 days.

Due to the growth and reproduction of cells, the blastula rapidly increases, the fertilization membrane on its surface is destroyed, and the gastrulation stage begins. In this case, a three-layer embryo is formed with 3 germ layers (ectoderm, endoderm, and mesoderm) - cells separated by slit-like spaces.

The epithelium of the skin, some sense organs, and the nervous system are subsequently formed from the ectoderm; from the endoderm - a significant part of the epithelium of the digestive canal, pancreas, liver, epithelium of the respiratory tract; from the mesoderm - pleura, pericardial sac, peritoneum, genital and urinary organs, skeletal and smooth muscles, blood, circulatory system, connective tissue, skeleton, and others. With the development of tissues from the mesodermal layer, the mesoderm cells are separated, and they have long processes that form the embryonic connective tissue - mesenchyme. Membranes provide nutrition, respiration, and protection for the embryo from part of the cells of all germ layers. are formed

On the 6-7th day, the embryo attaches to the mucous membrane of the uterus and begins to sink into its wall. The epithelium of the mucous membrane grows on the embryo and soon covers it completely. This process is called implantation. Implantation usually occurs in the upper part of the uterus on the anterior or posterior surface. During the implantation period, the embryo feeds on substances that penetrate from the fluid into the uterine cavity, as well as the remains of destroyed endometrial cells.

Several membranes develop around the embryo. The internal called the amnion forms a bladder filled with fluid in which the embryo develops. Amnion protects the fetus from mechanical shocks and creates a favorable environment for its development. On top of it, a chorion villous membrane develops, arising from the trophoblast. The chorionic villi contain blood vessels and form the fetal part of the placenta. With their help, the nutrition and respiration of the embryo are carried out (Figure 1).

Figure 1. The first few weeks of embryogenesis in humans. Beginning at the fertilized egg and ending with the closing of the neural tube (Zephyris, n / d)

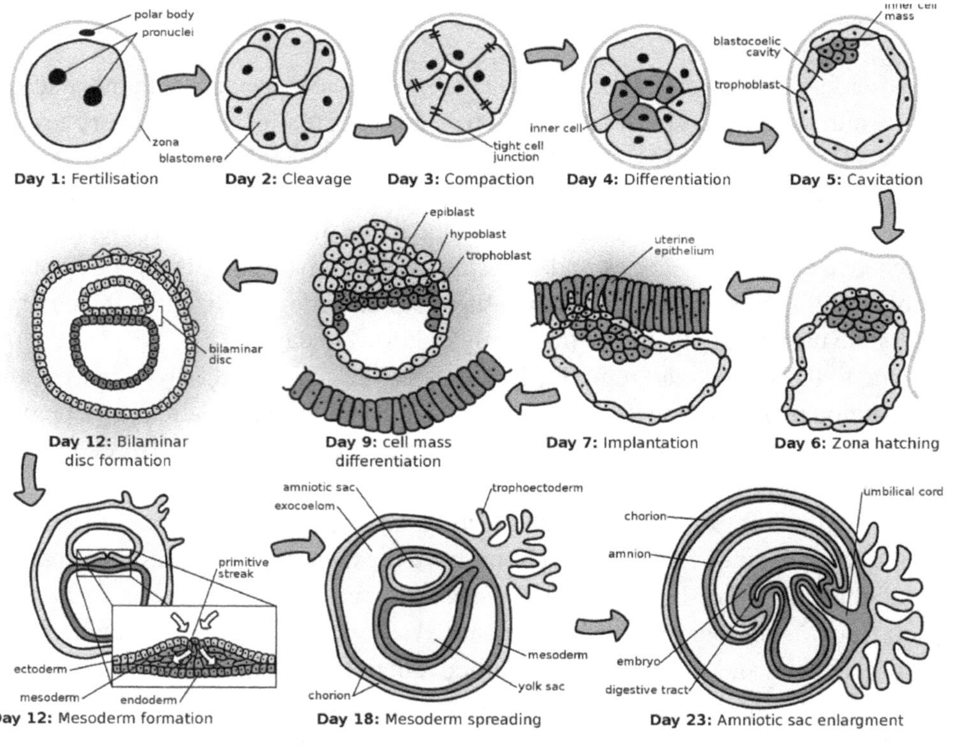

By the end of the second month of intrauterine development, the chorionic villi remain only on the side of the embryo where the placenta is formed. They grow and branch, plunging into the uterine mucosa and destroying its blood vessels. Depressions form in the mucous membrane of the uterus, which is filled with maternal blood flowing from the destroyed blood vessels and washing the chorionic villi. A placenta is formed, like a disk fixed in the uterine mucosa.

The blood of the mother and the fetus does not mix: the fetus's circulatory system from the very beginning is a closed system, isolated from the mother's circulatory system. Through the wall of the blood capillaries of the fetus and the sheath of the villi, gases, and nutrients are exchanged between the mother's body and the fetus. The wall of the villi - the placental barrier - consists of the trophoblast, the connective tissue between it and the blood vessels. The decay products emitted by the fetus (CO_2, water, electrolytes, urea, and other substances) penetrate the placental barrier from the fetus into the maternal blood. The fetus receives the substances from the mother necessary for nutrition, respiration, growth, maintaining water balance, and immunological protection (O_2, water, inorganic ions, sugars, amino acids, lipids, hormones, and antibodies). Maternal antibodies protect the fetus (and subsequently the fetus and infant) from infectious diseases until its immune system begins to function. The function of the placenta is that the contact of the fetus with the mother's body is established through it.

By the beginning of the eighth week, the embryo has a length of 15-20 mm, and by the end of it - 40 mm. The mass of the embryo is five grams. The main features of the eighth week are the intensive development and modification of the embryo (Figure 2):

- There is a straightening of the body, and it is increasingly divided into segments - limbs, head, and torso.
- The nervous system is actively developing, the brain is divided into sections, the hemispheres are more clearly outlined, convolutions are formed, and brain structures are responsible for muscle tone.
- The muscular system develops, which enables the embryo to make a variety of movements.
- Ears, nostrils, and eyes are formed - by the end of the 8th week the upper lip is fully connected, and the face looks quite formed.
- There is an ossification of the skull, arms, and legs, the bones harden, fingers are formed, and the large one becomes isolated and opposed to the palm, elbow, and knee joints are formed.
- The development of the digestive tract is almost completed; there is the innervation of the intestines and stomach, due to which the motor functions of

the gastrointestinal tract are ensured, and the cavities of the kidneys, heart, bladder, and ureters are formed.

Figure 2. 8 weeks (Marple, 2023)

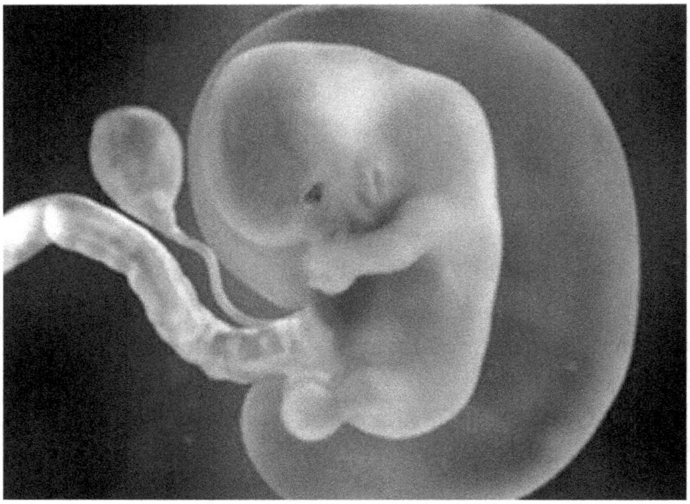

The fertile period begins in the third lunar month. Here is a chronology of the main events of this period (The course of pregnancy by weeks (n / d).

By the end of the 12th week, the length of the fetus is 90 mm, and its weight is approximately 20 g. There is an intensive development of the brain and the formation of connections between the spinal cord and the cerebral hemispheres. The structure of the brain already resembles an adult brain. The fetal blood includes leukocytes and erythrocytes. The liver begins to produce bile, and not only to provide hematopoiesis. The kidneys start to function. The intestine is growing and begins to fit into an adult's loops. The peristaltic movements occur. The formation of the placenta is completed.

16th week: the formation of facial muscles in the fetus. The fetus starts blinking. In place of the eyelashes and eyebrows, fluffy hairs can be seen. The auricles are already placed as in an adult. The reaction of the fetus to loud sounds due to bone conduction is traced. The formation of all joints and bones has been completed, but the process of compaction of bone tissue will continue until puberty. It is possible to determine the sex of the child reliably; his external genitalia have formed. The kidneys work intensively. The urinary canal is formed, and urination is performed every hour.

20th week: the body length of the fetus is 28 cm (with the lower limbs), and the weight is about 340 g. The fetus is active, making numerous movements, since it has enough space in the uterus for somersaulting, pushing away from the walls, and so on. The fetus can smile, grab the umbilical cord, feel itself, frown, and lose its eyes. His facial expressions are quite developed. The skin becomes dense, and actively covered with fluffy hairs and a special lubricant that prevents the risk of mechanical friction. Both externally and internally, the fetus already corresponds to the appearance and structure of a person, and in the future, quantitative rather than qualitative changes occur.

24th week: the fetus continues to grow, it weighs about 600 g, and its height is approximately 33 cm. The fetus occupies the entire uterine cavity, so sharp movements are replaced by smaller ones. The fetus can still change the position of the body but prefers to sort out the umbilical cord, and bend and unbend the arms and legs. Dramatically increases muscle mass. The sense organs are improved: the ability to distinguish tastes, hear sounds, and navigate in space develops; the skin is highly sensitive. The brain continues to develop, connections with the senses are formed, and the fetus reacts to external stimuli.

28th week: the weight of the fetus reaches 1,200 g, and the height is 35 cm. The body becomes more rounded, and it becomes more and more like a newborn. Tiny, tufted hairs begin to fall out. The hairs gradually darken. The fetus is getting better at distinguishing sounds and voices. The weight of the brain increases, and grooves appear on its surface. Taste develops: the fetus begins to get used to what the mother eats. The skeleton is already almost formed. Also, this week is important for the formation of the psyche and character of the baby.

36th week: the fetus begins to actively prepare for birth. All organs can function independently. The weight reaches 2,500 g, and the height is approximately 46 cm. The fetus is already in its final position in the uterus; the bones of the skull remain supple to facilitate passage through the birth canal.

40 weeks is a full-term pregnancy, most women give birth currently. The weight of a child who was born at this time ranges from 2,600 g to 4,400 g, and his body length is 48-53 cm.

Before childbirth, the child closely presses the arms and legs to the body, bends the head as much as possible, and presses against the exit from the uterus. This position allows you to make it possible to pass the birth canal with the narrowest part of the skull. During labor, with each contraction, the child gradually moves down, he makes helical-translational movements, screwing into the mother's birth canal. As the newborn advances, the cervix dilates. This is followed by uterine contractions that move the baby through the birth canal. Gradually, the head of the baby is shown, and after it - his torso. Childbirth is aimed not only at the safe passage of the birth canal by the child, protecting him from accidental injuries, but also at preventing ruptures of the woman's soft tissues.

The Prenatal Period Modeling

Let us consider the interpretation of the event of the prenatal cycle within the framework of the theory of self-organization.

The emergence of a new living organism because of sexual selection is shown in Figure 10, Chapter 6. The conditions for successful sexual reproduction are the presence of the reproductive age of both parents, the periodically arising readiness of the mother's reproductive apparatus (menstrual cycle), the father's spermatogenesis that has passed all phases, and a non-periodic event - sexual intercourse. We also note that the process of conception in a person is often accompanied by higher feelings - love and others, and the influence of the mother's psycho-emotional state on the development of the fetus is recognized by modern researchers (Blum, 1953). Following the assumption made in Chapter 6, this can be seen as the presence of external modulation and the occurrence of resonance.

We will interpret the chromosome component after the union of male and female chromosomes as a soliton pair (since the chromosomes are unchanged here, except for minor mutations). Let us also represent the body composition as a soliton, and the field component as a wave having the modality ϕ after the fusion of the gametes of the parents because of resonance modulated by external influences (explanations to Figure 10, Chapter 6).

This entire SWM system has the modality of the actual existence $\boldsymbol{A}$ as a cell and the potential existence $\boldsymbol{P}$ as a multicellular embryo that has yet to form (Figure 3).

Figure 3. SWM of the zygote with designations of modalities and external modulating influences

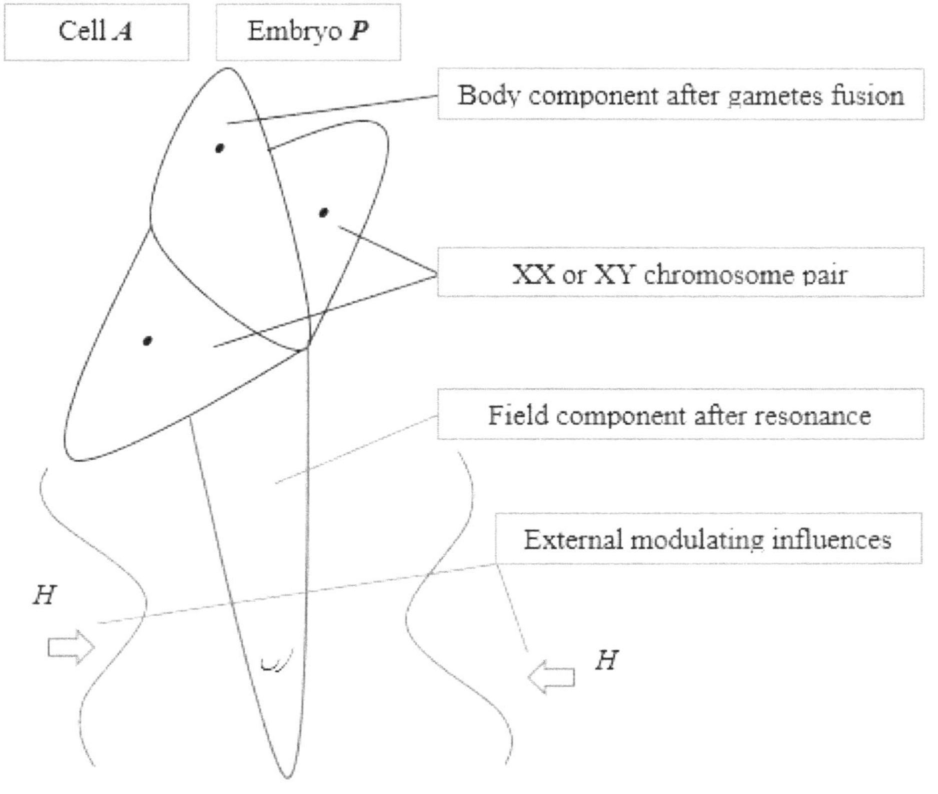

We note the asymmetry and imbalance of the system, the high potential (due to resonance) of its wave component, and its increased sensitivity to external influences.

Embryonic period. The following model changes take place: due to the potential of the field component, all soliton components are activated and acquire *ø* modality.

In this case, the chromosomes do not change, and hereditary information is used for the development of the body of the embryo (synthesis of Complex Organics, Chapter 2).

An increase in body size and morphological changes show how the potential increases and the transition of this component into a wave state.

Generally, the system acquires modality ***A*** as an embryo and ***P*** as a fetus (Figure 4).

Figure 4. Model of embryonic development with the designation of modalities. Waves are shown in blue; solitons are shown in black

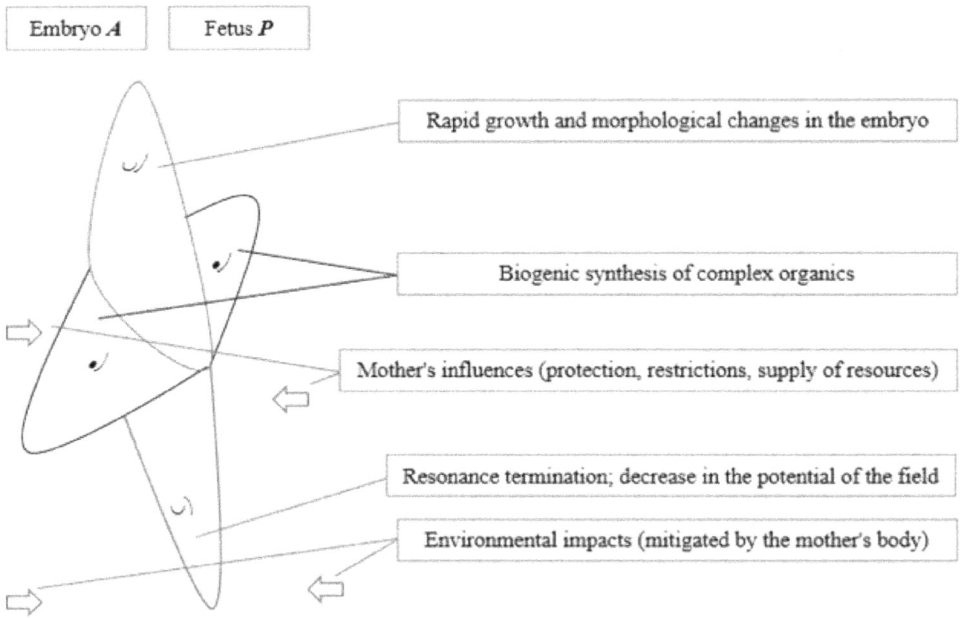

The dynamics of the processes are aimed at the formation of the embryo as an independent organism (development of membranes, isolation of blood circulation, etc.), as well as the formation of the system of the embryo and mother (attachment to the wall of the uterus, development of the placenta, etc.). Suppose this process is "cleared" of the mechanisms of heredity. In that case, an analogy with the behavior of a parasite becomes visible, where the parasite - the embryo - uses the resources of the donor organism.

External influences cease to correspond to the conditions of resonance, and the influences from the maternal organism can be characterized as protective and restrictive.

The fetal period entails such changes in the model (Figure 5):

- Return of the wave corresponding to the body component to the state of a soliton (conservative effect of the mother and a decrease in the potential of the field component).
- Decrease in the potential of the field component (environment impacts do not create resonance conditions, the potential is partially used up).

- The synthesis of organics is actively continuing with the use of the mother's resources and the program of heredity. Still, there are no qualitative changes, so there is no longer any reason to designate ϕ modality.
- The system acquires the modality ***A*** as a fetus and ***P*** as a child.
- The limits of adaptation of the donor mother to the parasitic fetus are exhausted by the end of pregnancy according to the most critical of the restrictions (the size of the birth canal) and its rejection (delivery) begins.

Figure 5. Model of fruit development. Color coding as in previous figures

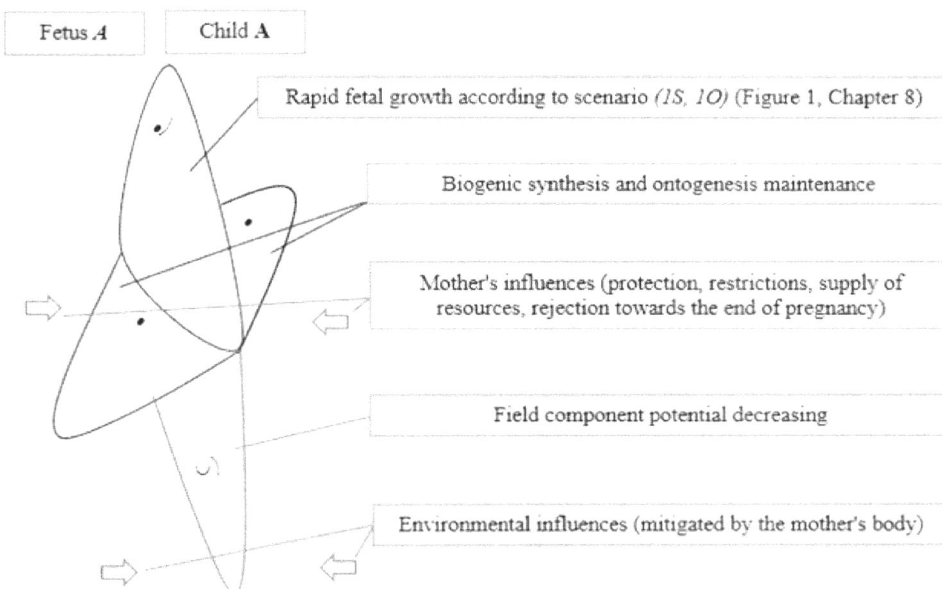

The system is gradually stabilizing.

It should be noted here that childbirth does not mean a break in all ties - mental ties (love, at the level of the intuitive channel) and partially physiological (the mechanism for feeding a child) remain and even intensify; in the latter case, the leading role is played by the complex mechanisms of heredity (with the connection of the humoral system, and so on). However, rejection by the mother marks the most important milestone: the child's transition to a new, rather hostile environment (albeit under the protection of parents). It also checks its viability as an independent organism: it needs great potential to adapt.

Critical events, time frames, systemic factors. The presence of wave-soliton models makes it possible to clarify the periodization of pregnancy and understand the meaning of ongoing events.

The most important of the events is the formation of a new living organism, which corresponds to the formation of a wave-soliton system in Figure 3. The system is open, strongly subject to external influence, and is in a non-equilibrium state.

But from the very beginning, it is already an independent system, different from both the parents and the external environment. It develops very quickly according to a given program and exists relatively autonomously until the start of implantation. These 6-7 days should be singled out as the second period.

Further, two processes occur in parallel: the embryo's independent development and the mother-embryo system's formation. The first of them unfolds according to the scenario *(1S, 1O)* and is qualitatively completed by the end of the fetal period by the end of the 8th week. The second lasts until the final formation of the placenta by the end of the 12th week.

Further, mainly quantitative, without the formation of new organs, but with partial changes in their morphology, the stable development of the fetus in the mother-fetus system continues, until the excess of the limits of self-regulation of the mother's body (and the corresponding hereditary mechanisms) begin preparation for rejection fetus, i.e., up to about 36 weeks in a normal pregnancy.

This preparation takes 36-40 weeks, at full term, and includes the preparation of both the mother and the fetus, culminating in childbirth.

Among other things, such periodization is important in addressing issues of bio-ethics. From the very beginning, the destruction of the embryo should be regarded as the murder of a living being, but when does potential homicide develop into actual?

Based on the logic of modeling, an abortion after implantation (that is, after the 1st week of pregnancy) should be regarded as an attempt to kill the mother, and abortion after the 8th and even more so the 12th week as an attempt to destroy the mother-and-child system. The legislative period of 20 weeks is overstated: the fetus is human in terms of morphology (Figure 2), homeostasis, and signs of nervous activity. In addition, in the 20th week, no important events would distinguish this milestone from the general series. The incompleteness of the processes of formation of the child's body and the imperfection of his nervous activity cannot be an argument. Firstly, the development of several systems, such as reproductive, ends much later than birth. Secondly, approximately 10% of the world's population is disabled - and in civilized countries, they are not discriminated against. Still, they are trying to create an inclusive environment where they would be comfortable like normal people.

The general plan for developing the embryo and fetus corresponds to the scenario (1S, 10) - and this correspondence is already inherited, not just direct. However, attempts to calculate the duration of the marked periods as a total, rather than individual events, and regularity based on simple ratios based on the "golden ratio" (Chapter 3) do not lead to success. Moreover, if, by estimating the duration of large evolutionary periods, it was possible to ascertain the acceleration of evolution (Chapter 4), then its deceleration is evident here.

This should not be surprising - the pace of development is determined by a combination of external factors acting in different directions.

First, if the emergence of a living organism is based on a single event - a resonance modulated by external factors - (this hypothesis was proposed in Chapter 4), then its consequences further influence. The analog here is the damping curve of oscillations, which demonstrates a decrease in the areas of "half-waves" - and based on this, the potential was determined in Chapter 2 - and, consequently, the "internal resource" for the ongoing changes, and, consequently, their rate.

Secondly, there is an adaptive need for accelerated development - by birth, the child must be ready to fight for existence.

Thirdly, during the fetal period, the mother's influence comes into play. In a stable environment created by her body, and with limited resources supplied by her, the rate of evolution decreases (Chapter 4).

There is another analogy to such a development of events - the evolution of the Universe itself, with its turbulent "first three minutes" (Weinberg, 1993) and much calmer evolution in the future - another similarity to the macro- and microcosm.

Finally, one can ask the following question: What is the mechanism of adaptation of a pregnant woman, mother, and child to the environment? Considered in isolation, of course, both pregnancy and the bondage of caring for a child reduce the mother's ability to both search for food and protect herself from danger. But in the system of family, clan, and society, mother and child are protected, and they contribute to the continuation of the existence of the family, clan, and culture, thus, everyone receives evolutionary advantages.

Another system factor is external modulations. The analysis of their composition and combinations of their effects is very complicated, and, in order not to stray too much from the topic of this study, we will only mention modulation - by the time of existence and the state of progress/regress of the species, which determines its evolutionary potential.

AGE PERIODS TERMS AND CHARACTERISTICS

Further life cycle events related to the postnatal period will be considered, including senility and death. However, all near-death events will be taken out in Chapter 10.

Studies of age-related changes were carried out with an emphasis on physiological, and psychosocial components, or considering the balance of all manifestations.

An example of such an approach is age periodization, which was adopted at the VII All-Union Conference on the problems of age morphology, physiology, and biochemistry (Table 1).

Table 1. Age periodization and a brief description of the periods

Periods	Age and gender		Some characteristic age-related events
	Male	Female	
Newborns	From birth to 10 days		The period of feeding a child with colostrum. The beginning of growth in length. Frequent drop in body weight
Breast age	From 10 days to 1 year		Start feeding "mature" milk. Eruption of the first milk teeth (from 6 months); the period of maximum intensity of the growth process, the beginning of the straightening of the body (formation of the curves of the spine), sitting, standing, and the first steps. The beginning of cognitive development (examination and recognition), imitation, and calling gestures. Pre-speech and the beginning of speech development
Early childhood	1–3 years		Completion of an eruption of the first generation of teeth. Decrease in growth rate. The choice of the object of affection and the selection of "I". Phrasal speech. Active subject activity. Recognizing pictures, fantasizing, and using substitute objects in the game
First childhood	4–7 years		The first growth spurt is often recorded. From the age of 6, the eruption of permanent teeth begins. At the end of the period, there are distinct manifestations of sexual dimorphism and the beginning of sexual identification (sex awareness). Visually symbolic thinking, space exploration, and the concept of the time sequence. Fundamentals of ethics and group behavior. Conscious tenacity. Leading activity - role-playing game
Second childhood	8–12 years	8–11 years	Eruption of permanent teeth (except for the third painter - "wisdom teeth"). The beginning of the development of secondary sexual characteristics (puberty) and the activation of the growth process. The transition from visual-figurative to logical-operational thinking, true voluntary acts, and social adaptation. Development of attention and arbitrary memory
Teenage	13–16 years	12–15 years	The puberty spurt is the second (true) growth spurt. Puberty and increased body growth in length. All this determines significant morphological and functional shifts affecting all body systems. Verbal expression of abstract thinking, intensive intellectual development, introspection, high gender identity, personal and emotional instability
Youthful	17–21 years	16–20 years	End of intensive growth and formation of the body (definitive state). At the beginning of the period of stabilization of the personality, self-determination, and the formation of a worldview, destructive manifestations of social activity are possible
The first mature age	22-35 years	21-35 years	Maturity itself is the relative stability of body parameters, the end of the formation of "typically female" and "typically male" features of the structure and psyche
The second mature age	36–60 years	36–55 years	Daily, weekly, seasonal, and other rhythmicity of physiological functions acquires leading importance. At the end of the period - the end of the female reproductive cycle - menopause and a climacteric complex of corresponding psychophysiological changes
Old age	61–74 years	56–74 years	A continued period of decline in social activity. The development of involutive changes in the body, including true negative growth, has also begun. Decreased adaptability. The disintegration of body functions at all levels of an organization. Usually, it is the end of the male reproductive period. Structural and functional changes in the central nervous system, and at the end of the stage, bright signs of "mental aging" are common
Senile age	75–90 years		

continued on following page

Table 1. Continued

Periods	Age and gender		Some characteristic age-related events
	Male	Female	
Long-livers	Older than 90 years		Those who lived up to this period are characterized by the relative stability of all parameters at the achieved qualitative and quantitative level, including due to compensatory (compensatory-senile) processes

(Nikityuk & Chtetsov, 1990)

A study of the postnatal development of a person with an emphasis on the formation of certain psychological characteristics was conducted by E. Erickson. (Erikson, 1950; Erikson & Erikson, 1987; Erikson et al., 1986). Its results are presented in Table 2.

A more detailed periodization, accompanied by detailed psychological characteristics, is given in (Rean et al., 2015).

Table 2. Life cycle stages according to Erickson

Stages of the cycle	Priority impact	Positive and negative formed qualities	The result of progressive development
Breast age, up to 1 year	Mother	Trust in the world - distrust in the world	Energy and the joy of life
Early childhood, 1 to 3 years	Parents	Independence - shame, doubts	Independence
Childhood, 3-6 years	Parents, brothers, sisters	Initiative - passivity, sense of guilt	Purposefulness
School-age, 6-12 years	School, neighbors	Competence - inferiority	Acquisition of knowledge and skills
Adolescence and youth, 12-20 years	Peers	Personal identity - non-recognition	Self-determination, devotion, fidelity
Early maturity, 20-25 years old	Friends, beloved	Closeness - isolation	Cooperation, love
Middle age, 25-65 years	Profession, family	Productivity - stagnation	Creativity and care
Late maturity, over 65	Humanity, beloved	Personality integrity – despair	Wisdom

Each of the age periods consists of several stages with characteristic features. For example, childhood includes the following stages:

- Newborns and infancy (from birth to 1 year).
- Early childhood (or first childhood, 1 to 3 years old) - the development period of functional independence and speech.
- Preschool childhood (or second childhood, 3 to 6 years old) is characterized by developing the child's personality and cognitive processes.

- Primary school childhood (or third childhood, from 6 to 11–12 years old) corresponds to the inclusion of the child in a social group and the development of intellectual skills and knowledge.

Adolescence is divided into two periods:

- Teenage (or puberty) corresponds to puberty and lasts from 11-12 to 14-15 years. Under the influence of constitutional changes, a teenager develops a new idea of himself.
- Youthful, from 16 to 20-23 years old, represents the transition to maturity - from a biological point of view, the young man is already an adult, but has not yet reached social maturity: youth is characterized by a sense of independence, although the person has not yet taken on social obligations. Adolescence acts as a period of making responsible decisions that determine the entire future life of a person: the choice of a profession and one's place in life, the selection of the meaning of life, the formation of a worldview and self-awareness, and the choice of a life partner.

During childhood, the law of uneven development manifests itself: each side of the child's psyche has its optimal period. According to L. S. Vygotsky, child development is subject not to biological but socio-historical laws and occurs through appropriating historically developed forms and activity methods. Education creates a zone of proximal development and sets in motion internal processes of development, which at first are possible for the child only in interaction with adults and cooperation with comrades, but then become the child's property. The zone of proximal development "defines functions that have not yet matured, but are in the process of maturation, characterizes mental development for tomorrow" (Vygotsky, 2009). No influence of an adult on the processes of mental development can be carried out without the real activity of the child himself. Leading activity is characterized by the fact that the main cognitive processes are rebuilt in it and changes in the psychological characteristics of the individual at a given stage of its development occur. The content and form of leading activity depend on the concrete historical conditions in which the child's development occurs. A change in the leading types of activity is prepared for a long time. It is associated with new motives that prompt the child to change his position in the relations system with other people.

D. Elkonin considers each age based on the following criteria (Elkonin, 2007):

- The social development situation - the relations system in which the child enters society.
- The child's main, or leading, type of activity during this period.

- The main neoplasms of development, and new achievements in development lead to the inevitability of change and the social situation, to a crisis.

During the transition from one age stage of development to another, critical periods are distinguished when the existing form of relations with the environment is destroyed and a new system of relationships is formed, accompanied by significant difficulties for a person and his social environment. There are small crises (1 year, 7 years, 17-18 years old) and large crises (newborns, 3 years old, adolescents 13-14 years old). During major crises, relations are radically rebuilt. Small crises are calmer; they are associated with the growth of human independence. During periods of crisis, children are difficult to educate, they show stubbornness, negativism, obstinacy, and disobedience.

E. Erikson's concept of personality development shows the connection between the human psyche and society. Each stage of human development corresponds to expectations, inherent in this society. A person can justify or not justify it, and then his personality development can go in a negative or positive direction.

So, in infancy, without close emotional contact, without love, attention, and care, the child's socialization is disrupted, mental retardation occurs, aggressiveness appears, and in the future various problems are associated with relationships with other people. A positive result is the formation of an infant's confidence in the world, people, and optimism, and the negative result is expressed in distrust of the world, people, pessimism, and aggressiveness.

The second stage of early childhood relates to autonomy and shame. The child begins to walk and learns to control himself when performing acts of defecation; society and parents teach the child to be tidy. Social disapproval through the possibility of punishment forms a sense of shame. At the end of this stage, a balance of autonomy and shame must be reached.

At the stage of childhood, the child knows how to speak, expands the area of mastering the world, develops a sense of enterprise, and takes initiative, which is laid down in the game. The game is significant for development because it forms initiative and creativity, and develops will, memory, and thinking. But if parents strongly suppress the child, then this negatively affects development and contributes to the formation of passivity, insecurity, and feelings of guilt.

The child has exhausted the possibilities of development within the family, and at primary school age, the school introduces him to knowledge about future activities, to the culture of society. Education is the leading activity. If a child successfully masters knowledge and skills, believes in himself, and is calm; failures lead to appearance feelings of inferiority, lack of confidence in their abilities, despair, and loss of interest in learning. If parents scold and punish the feeling of inferiority in the child is fixed.

Ego-identity is formed during adolescence. Rapid physiological growth, puberty, increased attention to how he looks in front of others, the need to find a professional vocation, and the demands of society are tough challenges. The leading activity influencing the development of a teenager's personality is his communication with peers. At this stage, the critical moments of the past reappear. If the child has formed independence, initiative, trust, and confidence in his significance, the teenager will successfully create a holistic form of ego-identity. Otherwise, identity diffusion occurs: a teenager cannot find himself, is not aware of his goals and desires, regression begins, a return to infantile, childish, dependent reactions, a steady feeling of anxiety, loneliness, emptiness appears, a constant expectation of something that can change a life, but the person himself does not actively undertake anything, a fear of personal communication and an inability to emotionally influence persons of the opposite sex, hostility, contempt for the surrounding society, a feeling of "non-recognition of oneself" on the part of the surrounding people are born.

The stage of maturity is divided into several stages and crises. The early maturity stage corresponds to entering a person into an intense personal life and professional activity. This is a period of formation, and self-assertion in love, sex, career, family, and society. The search for a life partner, close cooperation with people, and strengthening ties with one's social group becomes relevant, a person is not afraid of depersonalization, and a feeling of closeness, unity, cooperation, and intimacy with certain people appears. However, if identity diffusion passes to this age, the person closes in loneliness.

Next comes the adult stage. Positive symptoms: a person invests in his favorite work, cares for children, and is satisfied with himself and his life. The psychological and physiological regression are outlined if there is no favorite work, family, or children. Negative symptoms are strongly pronounced if the personality has always had negative choices at the stages of its development.

In mature years, crises are also distinguished (Hollis, 1993): 33-35 years - when, having reached a certain social and family status, a person begins to think: "Is this all that life can give me? "And some begin to change something: work, spouse, place of residence, hobbies, etc. Then comes a short period of stabilization from 35 to 40-43 years old, when a person consolidates everything that he has achieved, is confident in his professional skills, and authority, has an acceptable level of career success and material prosperity, health, position in the family, and sexual life are normalized.

Following a period of stability, a critical decade of 45-55 years comes, when a person begins to feel the approach of old age, health deteriorates, beauty and physical form are lost, alienation in the family and in relations with grown-up children grows, there comes a fear that you will not get anything in life, career, love. As a result, fatigue and depression arise, from which a person goes into dreams of new love victories, tries to move up the career ladder, and so on. The final maturity period lasts from 55 to 65 years - a period of physiological and psychological balance decreased sexual tension, and a gradual cessation of active work and social life.

The period of 65-75 years is defined as the first old age. After 75 years, age is considered advanced. In old age, a person overcomes three crises: reassessment of himself in addition to his professional role; awareness of the deterioration in health and aging of the body; and overcoming the fear of death (Krasnova & Leader, 2003).

Now we calculate the periodization of the life cycle, considering the change like external influences compared with the prenatal period. Assume that scenario (1S, 1O) takes place. Then the composition and duration of age periods are determined by the Fibonacci number series: 1-1-2-3-5-8-13-21-34-55-89, and in the first five stages the structure of levels and channels of human interaction with the environment is already fully formed environment (Table 3). Some researchers believe that the life cycle of women is formed not by the Fibonacci series, but by the related Luke series: 2, 1, 3, 4, 7, 11, 18, 29, 47, 76, 123; the counting of stages starts from the second number. If they are correct, the estimated age periods in Table 3 should be adjusted for women.

Table 3. Life cycle according to scenario (1S, 1O)

Cycle stages (age in years)	Level of interaction with the environment / basic needs
1. Infancy (< 1)	Intuition/need for love and protection
2. Early childhood (1-2)	Ego/awareness of one's difference from the environment
3. Preschool childhood (3-5)	Will, reason/interaction with the environment
4. Junior school childhood (5-8)	Nous/rationalization of relations with the environment
5. Teenage (8-13)	Passions/emotional attitude to the environment
6. Youthful (13-21)	Need for a feeling of adulthood
7. Adulthood (21-34)	Striving for success
8. Maturity (34-55)	Maintaining success
9. Old age (55-89)	Renunciation of life's blessings and life itself

Firstly, considering some disagreements between the cited sources, Table 3 correlates well with the facts presented.

Secondly, a definition of age periods, justified by the laws of self-organization and the sequence of development of channels of perception, is proposed.

Thirdly, there are numerous correlations between the degree of development of certain channels of human interaction with the environment and his psychological type, employment, belonging to a social stratum, needs, and so on. All this and comparisons with similar systems are discussed in detail in (Mkhitaryan et al., 2021). The discussion has an important practical goal: creating a comfortable and stimulating living environment for representatives of different psychotypes and ages (other criteria are also considered).

Fourthly, transitions at the age of 13, 21, and 55 are perceived subjectively as major crises, others are easier to endure. An adult does not remember the crises of a newborn (his ego has not yet been formed); the crises of one and three years are remembered poorly.

Fifth:

- By the age of 13, the structure of the system of interactions is fully formed; in the future, only redistribution of potentials between channels and their slight increase because of properly organized interactions with the environment is possible.
- By age 21, the body grows, and social connections and needs are formed, for which a certain potential is also spent.
- When growth stops, these costs also stop. The adaptive potential is used for external expansion - an active struggle begins to embody one's ideals and build a career, and family affairs, which is the main content of the adult stage.
- And one of the most important properties of open systems begins to appear - their finiteness, due, in this case, to the soliton-wave nature of the system. The potential is not restored, and gradually, instead of expanding, there is a desire to maintain the achieved positions, especially at the end of the maturity stage.
- The period after 55 years - old age - there is a time when the potential decreases. As a result, metabolic disorders and weight gain, problems with the musculoskeletal system, etc. The rejection of excessive needs helps to prolong life and, in the end, to renounce it. The period of old age will be discussed in more detail below.
- Accordingly, at stage 6 it is necessary to ensure the implementation of new views on life inherent in the younger generation, at stage 7 - to create conditions for self-realization, at stage 8 - to maintain achievements, and at stage 9 - for their painless renunciation.

As for the prenatal period, we have an increase in the duration of subsequent life cycle stages compared to the previous ones, while the duration of large evolutionary periods was reduced (Chapter 4). This confirms the crucial importance of the event of resonance modulated by external factors during the fusion of the parent's gametes on the events of the entire human life cycle.

Let us give one more comparison - with how the activity of the channels has changed during human evolution from the Upper Paleolithic to the present (the Upper Paleolithic is due to the appearance of symbolic activity, which made it possible to trace the activity of the channels (Figure 6).

Figure 6. Changes in the activity of channels and expressive means for presenting the myth of the World Tree in different historical eras (Kovalyov et al., 2020)

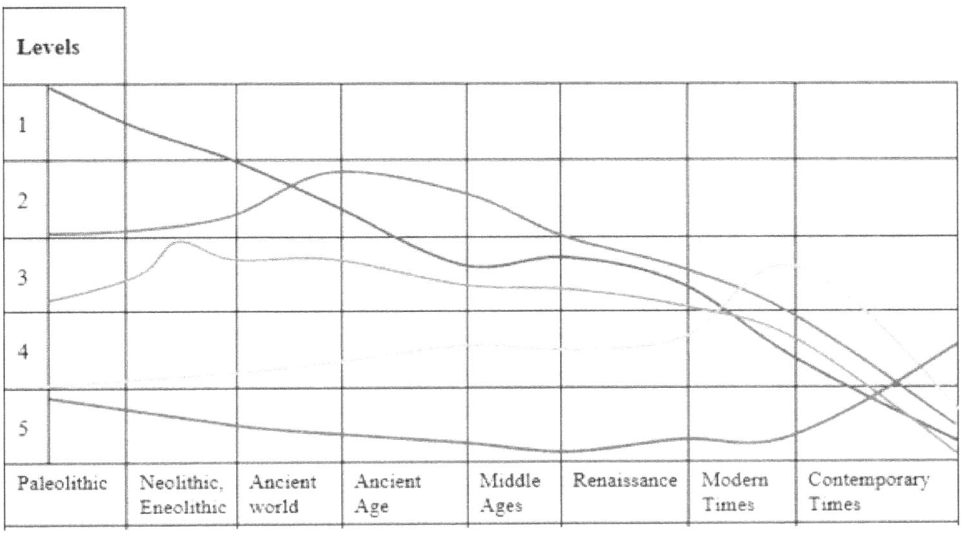

OLD AGE

Phenomenology

Advances in modern medicine distort the picture of natural aging; therefore, an earlier scientific source is needed. Such is (Mechnikov, 1964/1907); quoting his most expressive passages is expedient.

"Most of the hundred-year-olds I managed to see represented such a strong mental decrepitude that their study had to be reduced to observing purely physical manifestations. A few years ago, in the vicinity of Rouen, they celebrated the centenary of an old woman. Despite the relatively well-preserved physical side, her mental abilities turned out to be so weakened that there could be no question of any interesting study of them.

Of all the persons known to me who have reached the age of one hundred, the most interesting was the oldest woman, Madame Robineau, who died at the age of one hundred and seven. Studies of its physical functions point to several fascinating facts.

Dr. Ambar, on auscultation, found that the heart murmurs were normal, although perhaps a little exaggerated. The pulse is regular, 70 to 84 beats per minute, and intravascular pressure is slightly elevated. Her blood pressure is 17. Lungs are healthy. All these signs point to a good condition of the body. Particularly remarkable is the absence of arteriosclerosis at such an advanced age. This fact refutes the opinion that arteriosclerosis is one of the normal signs of old age. Urinalysis, which I examined in several sessions, indicates chronic, though not serious, kidney suffering. The sediment contained uric acid crystals, squamous epithelial cells, rare cells of the renal canals, several hyaline cylinders, and single white blood cells, and its density was 1019. The urine contained neither protein nor sugar. For 1 liter of urine, there were 11 g 50 cg of uric acid, 9 g of uric acid salts, and 1 g of 15 cg of phosphates.

Despite a significant decrease in taste, Ms. Robineau had a satisfactory appetite, although she ate and drank little. Her food was varied. She almost did not eat meat, but often ate eggs, fish, flour, vegetables, and boiled fruits. I drank sweetened water with white wine. After eating, she occasionally drank a little dessert wine. Her digestion and bowel movements were generally normal.

It is generally thought that the length of life results from heredity. But this rule does not apply to Ms. Robineau. Her parents died not old, and no one in her family reached 100 years old. Her advanced age is therefore an acquired property. She led a very modest life. She loved the hearth with her gentle and kind nature and led a family life, avoiding social entertainment.

After 106 years, Ms. Robineau's mental faculties declined sharply. She almost completely lost her memory, and her speech often became incoherent, but until recently she retained her soft and sweet character. Ms. Robineau passed away before reaching the full age of 107. For the past year, she has been wearying of life and often expressed a desire to die. One must think that her "sense of life" completely dried up in the end. After being ill for a week with influenza, which prevailed during the winter, she died quietly with symptoms of pneumonia.

In old age, growth decreases. Numerous measurements show that a man between 50 and 85 years of age decreases by more than 3 cm (3.166), and a woman even more - by 4 cm and 3 mm. Sometimes old people decrease by 6 cm and even by 7 cm. Weight also decreases in old age. According to Kettle, a man reaches his maximum weight at 40, and a woman at 50. From the age of 60, the weight begins to decrease and by the age of 80, on average, it falls by 6 kg.

A decrease in size and weight indicates general atrophy of the body of the elderly. Not only do the soft parts, like muscles and viscera, become lighter with age, but also the skeleton, due to the depletion of minerals in it. This loss of lime extends to all parts of the skeleton, causing old people's bones to become brittle and break easily, which is often the cause of their death.

Muscles in old age are also very susceptible to atrophy. They decrease in volume and become paler; the amount of fatty tissue between the muscle bundles decreases and sometimes almost completely disappears. Therefore, movements become slower and muscle strength is weaker. Measurements using a dynamometer of the strength of the arms and trunk indicate a gradual weakening of it with old age, which is more pronounced in men than women. The volume and weight of the viscera also decrease, although not to the same extent for the various organs.

The senile characteristics of mammals represent similar features to those of humans."

Along with the physiological signs of old age, I. Mechnikov also explores psychological ones, noting that at the end of life, a person first has an increased desire to live and fear of death, and then resignation with such an outcome - *"death becomes the same physiological need as sleep."*

Physiological, Psychological, and Medical Manifestations of Aging as a System

The modern understanding of aging is systemic and includes tracking processes at different levels of the body (Aging changes in organs, n / d; Wang et al., 2017; Frenzel et al., 2020; Tian et al., 2023). According to (Monakhova et al., 2018), *"Aging is a process of gradual inhibition of the basic functions of the body, including regenerative and reproductive ones, as a result of which the body becomes less adapted to environmental conditions (loses the ability to withstand stress, disease, and injury), which makes its death inevitable Aging is a complex process that affects various levels of biological organization, from the molecular to the organismic. At the molecular level, there is an accumulation of damage in the cell's macromolecules. At the cell level, this manifests as damage to the endoplasmic reticulum, accumulation of aberrant proteins, and active synthesis of pro-inflammatory cytokines. Active inflammatory processes cause tissue damage. At the level of the body, aging leads to the emergence and development of age-related diseases such as Alzheimer's disease, cataracts, and Parkinson's disease. Other age-related changes in the body include disruption of circadian rhythms and diseases of the cardiovascular system».*

Psychological and behavioral ones accompany physiological changes. Biological manifestations include changes caused by aging and the deterioration of general health. A person at a later age is vulnerable to diseases, which is associated with a decrease in the immune system's effectiveness. The diseases of old age are a combination of the symptoms of aging and diseases that the body can no longer fight against. For example, a young person can recover quickly from pneumonia, while for an older person, it can be fatal.

Changes in the body in the aging process are due to changes that occur in individual cells and entire organs. These changes cause changes in function and appearance (Aging changes in organs, tissue, and cells, n / d; Development: birth through death, n / d).

Cell aging. As we age, cells begin to function less well. Old cells must die, because the genes program a regulated process to destroy damaged, mutated, infected cells- apoptosis. As a result, the cell breaks down into bodies limited by the plasma membrane, which phagocytes destroy. Old cells die to make room for new cells. Old cells also die because they can only divide a limited number of times -Hayflick limit (Hayflick, 1998). When a cell can no longer divide, it grows, exists for a while, and then dies. The division mechanism involves telomeres, which are necessary to move the genetic material of cells in preparation for division (Chapter 2.) With each cell division, telomeres become shorter and finally, the cell can no longer divide, proceeding to senescence. Sometimes cell damage (radiation, sunlight veterinary, chemotherapy drugs, and so on) directly causes her death. Cells can be damaged by-products of their activity (free radicals).

The aging of organs depends on the functioning of their cells: old cells function worse, and some dead cells are not replaced by new ones, which leads to a reduction in the number of cells (in the testicles, ovaries, liver, and kidneys, it decreases significantly as the body ages). However, not all organs, such as the brain, lose many cells. A decrease in one organ's function can affect another's function. Some organs are more likely to malfunction under stress (the heart and blood vessels, the organs of the genitourinary system, especially the kidneys, and the brain).

The musculoskeletal system often shows signs of aging earlier than others. Bones are prone to loss of density (osteopenia and osteoporosis). Bones become less dense due to a decrease in calcium content, due to greater wear (the end of the femur in the thigh area, the ends of the radius and ulna in the wrist area, and vertebrae), in women after menopause due to the production of less estrogen. Changes in the vertebrae at the top of the spine cause the head to tilt forward, constricting the throat. As a result, swallowing is difficult, and the likelihood of suffocation increases. The vertebrae become less dense, and the discs between them lose fluid and become thinner, thereby shortening the spine and reducing a person's height. The cartilage lining the surfaces of the joints thins out due to wear and tear, often leading to osteoarthritis, one of the most common disorders of old age. Ligaments that connect joints and tendons that connect muscles and bones become less flexible, causing joint stiffness. Therefore, most people become less flexible. Ligaments and tendons tend to tear more easily and heal more slowly. These changes occur because their cells become less active. Muscle mass and muscle strength tend to decline from around age 30. This is partly due to decreased physical activity and a decrease in growth hormone and testosterone levels, which stimulate muscle development. In addition, muscles cannot contract as quickly because more fast-twitch muscle fibers are lost than slow-twitch ones. However, aging reduces muscle mass and strength by no more than 10-15% during adulthood. In the absence of disease, regular exercise can prevent most losses. More significant loss of muscle tissue (sarcopenia) occurs due to diseases or excessive immobilization, and not just aging. By age 75, the body's fatty tissue proportion typically doubles compared to young adulthood. Too much body fat can increase the risk of health problems like diabetes. Fat distribution in the body also changes, causing a change in the shape of the torso.

Vision. With age, the lens becomes stiffer (harder to focus on nearby objects), denser (harder to see in dim light), turns yellow (color perception changes), the pupil's reaction to light changes slows down, the number of nerve cells decreases (depth perception is disturbed), the eyes produce less fluid (feeling dry). Vision change is an undeniable sign of aging. The appearance of the eyes changes: the whites (sclera) of the eyes may become slightly yellowish or brownish (due to years of exposure to ultraviolet light, wind, and dust), irregular color spots may appear in the whites of the eyes, a grayish-white ring may appear on the surface of the eye, the lower eyelid may droop, the eyes can seem to fall into the skull.

Hearing is weakened by exposure to noise and aging. Hearing high-pitched sounds becomes more difficult with age (presbycusis). The consequence of presbycusis is that it becomes more difficult for older people to pronounce words. Many older people cannot hear in noisy places due to background noise. In addition, earwax, which impairs hearing, tends to accumulate in large quantities. Thick hairs may grow out of the ears.

The ability to distinguish tastes and smells over 50 begins to weaken. Decreased sensitivity of the taste buds on the tongue. The ability to distinguish smells is reduced because the nasal mucosa becomes thinner and dryer, and the nerve endings in the nose are destroyed. The mouth is more likely to feel dry, partly due to reduced saliva production. A dry mouth further reduces the ability to taste food. As we age, the gums recede slightly. As a result, the lower parts of the teeth are exposed to food particles and bacteria. In addition, tooth enamel tends to wear out. These changes, along with dry mouth, make teeth more susceptible to decay and cavities (cavities), increasing the chances of tooth loss. With age, the nose becomes longer and larger, and its tip begins to fall. Thick hairs may grow on the nose, upper lip, and chin.

The skin becomes thinner, loses its elasticity, becomes drier, and is covered with small wrinkles. Long-term exposure to sunlight contributes to the formation of wrinkles and roughness, as well as uneven skin pigmentation. Changes in the skin are partly caused by chemical changes and a decrease in the flexibility of collagen and elastin, in addition, with aging, their production decreases. As a result, the skin breaks more easily. The subcutaneous fat layer becomes thinner, as a result, the likelihood of wrinkles increases and cold tolerance decreases. The number of nerve endings in the skin is reduced. As a result, people become less sensitive to pain, temperature, and pressure, which increases the likelihood of injury. The number of sweat glands and blood vessels decreases, causing a decrease in blood circulation in the deep layers of the skin. Therefore, the body partially loses the ability to transfer heat from within the body through the blood vessels to the surface of the body. The number of melanocyte cells that produce pigment decreases. As a result, the skin is less protected from ultraviolet radiation and brown spots appear (also because the skin removes waste products worse). The ability of the skin to produce vitamin D in response to sunlight is reduced and the risk of vitamin D deficiency is increased.

Brain and nervous system. The number of nerve cells in the brain decreases. However, the brain can partially compensate for this loss: new connections are created, and new nerve cells can form even in old age; the brain initially contains more cells than it needs. The levels of chemicals involved in transmitting messages in the brain are reduced, but the levels of some may increase. Nerve cells may lose some of the receptors that convey these chemical messages reducing blood flow to the brain. Due to these age-related changes, the brain can work less well. Older people may react and complete tasks more slowly Some thinking abilities - active vocabulary, short-term memory, the ability to learn and remember words - may decrease after age 70. After age 60, the number of cells in the spinal cord begins to decline. As we age, nerves may transmit signals more slowly. The nerves are slower and do not fully recover, which impairs sensitivity.

The heart and blood vessels become stiffer. The heart fills with blood more slowly. Stiff arteries dilate worse when more blood is pumped through them. Therefore, blood pressure increases. The difference between a young and an old heart becomes apparent when the heart has to work harder and pump more blood. Regular aerobic exercise may improve athletic performance in older adults.

Lungs and muscles of the respiratory system. The diaphragm and intercostal muscles tend to weaken. The number of pulmonary alveoli and capillaries decreases. Therefore, slightly less oxygen is absorbed from the air than is inhaled. The lungs become less elastic. In people who do not smoke or have lung disease, these changes do not affect normal daily activities, but they do make themselves felt during exercise. It becomes more difficult to breathe at high altitudes. The ability of the lungs to fight infections decreases. The cough that clears the lungs subsides.

The digestive system is least affected by old age. The muscles of the esophagus contract with less force. Food is removed more slowly from the stomach, which cannot hold as much food as it does when young, because it becomes less elastic. The digestive tract produces less lactose needed to digest milk. As a result, older people develop intolerance to dairy products. In the large intestine, feces move more slowly, which leads to constipation. The liver decreases as the number of cells decreases. Less blood flows through it, enzymes that help metabolize drugs and other substances work less efficiently, and the effects of drugs last longer.

Kidneys and urinary tract. The kidneys are reduced in size due to a decrease in the number of cells. Less blood flows through them, and around 30, they begin to filter blood worse. Over the years, they may be less able to remove waste products from the blood. They can excrete too much water and too little salt, increasing the chance of dehydration. Certain changes in the urinary tract can make it difficult to control urination. In women, the urethra shortens, and its mucous membrane becomes thinner. The decrease in estrogen levels that occurs during menopause is one reason for the changes in the urinary tract. In men, the prostate gland enlarges, interfering with the passage of urine and preventing the bladder from emptying. As a result, older men tend to urinate more frequently and with less force or are unable to urinate despite a full bladder. This disorder requires urgent medical attention.

Reproductive organs. The effect of aging on sex hormone levels is more evident in women than in men. In women, it is associated with menopause, when the levels of female hormones are significantly reduced, and menstruation ends forever. Due to the decrease in the level of female hormones, the ovaries and uterus shrink. The tissues of the vagina become thinner and drier, and less elastic. The mammary glands become less hard and more fibrous and tend to sag. For most women, aging slightly reduces the enjoyment of sexual activity. The thought that you no longer have to worry about pregnancy can even improve your sex life and the pleasure you get from it. For men, changes in the level of sex hormones do not occur so suddenly. The level of the male hormone testosterone decreases, which leads to the production of less sperm and a decrease in libido, but this happens gradually. Most men can have erections and orgasms throughout their lives. However, erections may not last as long, maybe weaker, or require more stimulation. Impotence becomes more common with age and is often caused by a disorder, usually a disorder that affects the blood vessels or diabetes.

Endocrine system. The levels and activity of some hormones produced by the endocrine glands are reduced growth hormone (decrease in muscle mass), aldosterone (dehydration increases), and insulin (blood sugar control worsens).

Hematopoiesis. Active bone marrow decreases, and fewer blood cells are produced.

The immune system. Immune system cells work more slowly, which may partly explain some of the facts associated with aging (cancer is more common in older people; vaccines are less protective; pneumonia and flu are more common and more likely to die; allergy symptoms may become less severe; rare autoimmune disorders).

The nature of the decline in the efficiency of the heart, kidneys, brain, lungs, and other organs and systems in old age is shown in Figure 7.

Figure 7. Averaged aging curves of some human systems and organs (n / d)

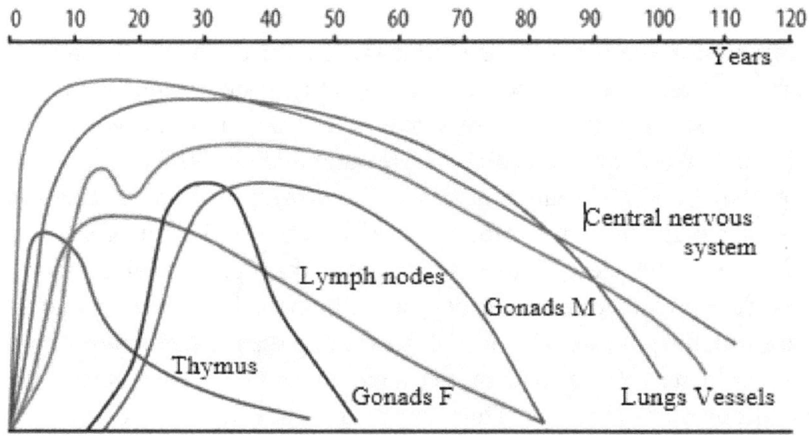

However, most functions remain sufficient, because the organs have a greater functional capacity than the body requires (functional reserve). So, even if half of the liver is destroyed, the remaining tissue will be more than enough to keep the liver functioning properly. Thus, disorders, rather than normal aging, are usually the main cause of loss of vital functions in old age.

Types and Assessments of Aging

Gerontology distinguishes three types of aging (Aging, n/d).

Natural aging. Age-related changes usually come on gradually and manifest themselves differently in different people. If we give a certain generalized picture, then at the age of about 40-50 years, the joints begin to lose flexibility and mobility, and it is more difficult for a person to move quickly, run, and jump; by the age of 50, physical strength falls by 7% compared to 45 years, and by 60 by about 35%. If a person constantly trains, these changes are less pronounced.

Premature aging is expressed in the early manifestation of signs of old age when a person's biological age exceeds his chronological age. In this case, changes in the body and psyche occur faster than usual. It is believed that the causes of this process are unfavorable living conditions, constant stress, and serious illnesses. Fatigue, weakness, and memory impairment are the main signs of premature aging.

Delayed aging is inherent in a small number of people, usually residents of high mountain areas. A low blood pressure level unites all centenarians, but the factors leading to longevity are still unknown.

Objective assessment of aging is carried out using various methods tracking up to 100 parameters, called biomarkers (Medvedev, Artyukhov, et al., n / d; Wang et al., 2017; Frenzel et al., 2020; Krut'ko et al., 2021). With the help of biomarkers, the characteristics of the type and speed of aging are determined, and individual risks are assessed, including the prognosis for developing several diseases. For example, under the leadership of V. Giampapa, the Biomarker Matrix Protocol was formulated, which determines the degree of aging at 4 levels:

1. At the physiological level or the level of general functioning (the ratio of lean body mass to fat mass, flexibility, aerobic endurance, bone density, tactile reaction time, forced expiratory volume, visual and auditory parameters).
2. At the cellular level (a biopsy and analysis of the skin condition is carried out, during which the change in the base membrane, the speed of circulation of the epidermis, the ratio of collagen, the architecture of the fatty glands, microvascular changes, the content of elastic fibers are determined).
3. At the molecular level (biochemical assays of human growth hormone, thyroid hormone, cellular enzyme Q10, determination of insulin sensitivity, analysis of heat shock proteins, blood tests for oncogenes, and serum levels of antioxidants).
4. At the chromosomal level (telomere position and DNA decay rate).

The interpretation of biomarker analysis results depends on the accepted theory of aging.

Theories of Aging

Consider the content of some theories of aging (Human aging, n/d; Finch, 1990; Mueller & Rose, 1996; Frolkis, 1988; Medvedev, 2011; Anisimov, 2015).

The energy theory or wear and tear hypothesis (Rubner, 1908) assumes that each species has a certain energy reserve, the depletion of which leads to aging and death.

The intoxication theory (Mechnikov, 1903) asserts that aging is caused by self-poisoning due to the accumulation of nitrogen metabolism products and decay products in the large intestine. Since the efficiency of systems and organs decreases with age (Figure 6), self-poisoning is seen, on the one hand, because of a decrease in efficiency, and on the other hand, as a factor that enhances this decrease ("vicious circle"). The reasons for the decline in efficiency lie deeper; as one of the options, the hormonal hypothesis (Brown-Sequard, 1889; Voronov, 1924) can be considered, which sees a decrease in the production of sex hormones as the cause of aging. In childhood, sex hormones are also not produced, which does not prevent the body from developing rapidly. However, the very idea of hormonal regulation has been developed in the modern hormonal-genetic theory (Dilman, 1972): during a person's life, starting from birth, there is an increase in the threshold of sensitivity of the hypothalamus, which ultimately leads to hormonal imbalance and progressive disorders of all types after 40 years of age - metabolism, including hypercholesterolemia. Overstrain of the central nervous system because of nervous shocks and constant stresses can also contribute to various imbalances, and ultimately, premature aging (hypothesis of overstrain of the central nervous system (Pavlov, 1912, Selye, 1936).

The mutation accumulation theory (Medawar, 1952) is based on the observation that animals rarely survive in old age. Therefore, alleles that appear later in life, resulting from mutations, are subject to little evolutionary pressure, even if they are negative due to their action. Thus, these mutations accumulate in the genome over many generations. However, an individual living for a long time experiences their effect, which manifests as aging. In 1957, D. Williams suggested the existence of pleiotropic genes that have different effects on the survival of organisms over various periods of life: they are useful at a young age when the impact of natural selection is strong, but harmful later when the effect of natural selection is weak. However, identifying such genes (for example, the telomerase gene in eukaryotes) has had limited success. Evidence for the accumulation of mutations is also controversial (Shaw et al., 1999).

Disposable soma theory (Kirkwood, 1977) solves an optimization problem: the distribution of body resources between body support - soma ¬ and other functions necessary for survival. Body maintenance must be carried out as needed during nature's normal lifetime. Thus, 90% of wild mice die during the first year of life, mainly from the cold, and using resources to combat old age will concern only 10% of the population, which is inefficient. Resources should be spent on improving heat retention or reproduction for the evolution of a species. Thus, life expectancy is best suited to environmental conditions. According to the theory, aging occurs due to non-ideal repair and maintenance functions of somatic cells adapted to meet environmental needs. Damage results from stochastic processes associated with the vital activity of cells. Longevity is controlled by the genes responsible for these functions, and the immortality of generative cells, unlike somatic ones, results from a large expenditure of resources and, possibly, the absence of some sources of damage.

Molecular mechanisms of aging. Several mechanisms of damage to macromolecules usually act in parallel or depend on each other (Kirkwood, 2005). In many of them, reactive oxygen species (in particular, free radicals) play an important role, the evidence for their influence was obtained quite a long time ago and is now known as the free radical theory of aging. However, there is evidence that free radicals do not cause natural aging. Another factor is increased somatic mutations and other forms of DNA damage with age, suggesting DNA repair (Chapter 2). DNA damage is caused by hard radiation and reactive oxygen species, so its integrity can only be maintained through repair. Indeed, a relationship exists between longevity and DNA repair, as demonstrated by the enzyme poly-ADP-ribose polymerase-1 (PARP-1), as a component of the cellular response to stress-induced DNA damage (Burkle, 2001). Higher levels of PARP-1 are associated with a longer lifespan.

Also, for the survival of cells, the circulation of proteins is important, and the appearance of damaged and excess proteins is critical. Damaged proteins accumulate with age and may be responsible for Alzheimer's, Parkinson's, and cataracts. Oxidized proteins result from the influence of reactive oxygen species, which are formed because of a few metabolic processes of the cell and often interfere with the correct functioning of the protein. Repair mechanisms cannot always recognize damaged proteins and become less efficient with age due to decreased proteasome activity (Carrard et al., 2002). In some cases, proteins are part of static structures, such as the cell wall, that cannot be easily destroyed. The turnover of proteins also depends on chaperone proteins, which help proteins obtain the necessary conformation (Chapter 2). With age, a decrease in reparative activity is observed.

The mitochondrial theory of aging (Bratic & Larsson, 2013) is that the slowdown in mitochondrial reproduction in highly differentiated cells due to a deficiency of mitochondrial proteins encoded in the nucleus creates conditions for the emergence and selective selection of defective mitochondrial DNA, an increase in the proportion of which gradually reduces the energy supply of cells. Such cells often have disturbances in ATP production and cellular energy balance.

Telomere theory (Olovnikov, 1973). In many human cells, losing their ability to divide is associated with losing telomeres (Chapter 2) at the ends of chromosomes after a certain number of divisions. This is due to the absence of the enzyme telomerase, which is normally only expressed in germ and stem cells. Telomerase allows them to divide, forming tissues and organs continuously. In adults, telomerase is expressed in cells that must divide frequently, but most somatic cells do not produce it.

Epigenetic theory. Cells slowly lose repressed chromatin markers over time, which may be associated with cell differentiation in the body. The loss of markers should lead to the depression of transposons and an increase in the DNA damage they cause, followed by the activation of the cellular DNA repair system. In addition to participating in DNA repair, the latter causes unauthorized recombination in telomeres. As a result, extended sections of telomeric DNA are converted into rings and lost, and the length of the lost circular DNA shortens telomeres. This process accelerates telomeric DNA loss tenfold, and most cells' subsequent apoptosis predetermines aging as a biological phenomenon. The proposed theory is an alternative to the hypothesis of genetically programmed aging and the hypothesis of aging because of the accumulation of errors and damage, explains the mechanism of the accelerated loss of telomeres in the case of oxidative stress and DNA damage, as well as the relationship between aging and the onset of tumors (Galitsky,

2009; Monakhova et al., 2018). Recently, DNA methylation has been considered an important factor in aging.

Numerous studies aim to find ways to preserve the integrity of DNA, particularly its telomeric regions. Oxidative stress may impact telomere loss, significantly accelerating this process in certain tissues (Von Zglinicki, 2002; Hachmo et al., 2020), and reversing aging. Future studies will show how justified these hopes are: there are publications with optimistic (Miklashevskaya, 2020) and pessimism (Khramov, 2017); Nelson & Masel, 2017) assessments.

Cellular response to aging. Due to the stochastic nature of the damage, individual cells age faster than other cells. Such cells threaten the health of the entire tissue, especially rapidly dividing stem cells (bone marrow cells, intestinal epithelium, and others), due to the great potential of such tissues to create mutant, possibly cancerous cells. The cells of these tissues quickly respond to injury by initiating apoptosis. Even low doses of radiation induce apoptosis in intestinal epithelial cells, and mild chemical stress induces apoptosis of stem cells in old mice. In these tissues, mass apoptosis is a sign of increased cell damage. In other tissues, the response to increased levels of damage may be to stop dividing. The balance between apoptosis and stoppage of division is a compromise between aging and cancer - the body must kill the damaged cells, or allow them to exist, increasing the risk of cancer.

Systemic mechanisms of aging. Whereas numerous theories of aging were previously considered competing, today it is believed that many mechanisms of cell damage operate in parallel, and cells must also expend resources to combat many adverse factors. The systems approach to aging attempts to simultaneously consider a large number of such mechanisms (Medvedev, et al. The System Theory of Aging Project, n / d) and separate the mechanisms that act at different stages of the life of an organism and on various types of cells and tissues of the body. For example, cells actively dividing are more likely to suffer from mutation accumulation and loss of telomeres than differentiated cells. In contrast, differentiated cells are more likely to suffer from protein damage.

Mathematical Modeling of Aging

The population approach and models of two types are mainly used: data models and system models (Novoseltsev et al, 2003). Data models are based on the statistical processing of datasets associated with aging; they do not use theories of aging. For example, the Gompertz-Makham mortality law is a statistical distribution that describes human and animal mortality as the sum of an age-independent Meikham term and an age-dependent Gompertz function that increases exponentially with age (Gompertz, 1825).

The Gompertz-Meikham mortality law satisfactorily describes the dynamics of human mortality in the range of 30-80 years; for older ages, there is a plateau in mortality instead of exponential growth. To explain this effect, the Strehler-Mildwan models (Mueller, Rose, 1996) and reliability theories (Masoro & Austad, 20060) were proposed - in terms of this theory, the Gompertz-Makham mortality law is the law of failures, where the risk rate is a combination of failures independent of age and failures associated with aging, with an exponential increase in the rate of these failures.

System models are built with data on the structure of body systems, ongoing processes, and theories of aging, which are verified during the simulation. These models consider aging as a balance and redistribution of resources during the life cycle of an individual organism and evolutionary aspects. For example, the balance between the resources associated with the birth of offspring and the resources necessary for the survival of parents is considered.

Peculiarities of the soliton-wave model. First, it is necessary to assess how much the potentials of the wave and soliton components have decreased and the adaptive potential (Figures 9-12). Even qualitative (without units of measurement and scaling) and incomplete data, presented in Figure 6, show that, compared with peak values, the potentials have decreased. Taking a very rough assumption that the decrease in the potential of systems and organs is proportional to the drop in their performance, we have the following estimates of the decline in potential: for the central nervous system, approximately 3 times; for the lungs, about 4 times; for the thymus approximately 7 times. Self-regulatory mechanisms cannot compensate for such a decrease (Chapter 8).

It follows that the scenario of self-organization in old age is convolution (Chapter 4). However, although severe violations of the functions of organs are observed, their physical disappearance does not occur. This fact can be explained by the operation of the inheritance mechanism that maintains the established structures. Therefore, it will be more correct to speak of a partial convolution. When this process covers more than one organ or organ system but will become complex and irreversible, it should be considered the beginning of old age. This usually occurs (Table 3) over 55 years of age.

Now let's try to set its parameters. Theoretically, three options are possible: pronounced degradation of the subject space (wave component), which can be expressed in senile dementia; pronounced degradation of body functions (soliton component), which can be expressed in the form of extreme weakness and systemic age-related diseases; pronounced degradation of the soliton-wave system components while maintaining their balance. All these cases are described in (Mechnikov, 1964/1907).

However, what is the most likely scenario? We can conclude that the soliton (substance) component most often degrades faster, creating the basis for destroying the unity of the system. Interpreting the degradation process as a partial convolution (the potential decreases, but the channels and levels remain), we obtain a model of this process shown in Figure 8. The question arises when the degradation becomes irreversible and leads to the system's destruction, that is, the organism's death. Let's defer the answer to Chapter 10.

Figure 8. Interpret old age as a convolution (1) and the distribution of potentials of the soliton and wave components in the later stages of life (2). The blue dotted line shows the partial convolution sequence. The designation of the components is the same as in Figures 3-5

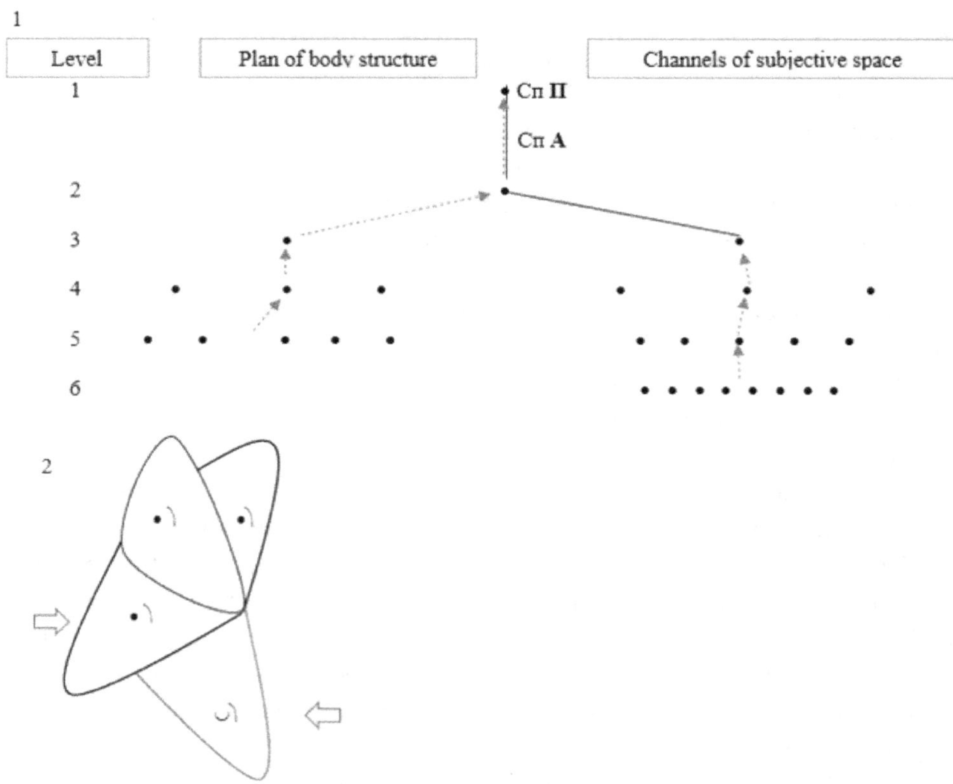

This is also connected with the emergence of wisdom that comes with age: the convolution of subjective space, weakening the channels of feelings and practical reason, increases the role of intuition and reason, and experience accumulates too.

Another factor is involuntary mental and physical asceticism (weakening of sexual desires due to a decrease in hormonal activity) - and, according to the unanimous opinion of ascetics of all times, it is abstinence that is an indispensable condition for spiritual progress.

A less pleasant consequence is disorientation in time: the convolution and the "arrow of time" point in opposite directions. Therefore, an aging person "falls into childhood", is prone to nostalgia, remembers the distant past better than recent events, tends to evaluate the past better than the present, and shows excessive conservatism and stubbornness.

We conclude this part with the following remark. There is a practical need to calculate the interdependence of biomarkers, determine the optimal number of them, display the results of biomarker analysis, and help diagnose the result. In all these cases, using HEM and SWM is advisable, since the models are non-additive, which makes it possible to combine indicators of different quality with different units of measurement within one model due to its calibration (Chapter 8).

LIFE CYCLE MODELING IN GENERAL

Modeling of the life cycle based on the relationship of wave and soliton is shown in Figure 4 of Chapter 4. Both options are suitable for modeling. We will use the option on the left for the image's convenience. Let us show a graph of the change in the value of the *SpA* and *Sp P* potential, which determines the course of the sweep scenario *(1S, 1O)* and its inverse convolution, including the theoretical life span (corollary 1 of property 4.9, chapter 4). Recall that the potential is a dimensionless quantity, calibrated depending on the conditions of the problem.

The total potential of *SpA* manifests itself in the formation of the human body and psyche, ensuring the functioning and resistance of the organism throughout life. It is interpreted as an area bounded by the graph of changes in the wave and soliton amplitude values, and the axis shows the lifetime. The current total value of the amplitudes is the potential that can be used at one or another moment of life. We accept the same signature •. Since the variety of factors, and the modeled characteristics are huge we will use the iteration method, moving from simpler models to more complex ones.

Iteration 1 (Figure 9). We will show the theoretical life expectancy and designate the age periods following Table 3.

Iteration 1 shows a rapid increase of the potential *SpA* during pregnancy and early childhood - it is enough for the formation of the new structure; a decrease in its growth rate to zero in maturity - here it is enough to maintain existing structures and ensure reproduction; a progressive reduction in old age when it is not enough to keep vital parameters and compensate the pathologies.

Iteration 2 (Figure 10). Let's consider the data on the aging of systems and organs, shown in Figure 6, and the actual life expectancy. Unfortunately, these data are not enough to accurately calibrate the model. Death can occur much earlier if one of the vital organs is affected. The same applies to the mental component - a distraught person in natural conditions is doomed to a quick death; in a humane society, of course, he can live much longer.

Following the progressive decrease in Sp A potential, signs of aging are observed: muscle atrophy, reduced growth, and so on. Death occurs due to a such disruption in functioning, which cannot be compensated. And that destroys the integrity of the SWM. The soliton component continues to exist after the destruction of the integral system - this explains the fact that the body does not disappear after death.

Iteration 3 (Figure 11). Let us introduce the next portion of refinements - we will show interactions with the environment. They can be beneficial or destructive. The former increases the potential and reduces it, demanding compensation. In addition, external influences are necessary to construct psychophysiological structures (Chapter 5, Figure 8). We also consider and show the following:

- During pregnancy and childhood, the most favorable external conditions are provided, and the potential increases rapidly.
- A person can form useful or bad habits, and fall into favorable or unfavorable conditions, which also affects potential changes.

Figure 9. Life cycle SWM, iteration 1. The value of the potential Sp **A** *is shown conditionally*

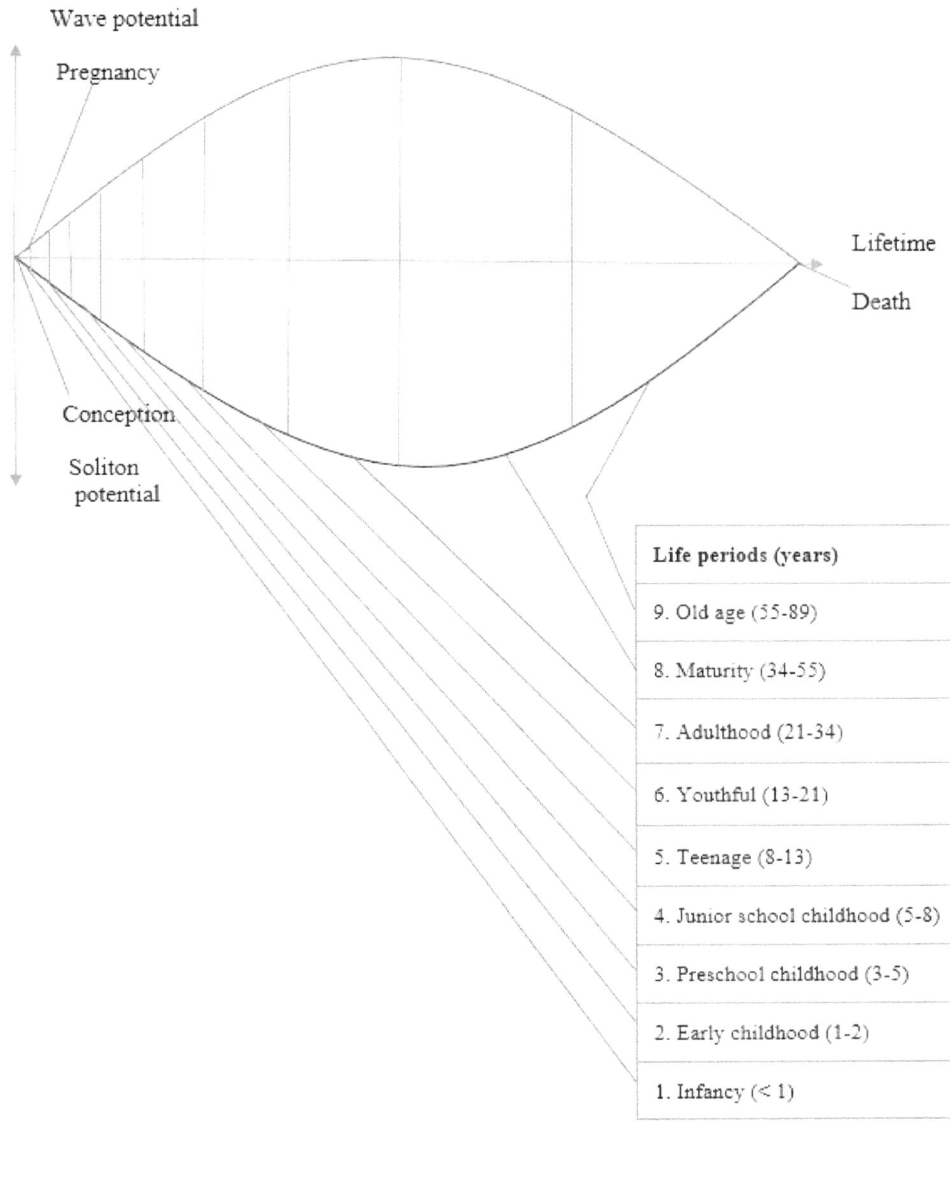

Figure 10. Life cycle SWM, iteration 2. The minimum potential of the Sp **A**, *which required the capacity to maintain functionality is shown in green*

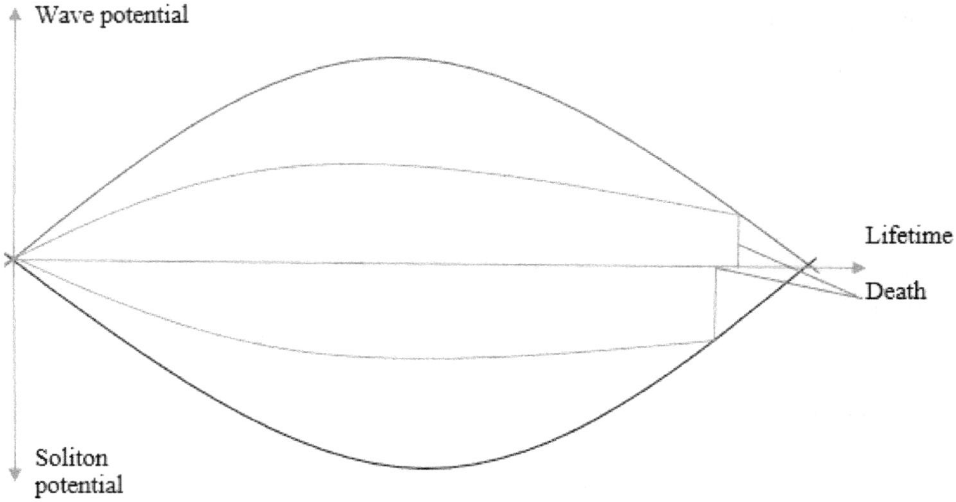

Figure 11. Life cycle SWM, iteration 3. The minimum potential of the Sp **A**, *which required the capacity to maintain functionality is shown in green. External influences are shown conditionally and are indicated by contour arrows.*

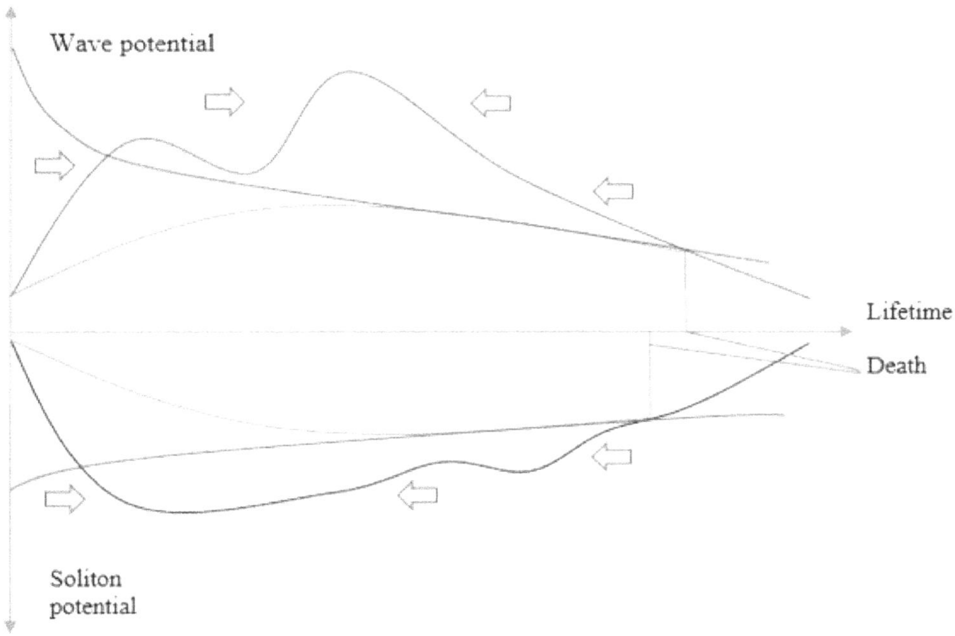

So, due to asymmetric external influences, various asymmetries of the soliton-wave pair arise, which, as noted above, is one of the properties of living systems.

Note that for the mental component, the correct organization of interactions with the environment is necessary - through the intuition, ego, etc channels; powerful intuitive interaction is manifested along the mother-child line.

Iteration 4 (Figure 12). Finally, let's consider the resonance associated with conception, as suggested in Chapter 6. This is the undistributed potential *SpP*, which has not yet expressed itself in forming any structures of the body or psyche but creates the potential possibility of their appearance. It can also be used to compensate for destructive external influences and pathologies. Let us draw a graph of its change separately for the wave and soliton components of the SWM, denoting them as a violet line.

The soliton and the wave already at the moment of gamete fusion have a non-zero potential, and they are consistent with a certain error; after birth, this inconsistency can lead to the destruction of the integrity of the system, which may explain the phenomenon of infant death, which has no apparent cause.

Figure 12. Life cycle SWM, iteration 4. The minimum potential of the Sp **A**, *which required the capacity to maintain functionality is shown in green. The purple line shows the change in the Sp* **P** *potential after the supposed resonance. Arrows indicate external influences.*

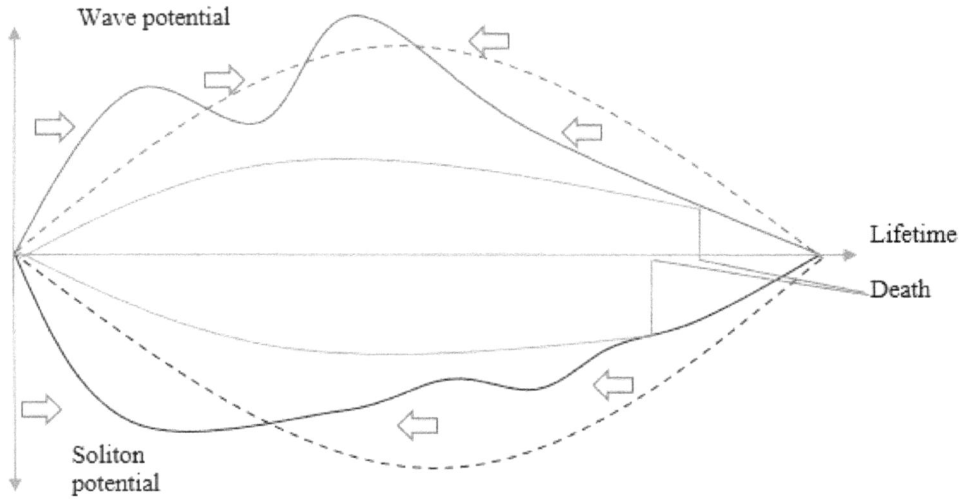

So, SWM iterations 1-4 correspond to the available data on the human life cycle. Their further refinement and development using the OM2 and OM3 models are possible.

THE METHODS OF LIFE EXTENSION

The presence of iterations makes it possible to evaluate the methods of life extension described earlier. They all come down to 3 strategies, many of which have ancient analogs.

Strategy 1: A healthy lifestyle is aimed at increasing the potential of the soliton component to move the date of death as close as possible to the theoretical limit. The goal is achieved through a balanced diet, special exercises, and favorable conditions. This also applies to medicine, especially preventive medicine. At the same time, much less attention is paid to the mental component, even though its degradation also leads to death.

Here is the time to recall the ancient ideal of beauty - "kallos kagatos" - when both components should have been perfect and harmonious, or about yoga, when physical development (including nutrition and the choice of favorable conditions) was a prerequisite for spiritual development.

Strategy 1 cannot change the wavelength or the soliton and push back the theoretical limit of life.

Strategy 2: Proper conception aims to increase the length of the soliton and its potential through the selection of parents and the time of conception.

According to the hypothesis proposed in Chapter 4, these activities are interpreted as creating conditions for the emergence of resonance.

In antiquity, this strategy was given much more attention than it is now, and very sophisticated means of achieving the goal were developed. The following may be noted:

- Caste division of society and marriages within the same caste ensured the selection of parents with similar potentials and other characteristics.
- Careful consideration of astrological data - an unscientific method with a completely rational goal - to choose the right time for conception.
- Carrying out special rituals and cleansing practices is also an unscientific method of increasing potential by attracting divine grace.

Modern eugenics studies (and genetic manipulations can be attributed to them) approach the problem from the other side, and, in addition to the selection of parents, do not consider other nuances.

Strategy 3: Creating a cyborg involves implanting or replacing vital organs with mechanical counterparts as they age and using various chips to help the brain work. With the progress of science and technology, theoretically, a mechanical analog can replace the entire body, and consciousness can be replaced by a kind of computer that stores data about the personality of the former person in memory and implements the algorithms of his behavior. Then the cyborg will turn into a robot.

It is curious how the most worldly and even technological strategies implicitly recognize the existence of a subjective space (programs, data) independent of the bodily (mechanical systems) component! Moreover, he believes that he will exist for an arbitrarily long time with the right substrate, retaining the main personality traits!

Some prototypes of such a strategy were known to Egyptian priests, and it seems they understood the problem's essence better. While mummifying and preserving the body, the priests, with the help of rituals, laid in it a "program" that made it possible to act and speak, but at the same time they created another body adapted to life in the afterlife. They also believed that in the afterlife, consciousness changes, gradually becoming no longer human, but one that, together with a new body, is worthy of being with the gods.

So both the cyborg and the robot, despite the initial human consciousness, will gradually cease to be human personalities.

The death of a person is the irreversible destruction of a soliton-wave pair; without their unity, a full life is impossible; none of the listed strategies provides for preserving such unity. Therefore, strategies 1-3 in their current form should be viewed skeptically.

A system **Strategy 4** is needed. Within the framework of the definition of life given in Chapter 4 and the understanding of human structure presented in Chapter 8, strategy 4 should ensure:

- The synergy of soliton (substance) and wave (field) components and the emergence of a system.
- Evolution of the emergent system with the formation of the ego level as the basis of personality.
- Modification and replacement of elements of lower levels within the lifetime of the ego.

This is how to reproduce the quality of life and preserve the personality foundations.

On the fractality of the SWM life cycle. We have repeatedly mentioned fractality as a principle for constructing complex systems[1]. Let's see if the self-similarity of SWM occurs when modeling species' life cycle, having sexual reproduction, and higher taxa, using iteration 4, Figure 11.

At the beginning of the cycle, the potential is sufficient to implement allogenesis, which leads to the emergence of a new species. The rate of evolution is very high, which can explain the "missing links problem" - there are few leftovers, and it is very problematic to find them. Further, the potential is no longer enough for significant changes, the rate of evolution decreases, and if the species maintains an equilibrium state with the environment, it can exist practically without changes for quite a long time. However, the magnitude of the potential decreases faster and faster. When the current potential is no longer enough to maintain equilibrium, the species dies out even for no apparent reason. As we can see, self-similarity for species exists.

Similarly, for higher taxa - at the beginning of their formation, the potential is sufficient for rapid arogenic changes, as well as the formation of lower taxa (here, the scenario of sweep *(1S, {O})*, Chapter 4) is implemented with the formation of a multilevel system, for example, class - families - genera - kinds). Further, the rate of evolution falls, and this whole system exists for a long time, adapting in the direction of an equilibrium existence with the environment. However, when the time of existence of the wave corresponding to the first level of the system ends, all subsequent levels also collapse (corollary 1 of property 4.9, Chapter 4).

Outwardly, this manifests itself as a mass extinction of species, genera, families, etc., which has repeatedly happened in the history of the Earth (Chapters 1, 5). External catastrophic impacts trigger another extinction mechanism, destroying ecological systems (Chapter 5).

Thus, the life cycle of higher taxa is also similar to the human life cycle.

FUTURE RESEARCH DIRECTIONS

An immediate goal will be to address near-death states (Chapter 10) using SWM. The detailing of the models is due to the current tasks of the work. Let's talk about research beyond the scope of this work. Soliton-wave models seem appropriate for detailed modeling of various structures and processes of the entire human life cycle in cases where it is impossible to neglect their system properties and interactions with the environment as open complex systems. Auspicious (but also technically difficult) is the use of OM2 and OM3, which were mentioned here.

CONCLUSION

The soliton-wave model is applicable for modeling the main life cycle from conception to senility; applicability is due to the nature of the human body as a complex open system and the correspondence of the S-space axiomatics to the properties of such systems. The accuracy of modeling corresponds to the tasks of the work. If necessary, it can be increased by using models of the OM2 and OM3 types, which provide better dynamics and the possibility of simulation.

We note the modeling of the aging process, which makes it possible, due to the consistency and non-additivity and the possibility of calibration, to track changes of different quality at different levels within the same model. This corresponds to modern ideas about the systematic nature of aging, and changes in energy consumption and other resources.

The theoretical significance of SWM lies in a better understanding of the aging process as a partial convolution, which is not the same for the soliton (substance) and wave (field, mental) components, which explains several phenomenological manifestations (uneven aging, pathologies, disorientation in time, and so on). The aging causes are also better understood the influence of the "age" of a species on the course of ontogenesis, the role of systemic factors, evolution as self-organization, environmental influences, and especially modulation during conception.

The practical use of SWM is:

- The age periods and their psychological and physiological characteristics clarification are important for designing a comfortable environment.
- The ability to calculate the correspondence between various parameters of the organism, including determining the necessary and sufficient set of biomarkers.
- Classification and refinement of life extension strategies and selection of appropriate means.
- Better understanding of the evolutionary process of taxa at different levels.

And finally, the last addition to the definition of a living system given in Chapter 4:

A living system is:

1. (Invariant part): an existing boundary Sp, with potential π and modality A, arousing under the influence of external impulses in a dissipative environment, corresponding to the abstractions and axioms of the S- space model, including soliton and wave components (which modalities are æ, ö or ↓), developing according to Sp self-organization scenarios, and
2. (Calibration part): calibrated by conditions specific to possible bases for the presence of substances and fields necessary for life, as well as conditions for the feasibility of reactions needed for life, energy sources, self-organization mechanisms, and
   3) (Life cycle part): being in one of the life cycle stages.

REFERENCES

Ageing. (2024, August 11). Wikipedia. https://en.wikipedia.org/wiki/Ageing

Aging changes in organs, tissue, and cells. (n.d.). *Medical Encyclopedia*. https://medlineplus.gov/ency/article/004012.htmDOI: 10.1098/rstb.1991.0035

Blum, G. S. (1953). *Psychoanalytic theories of personality*. McGraw-Hill Book Company.

Bratic, A., & Larsson, N.-G. (2013). The role of mitochondria in aging. *The Journal of Clinical Investigation*, 123(3), 951–957. DOI: 10.1172/JCI64125 PMID: 23454757

Burkle, A. (2001). Physiology and pathophysiology of poly(ADP-rebosyl)ation. *BioEssays*, 23(9), 795–806. DOI: 10.1002/bies.1115 PMID: 11536292

Carrard, G., Bulteau, A. L., Petropoulos, I., & Friguet, B. (2002). Impairment of proteasome structure and function in aging. *The International Journal of Biochemistry & Cell Biology*, 34(11), 1461–1474. DOI: 10.1016/S1357-2725(02)00085-7 PMID: 12200039

Drenos, F., & Kirkwood, T. B. L. (2005). Modelling the disposable soma theory of aging. *Mechanisms of Ageing and Development*, 126(1), 99–103. DOI: 10.1016/j.mad.2004.09.026 PMID: 15610767

Erikson, E. H. (1950). *Childhood and Society*. W. W. Norton & Co.

Erikson, E. H., & Erikson, J. M. (1987). *The Life Cycle Completed*. W. W. Norton & Co.

Erikson, E. H., Erikson, J. M., & Kivnick, H. Q. (1986). *Vital involvement in old age: The experience of old age in our time*. W. W. Norton & Company.

Finch, C. (1990). *Senescence, Longevity, and the Genome*. The University of Chicago Press.

Flatt, T., & Schmidt, P. S. (2009). Integrating evolutionary and molecular genetics of aging. *Biochimica et Biophysica Acta*, 1790(10), 951–962. DOI: 10.1016/j.bbagen.2009.07.010 PMID: 19619612

Frankel, S., & Rogina, B. (2021). Evolution, Chance, and Aging. *Frontiers in Genetics*, 12, 733184. Advance online publication. DOI: 10.3389/fgene.2021.733184 PMID: 34567083

Frenzel, A., Binder, H., Walter, N., Wirkner, K., Loeffler, M., & Loeffler-Wirth, H. (2020). The aging human body shape. *NPJ Aging and Mechanisms of Disease*, 6(1), 5. Advance online publication. DOI: 10.1038/s41514-020-0043-9 PMID: 32218988

Gavrilov, L. A., & Gavrilova, N. S. (2002). Evolutionary Theories of Aging and Longevity. *TheScientificWorldJournal*, 2, 339–356. DOI: 10.1100/tsw.2002.96 PMID: 12806021

Gompertz, B. (1825). On the Nature of the Function Expressive of the Law of Human Mortality, and on a New Mode of Determining the Value of Life Contingencies. *Philosophical Transactions of the Royal Society of London*, 115, 513–585. DOI: 10.1098/rstl.1825.0026

Hachmo, Y., Hadanny, A., Abu Hamed, R., Daniel-Kotovsky, M., & Catalogna, M. (2020). Hyperbaric oxygen therapy increases telomere length and decreases immunosenescence in isolated blood cells: a prospective trial. *Research Paper 12*(22), 22445-22456. DOI: 10.18632/aging.202188

Harman, D. (1972). The biologic clock: The mitochondria? *Journal of the American Geriatrics Society*, 20(4), 145–147. DOI: 10.1111/j.1532-5415.1972.tb00787.x PMID: 5016631

Harman, D. (1981). The aging process. *Proceedings of the National Academy of Sciences of the United States of America* 78(11), 7124-7128. DOI: 10.1073/pnas.78.11.7124

Hayflick, L. (1998). How and why do we age? *Experimental Gerontology*, 33(7–8), 639–653. DOI: 10.1016/S0531-5565(98)00023-0 PMID: 9951612

Hollis, J. (1993). *The Middle Passage: From Misery to Meaning in Midlife*. Inner City Books.

Human Physiology/Development: birth through death. (2022, August 15). Wikibooks, Open Books for an Open World. https://en.wikibooks.org/wiki/Human_Physiology/Development:_birth_through_death

Kirkwood, T. B. L. (1977). Evolution of aging. *Nature*, 270(5635), 301–304. DOI: 10.1038/270301a0 PMID: 593350

Kirkwood, T. B. L. (2005). Understanding the Odd Science of Aging. *Cell*, 120(4), 437–447. DOI: 10.1016/j.cell.2005.01.027 PMID: 15734677

Kovalyov, Y., Mkhitaryan, N., & Nitsyn, A. (2020). *Self-organization of the human mind and the transition from paleolithic to behavioral modernity*. IGI Global. DOI: 10.4018/978-1-7998-1706-2

Krut'ko, V. N., Dontsov, V. I., Ermakova, N. A., Makarova, V. V., Mitrokhin, O. V., Shashina, E. A., & Shcherbakov, D. V. (2021). Method and Computer System for Dialog Optimization of Aging Biomarker Panels for Biological Age Assessment. *Frontiers in Genetics*, 12, 634734. Advance online publication. DOI: 10.3389/fgene.2021.634734 PMID: 33747048

Le Bourg, E. (2001). A mini-review of the evolutionary theories of aging. *Demographic Research*, 4, 1–28. DOI: 10.4054/DemRes.2001.4.1

Magee, P. T. (1976). Meiosis in protists. Some structural and physiological aspects of meiosis in algae, fungi and protozoa. *Bacteriological Reviews*, 40(1), 190–240. DOI: 10.1128/br.40.1.190-240.1976 PMID: 773364

Marple, K. (2024, February 20). *Fetal development week by week*. BabyCenter. https://www.babycenter.com/pregnancy/your-baby/fetal-development-week-by-week_10406730

Masoro, E. J., & Austad, S. N. (Eds.). (2006). *Handbook of the Biology of Aging* (6th ed.). Academic Press.

Medawar, P. B. (1952). *An unsolved problem of Biology*. H.K Lewis.

Mueller, L. D., & Rose, M. R. (1996). Evolutionary theory predicts late-life mortality plateaus. *Proceedings of the National Academy of Sciences of the United States of America*, 93(26), 15249–15253. DOI: 10.1073/pnas.93.26.15249 PMID: 8986796

Nelson, P., & Masel, J. (2017). Intercellular competition and the inevitability of multicellular aging. *Proceedings of the National Academy of Sciences of the United States of America*, 114(49), 12982–12987. DOI: 10.1073/pnas.1618854114 PMID: 29087299

O'Rahilly, R., & Müller, F. (2010). Developmental stages in human embryos: revised and new measurements. *Cells Tissue Organs, 192*(2), 73-84. DOI: 10.1159/000289817

Olovnikov, A. M. (1973). A theory of marginotomy: The incomplete copying of template margin in enzymic synthesis of polynucleotides and biological significance of the phenomenon. *Journal of Theoretical Biology*, 41(1), 181–190. DOI: 10.1016/0022-5193(73)90198-7 PMID: 4754905

Shaw, F. H., Promislow, D. E. L., Tatar, M., Huges K. A., & Geyes, C. J. (1999). Toward reconsiling inferences concerning geetic variations in Drosophila melamogaster. *Genetics - Genetics Society of America, 152,* 553—566.

Theories of aging. (n.d.). An introduction to aging science brought to you by the American Federation for Aging Research. https://www.afar.org/imported/AFAR_INFOAGING_GUIDE_THEORIES_OF_AGING_2016.pdf

Tian, Y. E., Cropley, V., Maier, A. B., Lautenschlager, N. T., Breakspear, M., & Zalesky, A. (2023). Biological aging of the human body and brain systems. *Nature Medicine*, 29(5), 1221–1231. DOI: 10.1038/s41591-023-02296-6 PMID: 37024597

Valero, M., Richerd, S., Perrot, V., & Destombe, C. (1992). Evolution of the alternation of haploid and diploid phases in life cycles. *Trends in Ecology & Evolution*, 7(1), 25–29. DOI: 10.1016/0169-5347(92)90195-H PMID: 21235940

Vina, J., Borra's, C., & Miquel, J. (2022). Theories of Ageing: Critical Review. *IUBMB Life 59*(4 – 5), 249 – 254. https://iubmb.onlinelibrary.wiley.com/doi/pdf/10.1080/15216540601178067

Von Zglinicki, T. (2002). Cell Press. *Trends in Biochemical Sciences*, 27(7), 339–344. DOI: 10.1016/S0968-0004(02)02110-2 PMID: 12114022

Wang, C., Liu, C., Gao, H., & Liu, H. (2017). Order of aging of major human organs or systems and evaluation of health status based on aging. *Biomarkers*, 22(2), 145–148. DOI: 10.1080/1354750X.2016.1252951 PMID: 27800690

Weinberg, S. (1993). *The First Three Minutes: A Modern View of the Origin of the Universe*. Basic Books.

Zephyris. (n.d.). The first few weeks of embryogenesis in humans. Beginning at the fertilized egg, and ending with the closing of the neural tube. https://commons.wikimedia.org/wiki/File:HumanEmbryogenesis.svg

ADDITIONAL READING

Dilman, V. M. (1972). Pochemu nastupaet smertj [Why death occurs]. *Medicine*.

Elkonin, D. B. (2007). *Detskaja psykhologhyja* [Child psychology]. Publishing Center "Academy".

Frolkis, V. V. (1988). Starenye y uvelychenye prodolzhyteljnosty zhyzny [Aging and increase in life expectancy]. *Science*.

Galitsky, V. A. (2009). Epyghenetycheskaja pryroda starenyja [Epigenetic nature of aging]. *Tsitologia*, 51, 388–397.

Gamova, L. G. (2010). *Vozrastnaja anatomyja y fyzyologhyja rebenka* [Age-related anatomy and physiology of a child]. YSU im. I. A. Bunina.

Granovich, A. I., Ostrovsky, A. N., & Dobrovolsky, A. A. (2010). Morfoprocess y zhyznennыe cykly orghanyzmov [Morphoprocess and life cycles of organisms]. *Journal of General Biology*, 71(6). PMID: 21268862

Khramov, A. (2017). Uchenye zajavyly o nepobedymosty starenyja [Scientists have declared the invincibility of aging]. *INFOX.ru*. https://m.infox.ru/news/232/186580-ucenye-zaavili-o-nepobedimosti-starenia

Khukhlaeva, O. V. (2006). Psykhologhyja razvytyja. Molodostj, zrelostj, starostj [Psychology of development. Youth, maturity, old age]. *Academia (Caracas)*.

Krasnova, O. V., & Lidere, A. G. (2003). *Psykhologhyja starosty y starenyja* [Psychology of old age and aging]. Publishing Center "Academy."

Lysova, N. F., & Aizman, R. I. (2018). *Vozrastnaja anatomyja y fyzyologhyja* [Age anatomy and physiology]. INFRA-M.

Lyubimova, Z. V., & Nikitina, A. A. (2019). *Vozrastnaja anatomyja y fyzyologhyja* [Age anatomy and physiology]. Yurayt Publishing House.

Mechnikov, I. I. (1964). *Etjudy optymyzma* [Etudes of optimism]. Nauka. (Original work published 1907)

Medvedev, D. A., Artyukhov, I. V., & Evfratov, S. A. (n.d.). Proekt «Systemnaja teoryja starenyja» [System Theory of Aging Project]. http://sciencevsaging.org

Medvedev, D. A., Artyukhov, I. V., & Evfratov, S. A. (n.d.). Proekt «Systemnaja teoryja starenyja». Byomarkery starenyja [System Theory of Aging Project. Biomarkers of aging]. https://sciencevsaging.org/content/popular/b

Mkhitaryan, N. M., Badeyan, G. V., & Kovalyov, Y. N. (2004). *Erghonomycheskye aspekti slozhnikh system* [Ergonomic Aspects of Complex Systems]. Naukova Dumka.

Mkhitaryan, N. M., Kovalyov, Y. M., Malik, T. V., Safronov, V. K., & Safronova, O. O. (2021). *Dyzajn seredovyshha mista: Baghatokryterialjna optymizacija ta rozumni tekhnologhiji* [Design of the City Environment: Multi-Criteria Optimization and Smart Technologies]. Naukova dumka.

Monakhova, M. A., Akimova, N. I., & Kokaeva, Z. G. (2018). Ghenetycheskye y epyghenetycheskye mekhanyzmy starenyja [Genetic and epigenetic mechanisms of aging]. *ResearchGate*. https://www.researchgate.net/publication/329101904_geneticeskie_i_epigeneticeskie_mehanizmy_aging

Nikityuk, B. A., & Chtetsov, V. P. (Eds.). (1990). *Morfologhyja cheloveka* [Human morphology]. 2nd ed.). Publishing House of Moscow State University.

Novoseltsev, V. N., Novoseltseva, Zh. A., & Yashin, A. I. (2003). Matematycheskoe modelyrovanye v gherontologhyy - strateghycheskye perspektyvy [Mathematical modeling in gerontology - strategic perspectives]. *Uspekhi Gerontologii*, 12, 149–165. PMID: 14743613

Sergeev, I. Yu., Dubynin, V. A., & Kamensky, A. A. (2020). *Fyzyologhyja cheloveka y zhyvotnykh* [Physiology of man and animals]. Yurayt Publishing House.

Tyurikova, G. N., & Tyurikov, Yu. B. (2018). *Anatomyja y vozrastnaja fyzyologhyja* [Anatomy and age-related physiology]. INFRA-M.

ENDNOTE

1 (Bryusov, 1922). (trans. Timeyev, A.):

What if these things we call electrons
Are worlds with continents and seas,
And art and craft, and war and peace,
And scrolls and scrolls of memories?
And Universe in every atom,
With fivescore planets circling round -
With all the things that here abound,
But also things we've yet to fathom?
Minute in size, their dwellers follow
Same endless prospects, starry-eyed,
Same passions share and griefs, and wallow
To same degrees in worldly pride.
Their scholars set the whole creation
Centered within their native shores,
And strive to find the true equation,

And muse aloud, like truly yours.
And when through self-destroying fusion
Their tiny world new force emits,
They see, in cruel self-delusion,
Lord's lantern shattered into bits.

Chapter 10
Near the Death

ABSTRACT

The events preceding the onset of death and occurring after its ascertainment, the medical definition of death, the causes of its onset, and the problem of biological immortality are considered. A soliton-wave model of near-death processes has been constructed. Within the framework of the model and from the standpoint of the theory of self-organization, the definition of the death of the system, its causes, and the scenarios of subsequent events for the components of the system are considered. The results obtained are practically important for medicine and a comprehensive understanding of the problem of biological immortality. Simulation results were compared with known data; their correlations are shown.

BACKGROUND

For subsequent modeling, the following questions are of interest.

How is the onset of death determined in official documents and what are its causes? What is the significance of near-death experiences, and how is this related to meditative states? What is the modern view on the problem of biological immortality? Can the symptoms of dying be linked to the aging scenario (Figure 7, Chapter 9)?

Extensive literature is devoted to the problem of near-death states.

The symptoms of the dying process are well represented in the literature published by palliative care organizations (*The Dying Process - Palliative Care Australia*, n.d.; *The Process of Dying*, n.d.).

From a physiological point of view, near-death states have been studied for a long time, and the necessary data are available in many modern textbooks on pathological anatomy and pathological physiology (Strukov & Serov, 2010; Paltsev & Anichkov, 2001; Paukov et al., 2015; Lyubimova & Nikitina, 2019; Ado et al., 2020; Voinov, 2004). In animals, many pathological processes are similar to those in humans (Zharov, 2021).

DOI: 10.4018/978-1-6684-8509-5.ch010

In the medical literature, two topics are of interest: the immediate causes of death (*The Top 10 Causes of Death*, 2024; Leonov, 2008), as well as signs of death and protocols for its ascertainment (Sorokin, 2004; Walker, 1978; Lizza, 2018; Dachet et al., 2021); Bernat, 2013; Clinical Criteria, 2014; Instruction Nº 73, 2003).

The psychological aspect of dying is represented in (Kübler-Ross, 2008; Twycross, & Wilcock, 2016; Shermer, 2017), among which we note works devoted to the analysis of near-death experiences (Alcock, 1981; Moody, 1988; Kurtz, 1991), especially the point of view of practicing reanimators (Gubin, 2002).

It also seems natural to compare near-death experiences with meditative states (Ebert, 1986).

It should be noted that a special science, thanatology, deals with the scientific analysis of various aspects of death (Ryazantsev, 1994).

Biological immortality corresponds to two facts - the number of divisions of some cells is not limited by the Hayflick limit; as a result of asexual reproduction, the daughter cell is genetically identical to the mother cell. Most species are biologically immortal - which does not mean physical immortality: they die when eaten by other organisms, from adverse environmental factors, and so on. These are all prokaryotes and unicellular eukaryotes; colonial organisms (Martinez, 1998), and some sea urchins (Pristavka, 2021). The absence of biological aging is expressed in the absence of an increase in mortality, starting from a certain age. Complex organisms, including humans, can die of old age. Hovever, a human death from old age is not indicated in the medical lists of causes of death. Biological immortality remains controversial in terms of its existence (Hallyday, 2009; Olshansk et al., 2004) and implementation mechanisms (Landecker, 2000; Agarwal & Rimm, 2012).

BEFORE THE DEATH

The Signs of Natural Death

Interest in signs that make it possible to predict the time of death appeared in ancient times; the current data noted in the review are related to palliative care and hospice care. The statistical sample is large enough and the symptoms should be considered reliable.

3 stages of approaching death:

- Terminal stage (up to 1-2 years): a gradual decrease in the functioning of systems and individual internal organs, a feeling of constant fatigue.
- Pre-final stage (several weeks or months): the appearance of shortness of breath and pain.

- Final stage (last days): the dying person stops eating and drinking and slowly dies away.

However, individual variations are possible depending on the pathological picture (one example will be discussed below) and in the case of injuries incompatible with life. The latter case will not be analyzed here.

Predicting the exact time of death is almost impossible. According to (The Dying Process, 2020; Harbingers of imminent death[REMOVED HYPERLINK00 FIELD], 2019; Signs of impending death, n/d), such signs are observed near death:

- Change in lifestyle, daily routine, and behavior - there are early signs (several months before death).
- Impaired perception (usually 3-4 weeks before death).
- The person eats little or no food (usually 3-4 weeks before departure).
- Unable to swallow (a few days before death).
- A person is awake only 5-6 hours a day, the rest of the time he sleeps; he sleeps for days a few days before death.
- The person does not want to communicate with anyone.
- A person constantly complains of weakness, at first refusing to walk around the room, then sits down with difficulty, then does not want to move at all.
- The pattern of behavior changes drastically - unreasonable aggression, pity, and so on appear; the very fact of change is important here.
- A person begins to talk to someone not seen by others, usually long-departed parents, wife (husband), or other close people.
- A person's consciousness is clouded - at times he does not recognize others, or the place where he is, and speech resembles delirium (60-72 hours before death).
- Changes in breathing, frequency of urination and defecation, and urine color.

Shortly before death are observed:

- Black vomit.
- Emptying of the bladder and intestines.
- Clouding of the cornea.
- Drooping of the lower jaw, the mouth is open.
- The pulse is too slow or cannot be felt.
- Pressure is minimal.
- Jumping temperature indicators.
- There is noisy breathing and wheezing.

There are also secondary signs of imminent death: the nose becomes sharp, and eyes and temples fall. The ears become cold and flaccid, the tips turn forward, the face becomes symmetrical, the skin becomes grayish or yellowish, and the skin of the forehead becomes tight and rough. Before death, a person loses the ability to hear, due to a sharp pressure drop. Instead of the usual sounds, he hears a squeak, a strong ringing, extraneous sounds. A dying person hides his gaze from the light, his eyes are very watery, mucus accumulates in the corners, the proteins turn red, and the vessels in them turn white, often the right eye differs in size from the left.

According to surveys of hospice staff (Harbingers of imminent death, 2019) there are observed:

- Detached look turned inward (seen by 90% of respondents).
- The face of a person about to die became absolutely symmetrical (82% of respondents).
- Gray plaque on the face, which appears and disappears (45% of respondents).

Less reliable are:

- Dim or disappeared glow of the aura (seen by 7% of respondents).
- Blurring of facial features - only those who know the person well notice.
- The smell of death – a vague scent identified by some interviewees in this way and usually predicts death within 1-2 weeks.
- A premonition of death - sometimes a direct prediction of the date of one's death, occasionally conscious actions aimed at preparing for death - putting finances in order, saying goodbye to friends, performing religious rituals, meditation, and the like.

Let's name a few more individual signs:

- The desire of dying people to straighten their blankets and clothes, straighten them, and move their hands around themselves, as if clearing the body and space from non-existent straws and threads. Some try to throw off the veil or use gestures to ask others to take off their clothes.
- A temporary improvement in the condition (pre-mortem improvement or neurochemical oscillation) may indicate the approach of death. More common in cancer patients. The cause of this condition is unclear; perhaps it lies in the fact that the body fights the disease to the last, mobilizing forces and resources at the expense of other functions - motor, motor, etc. - but up to a certain limit. When it comes, the protection is turned off. There is a kind of

surge of strength, a person becomes active, mobile, and talkative, but after a few hours' death occurs.

- Hallucinations - when dying, brain cells begin to gradually die due to oxygen starvation (hypoxia) and a person has visual, auditory, gustatory, olfactory, and tactile hallucinations. Most often, visions are associated with religious concepts. Others see beautiful landscapes, rare fauna, and flora. Still, others talk to deceased relatives. It is believed that the nature of hallucinations is not associated with the form of the disease, age; religious preferences; individual characteristics; education; or level of intelligence[1].
- Venous or cadaveric spots - areas of the body soaked in blood - occur before death, when dying, and within a few hours after death due to blockage of blood circulation.
- Edema appears on the lower and upper limbs due to a global disruption of kidney function.
- Death rattle resembles crackling, murmuring, and blowing air out of the lungs through a straw to the bottom of a mug with water. On average, 16 hours pass from the onset of this phenomenon to the onset of death. This is a sign of impaired swallowing function.

Psychological and emotional states. The classic sequence of psychological reactions to death was described by E. Kubler-Ross in 1969 (Kubler-Ross, 2001). She identifies five emotional states through which a person passes after a fatal diagnosis.

The first of these is the stage of denial. "No, it's a mistake, it's not me!" – is the most common reaction of a person to the announcement of a fatal diagnosis.

Then comes the stage of anger, engulfing a person who asks questions: "Why me? Who is to blame here?" Anger is directed at people, who care about him, and any healthy person. To complete this stage, the dying person must have the opportunity to pour his feelings outward.

The next stage is "bargaining" - a person tries to prolong his life, promising to fulfill all the recommendations of doctors or bargaining for his life with the help of repentance before God for his sins. These three phases constitute a period of crisis and develop either sequentially or iteratively.

After this crisis, the dying person enters the stage of depression, realizing: "Yes, I will die," withdraws into himself, and often feels the need to cry out of pity for himself. This is the preparation for realizing the need to renounce life and readiness to meet death (the state of "social death").

The fifth stage is the "acceptance of death": a person realizes and comes to terms with the inevitability of imminent death (the state of "mental death"). Note that the state of humility, from the point of view of Christianity, is the most important condition for the forgiveness of sins and spiritual progress.

Kübler-Ross considered all five stages to be protective mechanisms of the psyche. In addition to the five main ones, she identified intermediate states - the first shock, preliminary grief, hope, and so on - from 10 to 13 types. However, the research stage that the authors are currently undergoing is missing. More seriously, the incompleteness of this scheme is pointed out by both followers (suggesting other stages) and critics (noting excessive generalization) (Bonanno & Trauma, 2004).

Based on the practice of observations in hospices (Harbingers of Imminent Death, 2019), we can distinguish the following emotional changes:

- Loss of interest in the outside world. It manifests itself as a limitation of the circle of communication, a reassessment of things and events that previously seemed important.
- Loss of spatial and temporal orientation, for example, events of the recent past are combined with very old ones, or very old events are remembered in the smallest detail, but the person may not remember what happened an hour ago; a person outside the context of the situation says that he needs to go home or go somewhere far away.
- Talking about people who have already died, as if they were alive - that the dying person heard or saw them recently, right here, conversations with them.
- The older the person, the less he is afraid to die. Some people want to bring death closer so that their loved ones do not see how they suffer.
- Before death, the elderly have indifference, discomfort, and sometimes pain.
- However, every 20th person has a spiritual upsurge.

Note that medications such as morphine, sedatives, and painkillers, an overdose of drugs, and metabolic changes associated with high temperature, dehydration, and self-poisoning are superimposed on this picture and their influence can be difficult to distinguish.

What conclusions follow from the described picture?

1. The system of the human body is no longer simply degrading, (Chapter 6), but is being destroyed;
2. Three scenarios (priority degradation of the material or field components or their relatively symmetrical degradation) take place, but they differ worse - they all approach a common finale - death - and then its symptoms prevail, which are different from the signs of old age;
3. An increase in space-time disorientation can be interpreted as a rapidly growing asymmetry of the substance and field components and confirmation of the different rates of their convolution, which was also noted in old age (Chapter 6);

4. Destruction is accelerating: the terminal stage lasts 1-2 years, the pre-final stage - months and weeks, the final stage - days;
5. One of the mechanisms of accelerated destruction is negative feedback that forms "vicious circles", for example, general weakness causes a lack of movement, and a lack of movement causes degradation of the musculoskeletal system (as well as respiratory, circulatory, and digestion and many others), which, in turn, increases the general weakness. All of this is consistent with Axiom 2.1 (Chapter 2).

Death Due to Illness

Now let's see how they affect the picture of the dying of the disease. Their number and variety are so great that medical reference books contain hundreds and thousands of pages (Chirkin et al., 1993) and cannot be presented here in any complete form.

Therefore, we will choose one example, but it would be as different as possible from the picture of natural death. To do this, you need to take any specific and severe disease with a fatal outcome. These conditions correspond to lung cancer (Lung cancer, n/d; Gardashnikov, n/d; Kiryanov et al., 2019).

General information. Lung cancer is a group of malignant tumors. A neoplasm develops due to a mutation in the cells that line the bronchi or lungs. The disease affects the right lung more often than the left, but also the upper lobes of the organ. The specificity of the disease consists of the fact that men get sick 6-10 times more often than women. There is a direct relationship with age: men aged 60-69 are several times more likely to suffer from this disease than people aged 30-39. In terms of prevalence, it is in 2nd place after skin cancer; among cancers of various organs, it accounts for up to 35-45%, and in terms of mortality, it is in first place. Distinguish radical (central) cancer, developing from the epithelium of the stem, lobar (Figure 1), or the proximal part of the segmental bronchus; peripheral cancer arising from the epithelium of the distal part of the segmental bronchus, bronchioles, as well as from the alveolar and bronchoalveolar epithelium; massive (mixed) cancer, occupying a lobe or all of the lung.

Figure 1. Lung cancer: 1 - radical from the upper lobe bronchus; 2- peripheral (nodular peribronchial form) (Paukov et al., 2015)

There are also several forms of lung cancer, depending on the nature of exophytic or endophytic growth; microscopic forms are represented by: squamous (epidermoid) cancer with keratinization or without keratinization; adenocarcinoma of varying degrees of differentiation; undifferentiated small or large cell carcinoma.

Lung cancer progresses over many years, passing through 4 stages (Figure 2).

Figure 2. Stages of lung cancer (FREE LDCT Lung Screening Program., n/d)

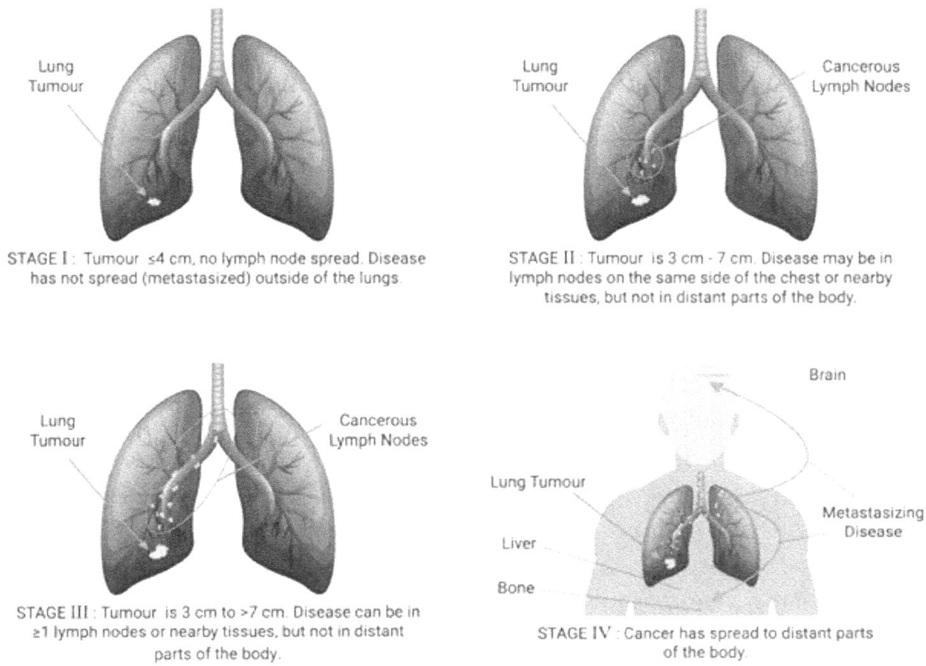

Lung cancer often (in 70% of cases) and early metastasizes:

- Lymphogenous way to the bifurcation and supraclavicular lymph nodes.
- By hematogenous route (to the liver, adrenal glands, bones, brain, etc.).
- Bronchogenic way (in different departments of the lung).

Lung cancer causes various complications:

- Blockage (obturation) of the bronchus with the development of lung tissue collapse (atelectasis).
- Disintegration (necrosis) of the tumor with bleeding.
- Cancer cachexia.
- Mechanical asphyxia.
- Cancerous pleurisy and pleural empyema.
- Gangrene of the lung.
- Insufficiency of "cor pulmonale".

The exact causes that can provoke the development of a cancerous tumor have not been determined; risk factors include sclerotic (cicatricial) and metaplastic processes in the bronchopulmonary tissue, smoking, dust, and gas pollution, especially characteristic of large cities.

The immediate causes of death from lung cancer are the destruction of the respiratory organs, bleeding and tissue breakdown, as well as the failure of other organs due to metastasis.

Symptoms Before Death

- Fatigue. Patients are often tormented by severe weakness and constant drowsiness. Every day they communicate less with loved ones, sleep a lot, refusing to do any physical activity. This is due to a slowdown in blood circulation and the extinction of vital processes.
- Refusal to eat. By the end of life, patients are severely emaciated, as they refuse to eat. This happens to almost everyone due to an appetite decrease, since the body does not need calories because a person does not perform any physical activity.
- Respiratory failure is the main symptom of lung cancer. In the terminal stage (4th), patients cannot breathe and are on artificial lung ventilation. Due to the breakdown of lung tissue and the accumulation of fluid in them (pleurisy), the body does not receive a normal amount of oxygen, and all the body tissues experience its deficiency. Carbon dioxide accumulates in the body. Metabolic processes in cells are disturbed, and some chemical processes become impossible.
- Development of physiological changes. There is a decrease in the amount of blood in the periphery and an increase in the flow to vital organs (lungs, heart, brain, liver). That is why the patient's arms and legs turn blue and often acquire a slightly purple hue.
- Depression of the functions of the nervous on the eve of death system. There is a decline in physical and emotional functions. Before death, the patient is in the stage of stunning.
- Change of consciousness. This leads to disorientation in place, time, and even self. Patients often cannot tell who they are and do not recognize relatives. As a rule, the closer death is, the stronger the mental state is depressed. There are hallucinations.

Psychological state before death. Personality changes depend on age, character, and temperament.

Before death, a person tries to remember his life and rethink it.

Gradually, the patient goes more and more into his thoughts and experiences, losing interest in everything around him. Patients become isolated because they understand that the end is inevitable, and no one can help them. Refusal to communicate is also associated with a special depressive state (the state of a martyr) - people are afraid of severe physical suffering (and experience it).

So, even though lung cancer is mainly a male disease, death can occur at a fairly young age and is accompanied by severe pain and the action of drugs and drugs that stop it, the overall picture is quite close to the symptoms of natural death. There is even a near-death improvement in the condition. This testifies to the commonality of the mechanisms of destruction of the system both in the case of natural death and in death caused by pathology. Special attention should be paid to the interpretation of disorientation in space, time, and personal identification from the standpoint of the theory of self-organization.

DEATH EVENTS

The next is the process of dying, which does not take place at once, but goes through several stages, collectively called terminal (Sorokin, 2004; Negovsky et al, 1987; Gubin, n/d).

The most significant is the extinction of the functions of the central nervous system.

Increasing hypoxia and subsequent anoxia in the brain cells (primarily the cerebral cortex) lead to the replacement of the oxidative type of metabolism by glycolytic.

In this case, there is a violation of the synthesis of ATP, ionic gradients, and free radicals are formed, and the permeability of cell membranes increases.

These metabolic changes entail destructive changes in cells and manifest themselves in the form of cloudy swelling, and hydropic degeneration. The brain tissues and spinal cord are the least resistant to the glycolytic type of metabolism.

These changes are reversible and, when normal tissue oxygen supply is restored, do not lead to life-threatening conditions.

With continued anoxia, they turn into irreversible degenerative changes, accompanied by protein hydrolysis, and autolysis develops. Just 4–6 minutes of anoxia is enough for irreversible changes in the cerebral cortex to occur. The subcortical region and the spinal cord can function somewhat longer with the glycolytic type of metabolism. The severity of terminal states and their duration depend on the severity and speed of development of hypoxia and anoxia.

In the process of dying, there are pre-agony, agonal pause, agony, clinical death, brain death, and biological death.

The pre-agonal state is a protective reaction of the body's vital systems. Characterized by:

- Disruption of the central nervous system (stupor or coma).
- Low blood pressure.
- An increase in heart rate from 90 beats per minute (tachycardia), followed by a slowing of the heart rate to less than 60 beats per minute (bradycardia).
- Deep and frequent breathing, alternating with rare and superficial, which contributes to the development of oxygen starvation of tissues and a decrease in pH (tissue acidosis).
- Increased heart rate.
- The acquisition of skin of various shades - at first it turns pale, turns yellow, then turns blue.
- The appearance of convulsions, convulsions.
- Confusion of consciousness, inhibition of reaction.
- Oxidative metabolism is the main one.

This state can last from several hours to one day.

A terminal or agonal pause does not always occur, for example, it is absent during electric shocks but is pronounced with blood loss. Characterized by:

- Respiratory arrest due to increased influence of the vagus nerve.
- Cessation of the heart activity with the disappearance of bioelectrical activity (asystole) lasting from 1–2 to 10–15 seconds.

The terminal pause lasts from a few seconds to 2-4 minutes.

The agonal state (Figure 3) begins with the first breath after the terminal pause.

Figure 3. Ivan Myasoedov. Portrait of a dying father. Agony. 1911. (Myasoedov, 1911)

Agony has a clear clinical picture and is always displayed on the ECG and EEG. The agonal state is characterized by:

- Sharp pallor of the skin.
- Dilated pupils.
- Inhibition of the function of the cerebral cortex, because of which its inhibitory effect decreases, which leads to increased excitation of the brain stem and is the cause of a temporary resumption of vital functions. The role of regulators of vital functions passes to the medulla oblongata and spinal cord. Regulation is aimed at mobilizing the last possibilities of the organism to save a life.
- The appearance of breathing. The amplitude of breathing increases, and its structure is disturbed - the muscles that carry out both inhalation and exhalation are simultaneously excited, which leads to an almost complete cessation of lung ventilation. Having reached a certain maximum, the respiratory movements decrease and quickly stop; towards the end of the agony, breathing resembles the act of swallowing.
- Increased blood pressure and heart rate.

- Appearance of pulsation of large arteries; the correct rhythm and blood flow can lead to the pupillary reflex and even consciousness restoration.

However, the fight against death is no longer effective, since the body's energy at this stage is based on anaerobic (oxygen-free) metabolism and not only becomes insufficient in quantitative terms but also leads to a rapid accumulation of under-oxidized metabolic products. During the agony the body loses, due to the complete burning of ATP and the depletion of cellular reserves, those few tens of grams, which D. McDougall considered the weight of the soul that left the body.

The agonal state lasts several minutes to several hours, usually up to half an hour.

Clinical death occurs from the moment the heart and breathing stop working. However, within 10–20 minutes it is still possible to bring a person back to life with medical efforts: pathological changes in all organs and systems are reversible.

This transitional state between life and death begins with the cessation of the activity of the central nervous system, blood circulation, and respiration and continues for a short time until irreversible changes in the brain develop. During clinical death, respiration, circulation, and reflexes are absent, but cellular metabolism continues anaerobically. Gradually, the reserves of energy drinks in the brain are depleted, and the nervous tissue dies. It is generally accepted that under normal conditions the term of clinical death in humans is 3-6 minutes; when the body cools (for example, when freezing or drowning in winter), this time can increase significantly.

Brain death is manifested by the development of irreversible changes in the brain, and in other organs and systems partially or completely reversible.

Its main diagnostic features (Walker, 1978):

- Lack of activity of the hemispheres (lack of reactions to stimuli).
- Lack of brain reflexes.
- EEG - silence (even with artificial stimulation).
- No signs of intracranial circulation.

Biological death. In ancient times, when ideas about humans and life differed from modern ones, the most important signs were cardiac and respiratory arrest.

In modern society, the most important sign of death is brain death (Lizza, 2018).

However, let us turn to official medical protocols that record both public ideas about death and the practice of resuscitation (Determination of Death, 2014.; Instruction N° 73, 2003.). According to these, the death of a person occurs as a result of the death of the organism as a whole.

Ascertaining the death of a person occurs with brain death or irreversible death of a person. The diagnosis of brain death is established in healthcare institutions that have the necessary conditions for ascertaining brain death.

Biological death is expressed by permanent, irreversible, and cadaveric (early signs, late signs) post-mortem changes in all organs and systems.

Post-mortem changes are functional (lack of consciousness, respiration, pulse, blood pressure, reflex responses to all types of stimuli), instrumental (electroencephalographic, angiographic), biological (maximum pupil dilation, pallor and/or cyanosis, and/or marbling (spotting) of skin covers, decrease in body temperature) and cadaveric (early, late) signs.

Resuscitation measures are terminated upon ascertaining the death of a person based on brain death, including against the background of ineffective use of the full range of measures aimed at maintaining life, with the ineffectiveness of resuscitation measures aimed at restoring vital functions within 30 minutes.

So, officially, death is considered the death of an organism as a system, and its diagnosis is made by a set of signs, and not only by brain death, cardiac or respiratory arrest.

To complete the picture, we should consider the events after biological death.

AFTER THE DEATH

The cells of the body do not die at the same time. After the heart stops, the body's cells stop receiving oxygen and nutrients, release carbon dioxide, and die. The duration of the process depends on the sensitivity to oxygen starvation.

It is believed that irreversible changes in the cerebral cortex occur much faster than in the more ancient brain stem and medulla oblongata. In the absence of oxygen in the cortex and cerebellum, necrosis foci appear in 2–2.5 min, and only single cells die in the medulla oblongata even after 10–15 min (Ado et al., 2000).

However, new research shows a more complex picture: while neurons degenerated, outgrowths of glial cells (performing a protective function) grew; also, the genes present in neurons degraded in the first hours after death, and the genes associated with glial cells increased their activity in the postmortem period for at least 24 hours (Dachet et al., 2021). Other organs show much greater vitality: the heart muscle remains viable for 1.5-2 hours, the kidneys and liver – for 3-4 hours, muscle tissue and skin – for up to 6 hours, and bone tissue – for up to several days.

Thus, death is, first of all, the destruction of a complex system of the body as such, and then the uneven destruction of its subsystems and organs.

On the other hand, if the system functions normally, individual cells' deaths do not lead to destruction. Thus, during regeneration, 80 ± 20 grams, or 0.33 ± 0.02 × 10^{12} cells, mainly blood and intestinal epithelial (Sender & Milo, 2021), die per day and are replaced by new ones.

Subjective space. Almost all ancient and modern religions recognized the life of consciousness outside the body. This thesis is reflected in myths, fairy tales, and literary creativity, and underlies occult and meditative practices.

These sources are not natural science. But you can study and use them, realizing that other, non-scientific methods, for example, characteristics of philosophy, religious studies, and the like, must be used.

While the scientific methods in this area develop, let us consider available-to-date observations.

The near-death experience (NDE): after the death or before? This name was introduced by R. Moody (Moody, 1988). He found out, by interviewing approximately 150 people who survived clinical death, their experiences, which he then systematized into 9 groups (subsequently 15), noting that no one had experiences related to all groups at once. After clarifying the picture by R. Moody, himself, and his followers, the most frequent experiences are grouped as follows (*Near-death Experience*, 2024):

- First impression - very unpleasant sound (noise).
- Understanding that he died (conscious death).
- Pleasant emotions – peace, and tranquility (Figure 4, panel 1).
- Frightening and even hellish images of fire and the presence of other suffering people (about 15%) according to (Blackmore, 1996) (Figure 4, panels 3, 4).
- The blind can see what is nearby.
- The deaf can hear sounds from nearby sources (Ring, 1985 Harper Perennial).
- Sensation of leaving the body, hovering over the body, the ability to move around in space.
- Sensation of deformation of space and time (figure 4, panel 3).
- The inability to influence the events that are taking place (to say something to relatives or resuscitators, etc.).
- A feeling of moving upward through a bright light tunnel or a narrow passage (Figure 4, panel 2).
- Meeting with deceased relatives or clerics.
- An encounter with a being of light (often interpreted as a deity).
- Consideration of episodes of the past life.
- Reaching a border or barrier.
- Unwillingness to return to the body.

- Anxiety or fear of ending their life in its current state.

Figure 4. Bosch H. Visions of the Hereafter. Between circa 1490 and 1516. Galleria dell'Accademia Venezia, Venice (Visions of the Hereafter, n / d)

NDEs include cardiac arrest, massive blood loss, anaphylactic shock, electrocution, coma, intracerebral hemorrhage or cerebral infarction, suicide attempt, drowning or asphyxia, respiratory arrest, or severe depression. Suicide attempts do not increase the likelihood of unpleasant experiences compared to other situations (Ring, 1985).

Criticism. R. Moody has been repeatedly criticized for various methodological errors (Gubin, n/d). This author is a practicing resuscitator. So, N. Gubin considers the methodological mistake to be the lack of data on when by whom, and how the fact of clinical death was recorded (it is necessary to determine the absence of breathing; circulatory failure; and the complete lack of reflexes). If a respondent does not have a near-death medical record, he cannot be included in the sample of patients demonstrating "life after death". Incorrect samples are not scientific data. data. It is not taken into account that patients who have undergone clinical death and post-resuscitation diseases have increased suggestibility; the patient "sees" what the doctor tells him. Table 1 presents survival statistics 3 months after clinical death.

Table 1. The outcome of clinical death

Case	Number of cases	Percentage
Death from cardiac arrest	345	68,59
Brain death	114	22,66
Incomplete Recovery	18	3,58
Full recovery	26	5,17
Total	503	100

(Gubin, n/d)

After 3 months, 41 patients were alive (8.15%). Of these, 18 (3.58%) had certain disorders of higher nervous activity and required outside care. Only in 5.17% of patients who underwent clinical death, subsequently, there was complete recovery of impaired functions. Of this number, 14 cases were initially diagnosed with drug poisoning, and the patients were in a coma for more than 30 minutes.

Therefore, if the publications provide data that more than 4.4-6.0% of patients who have experienced clinical death have detailed post-mortem memories, then the group of patients under study is incorrectly defined, the survey methods are incorrect, etc. Usually, they talk about 10-35% of people who "could clearly remember what they experienced" - probably a significant part of these memories are not connected with clinical death or the memories were forced on them by those who conducted the survey.

Finally, the above description of near-death experiences is compared with the description of the action of ketamine on the central nervous system (Checkman, 1980), which causes difficulty in expressing seen images in symbols of the second signaling system; strengthening the pathological dominant (center of excitation of the cortex) by an auditory stimulus; inhibition of the parts of the brain that stimulate the work of the cortex; stronger energy supply of central vision mechanisms; various violations of the integral work of the central nervous system; greater resistance of deep foci of excitation, which have pronounced connections with each other; complete cessation of the integration (unifying) function of the cortex. A distinctive feature of ketamine is the stimulator of changes in the excitation processes and inhibition of the cerebral cortex (dissociative anesthesia). As a result, patients do not feel irritation (pain, a feeling of pressure and stretching), but they hear, see (and quite often a tunnel or "pipe"), "leave", "ascend", meet relatives, etc.

Here are other conditions accompanied by the same symptoms: poisoning (often overdoses of certain medicinal substances, accompanied by a deep coma, are mistaken for death, and if the patient is taken out of this state, he describes almost all of the above symptoms, and sometimes he is even sure that he has been on that light (Thienes & Haley, 1972); effects of drugs other than ketamine, such as LSD, that affect the psyche (without overdose); coma caused by low blood sugar; coma when freezing.

Hence, it follows that near-death experiences manifest themselves at the first stage of dying, when the cortex has not yet turned off, and after revival if the brain has not died.

New scientific data. To date, results have been published that refute the allegation of complete brain shutdown after cardiac arrest (Li et al., 2015).

Moreover, there is a surge of electrical activity in the brain during cardiac arrest, which may be the reason for the vivid visions of people who have experienced a state of clinical death (Chawla et al., 2009).

There is also a release of dopamine, which plays an important role in arousal, attention, cognition, and emotional experiences, in the first minutes of asphyxia increased more than 12 times, as well as serotonin, which activates receptors that cause visual hallucinations in people - more than 20 times (Rolland et al., 2014).

All this also indicates that near-death experiences have a "pre-death", but not a "mortal" or "after-death" character.

Comparison with sensations during meditation. The classic source on yoga is the Yoga Sutra of Patanjali (the text in Sanskrit is given in (Yoga Sutra of Patanjali, 2006); modern presentations and comments are given in (Smirnov, 1981).

First, we note that many ancient and modern meditation techniques are based on the assimilation of the dead. In the Yoga Sutra, the stages of achieving expanded consciousness (samadhi) are described in the following sequence.

The first stage is the isolation of the sense organs from external stimuli (pratyahara).

Next comes the subsidence of the internal dialogue (nirvitarka, nirvichara).

Then the stage of consciousness purification from past impressions, habits, and complexes (vasanas and samskaras) begins. At this time, a person experiences visual, and also "replays" various life situations, choosing a different way than reality, a way to resolve them. Probably, "siddhis" belongs to this stage - the ability to become infinitely small or large, light or heavy, travel in space and time, meet different beings, and so on. It is also known about the appearance of various smells and sounds; there is a special technique for meditating on sounds - nada yoga (Russill, 2006). All these manifestations, very vivid, are very reminiscent of the described near-death experiences. Although they are signs of some progress, Patanjali warns against getting carried away with them, as they are not the end goal.

Overcoming these "temptations" gradually leads to the direct perception development of things (samprajnata).

The next stage is to dive into your ego and experience a kind of light (adhya atma prasadanah), in which the ego seems to dissolve.

After overcoming the ego, an all-consuming feeling of unity, love, grace, life, and "inspiration" (nirbija samadhi) arises, which is the goal.

All this is accompanied by external signs of death - cessation of breathing (kevala kumbhaka) and pulse, immobility of the body, etc.

Physiology of various meditation techniques. A review of such studies is given in (Ebert, 1986). Here is data on the measured indicators: a drop in oxygen consumption, changes in blood circulation, a change in the frequency of breathing, and characteristics of brain rhythms (Table 2). Note methodological inaccuracy: the state of samadhi was recorded according to the words of yogis, and it is not clear at what stage they were and whether they reached samadhi at all; the same applies to transcendental meditation.

Table 2. Decrease in energy exchange and oxygen consumption (assuming the respiratory coefficient is unchanged) under various conditions of rest)

Resting situation	Decrease in energy exchange, %	Number of subjects	Authors
Sleep (versus being awake)	8,73	9	Jana 1965
Hypnosis	-	14	
Deep sleep (versus REM sleep)	13		Brebbia, Altschuber 1965
TM (compared to being awake)	17	20	Wallage, Benson, Wilson 1971
Za Zen	19 (14-25)	4	Hirai, Sugi 1975
Yoga meditation	15,5	7	Dhanaraj 1975
	50	1	Anand, Chhina, Singh 1961
	30	1	Karambelkar, Vinekar, Bhole 1968
	25	1	Gharote 1977

(Ebert, 1986. Table 20)

The heart rate is often used to characterize the power of blood circulation, which changes with the excitation of the vagus and sympathetic nerves and reflects the sympathetic and parasympathetic systems' overall activity. There are two trends in the dynamics of heart rate:

1. The heart rate drops by 3-5 / min. in transcendental meditation,
2. The heart rate growth is noted by 1-2 / min. or by 15-20 / min., according to various studies with meditative practices of yoga and Zen.

Arterial pressure almost does not change under the influence of meditation; note a slight and uncharacteristic increase in blood pressure by the end of meditation.

Skin temperature can be used as an indicator of the amount of peripheral blood flow. During concentration, peripheral blood supply increases, which can be assessed as a sign of an increase in the tone of the parasympathetic nervous system. Various studies describe a slight increase in finger temperature by 0.5 K in yogis and its decrease during transcendental meditation.

The minute breathing volume measurement was carried out for transcendental meditation, and a clear decrease was found.

A decrease in the breathing frequency was recorded during meditation (Table 3).

Electroencephalographic studies of brain rhythms during meditation have yielded mixed results. There is evidence of a β-rhythm increasing in time, which from Sulcus Rolandi (Roland's furrow) spread throughout the cortex, and the skull. At the same time, an increase in the amplitude of the β-rhythm was also recorded. Most studies state an increase in the amplitude of the α-rhythm with a slight decrease in its frequency. During transcendental meditation, a high phase coherence of α-waves is found, primarily in the frontal regions of the skull, which is believed to be correlated with some mental phenomena (creativity, stream of consciousness, etc.).

Table 3. Decrease in respiratory rate with meditation techniques

Meditation	Breathing frequency [min^{-1}]	Breathing frequency [min^{-1}]	Number of examinees	Authors
	before	during		
	meditation			
Yoga (students	14	6-7	2	Wenger, Baghhi 1961
Yoga (masters)	26	1-2	2	
Za Zen	?	4-5	4	Hirai 1975
TM	13	11	20	Wallage, Benson, Wilson 1971
	10-15	4-6		Allison 1971
	18	15		Parulkar et.al. 1974

(Ebert, 1986. Table 21)

In contrast to the described patterns, in any state, associated with clouding of consciousness (with narcolepsy, hypersomnia, trauma, inflammation, or neoplasms of the brain), there is a decrease in frequencies in the ranges of θ- and δ-rhythms.

The death meditation (tukdam) practiced by Tibetan Buddhist monks was studied (its meaning will be discussed at the end of the chapter). According to the monks, this meditation can continue for a long time after death, preventing the appearance of cadaveric phenomena. Verification of this statement (Lott, et al., 2021) showed that the EEG of a thukdam monk in the state coincided with the EEG of a dead person (in Figure 5 below) and differed significantly from the EEG of a meditating healthy person (in Figure 5 above); the authors tried to see "life" through the auditory response of the brainstem (ARP).

Thus, signs of the activity of post-mortem consciousness were not recorded. Summarizing, we come to the following conclusions:

- NDEs and meditative states are pre-, not post-death experiences.
- After death, indicators of brain activity are not fixed.
- During meditation, there is a drop in oxygen consumption, changes in blood circulation, a decrease in the frequency of breathing, and changes in the characteristics of brain rhythms. However, these indicators are not reset; the number of test subjects is small.
- The state of samadhi is recorded only from the words of the yogis themselves; what stage has been reached is unknown; the number of test subjects is small.
- According to physiological indicators, the tukdam is identical to death. At the same time, cadaveric phenomena are weakened than during normal decomposition. The number of cases studied is too small.

How is it possible to fix the wave component using the described methods? As already noted in Chapter 9, the devices record the human body vibrations and electromagnetic fields, that is, the soliton component - and ascertain the body's death; in this case, the wave component is not measured by existing methods.

Figure 5. EEG of meditating monks (top) and monks in the state of tukdam (bottom) - the potential is not recorded (Lott et al., 2021)

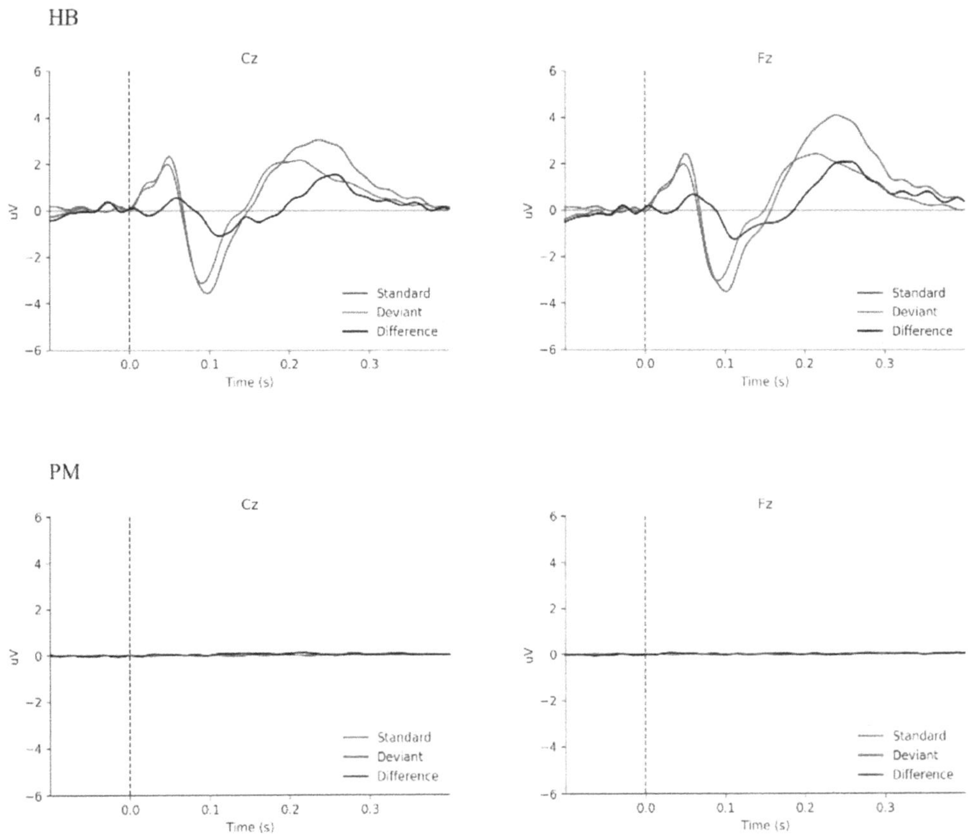

BIOLOGICAL IMMORTALITY

Speaking of immortality, one should separate the concepts of physical and biological immortality and consider separately the cells of the body and the body as a whole. Immortality in the social or religious sense is not considered here but will be given due attention in later chapters. So, a cell is called immortal if it is not limited in the number of divisions (it does not have a Hayflick limit). The biological immortality concept is not identical to physical immortality: sooner or later, cells die from natural causes (they are eaten by other living organisms, poisoned, frozen, and so on). It is more correct to speak of age-dependent (aging) and age-independent (random) cell death in a population or organ.

In this sense, one can also discuss the immortality of multicellular organisms. For example, a four-year experiment by D. Martinez (Martinez, 1998) proved the absence of age-related mortality among three groups of hydras. The immortality of hydras is likely associated with their high regenerative capacity.

In the database on aging and lifespan of animals (AnAge Database, n/d), 7 species of practically ageless multicellular organisms are recorded - Sebastes aleutianus, Proteus anguinus, Emydoidea blandingii, Terrapene carolina, Strongylocentrotus franciscanus, Arctica islandica, Pinus longaeva, among which there are 2 turtles, a sea urchin, and one of the pine species.

Multicellular organisms, that do not have age-related mortality include organisms, that reproduce asexually. Such immortality is based on the mechanisms of somatic embryogenesis and regeneration (Elchaninov et al., 2021).

A few works do not recognize biological immortality as such or dispute the proposed mechanisms for its maintenance (Holliday, 2008; Olshansky et al., 2004; Landecker, 2000; Agarwal & Rimm, 2012).

Limited immortality can probably be considered an evolutionary mechanism that ensures the reproduction and survival of the species in conditions of high mortality from external factors.

On the other hand, the absence of age-dependent mortality would be a disadvantage for sexually reproducing living beings with complex organization and high adaptive potential: they would win the competition from the next generations, depriving them of resources and opportunities for survival, as was demonstrated in the famous experiment of D. Kelhoun Universe 25 (Calhoun, 1973), when alpha male mice deprived other males of free access to food and water.

As for humans, the inevitability of death has always been one of the most difficult psychological problems. Let's leave the question of religious, psychological, historical, social, and other aspects of this problem for subsequent chapters and consider only the possibility of biological immortality.

All that has been said above regarding complexly organized animals also applies to humans, therefore, the prerequisites and mechanisms for the death of a person as an organism do exist.

At the cellular level, there are cells for which the Hayflick limit exists and is equal to 52 (they are the majority), and those that do not have a limit, which include stem and primary germ cells. In special environments, an unlimited number of divisions of cancer cells is observed. The most famous cell lines are HeLa, obtained in 1951, and en:Jurkat, obtained in 1970 (Biological immortality, n/d). Both lines are being reproduced at present. The lifespan of human cells is different (Table 4).

Table 4. The lifetime of some cells and organs of the human body

Cell / Organ	Description	Lifespan
Hair	The filamentous protein of hair follicles in mammalian skin	Four years
Leukocytes	A heterogeneous group of human and animal blood cells, different in appearance and function, distinguished by signs of the nucleus presence and the absence of independent coloring.	From several hours to several years (the main function of leukocytes is to protect the body from pathogens and remove tissue destruction products)
Lymphocytes	Cells of the immune system, are a leukocyte type of the agranulocyte group.	For NK cells it is 7-10 days, for B-lymphocytes - several weeks, for T-lymphocytes (in humans) - 4-6 years
Brain[2]	The central part of the nervous system	While maintaining brain health indefinitely (Magrassi, Leto, Rossi, 2013)
Neutrophils	The most numerous groups of granulocytes, neutrophils account for 40% to 70% of all leukocytes in humans.	For non-activated neutrophils in the bloodstream from 5 to 135 hours. When activated, neutrophils acquire the ability to squeeze through the vascular endothelium and migrate to tissues, where they live for 1 to 2 days
Spermatozoa	Male reproductive cell	After maturation - about 64 days - it can be stored in a man's body for up to a month. Depending on environmental conditions, they survive in the ejaculate for up to 24 hours. They die in the vagina in a few hours. In the cervix, uterus, and fallopian tubes, they remain alive for up to 3 days
Ovum	Female gamete (of humans, animals, higher plants, as well as many algae and other living organisms)	After ovulation, the egg remains viable, usually for 12-24 hours, in some cases up to 48 hours
Red blood cells	Blood cells that take in oxygen in the lungs and then carry it (carbon) around the body	125 days (every second, about 2.5 million erythrocytes are formed and many are destroyed).
Platelets	Cells involved in the process of blood clotting; their main function is to ensure the integrity of blood vessels	5-10 days; they must be constantly developed

(Cell Lifespan, 2020)

Some cells can renew themselves: if such a cell is removed, it will be replaced by another, and if left untouched, it can exist for a long time without changes. The same cell type can completely change in one place and will exist without changes in another (Arrojo e Drigo, et al., 2019).

This is a list of the most famous areas of search for biological immortality:

1. Medications – an extension of life due to the drug's effect on the state of biomarkers. From the "elixir of immortality", which Qin Shi Huang was looking for, to trendy bioactive supplements.
2. Transplantation and implantation - replacement of diseased or inoperable organs with biological or artificial analogs.
3. Genetic manipulations - there are several options (sometimes combined) - based on mitochondrial, telomeric, and epigenetic theories of aging (Chapter 6). For example, in 2017, it was reported that D. Sinclair's group tested the NAD+ compound in mice, changing the cellular aging process by protecting

DNA from damage (Sinclair, 2017), in 2020, Israeli scientists achieved success (Miklashevskaya, 2020) and so on.

4. Fetal therapy - the use of embryonic stem cells extracted from abortive material for rejuvenation. A variation is fetal gene therapy (Mattar et al., 2011; Wagner et al., 2009). It is considered promising for the treatment of several hereditary diseases. However, one should not forget about the ethical and historical sides: the method goes back to the ancient and cruel methods of rejuvenation, accompanied by occult practices and used by such historical figures as Báthory Erzsébet or Gilles de Retz.
5. Preservation of personality as an information structure not related to the physical body - the use of electronic devices for reading, writing, and storing, and in the future - cloning all information about a person's personality (or about the whole person, including his genome). Various techniques for preserving memories of past lives during reincarnations (according to legend, for example, Buddha and Krishna remembered them) serve as an ancient analog of this approach.
6. Anabiosis - postponing the solution of the problem of biological immortality or cure of fatal diseases to the future, with the hope that the progress of science and technology will make it possible, by immersing a person in a state of suspended animation in various ways, including by freezing (cryonics).

All these and similar efforts, which have not been crowned with proven success, face a complex of natural science, technical, ethical, and religious problems - an overview is given by (Derikot, 2018).

THE DEATH AS THE DESTRUCTION OF A SYSTEM

Now let us consider the dying processes from the theory of self-organization of systems standpoint. We stipulate that we will consider scenarios of events closest to death and for now, we will not use convolution scenarios (Chapter 4), leaving them to predict the distant afterlife.

Recall that dying is not instantaneous; that during clinical death as a result of external intervention, it is still possible to restore vital functions; and that even after the death of a biological cell, bodies continue to exist for some time and carry out both internal metabolism and exchange of substances with the external environment, even in the “perverted” forms of cadaveric phenomena.

On the other hand, the psychic component retains a certain activity potential, expressed in near-death (pre-death, as was found out above) experiences. At the same time, if the reflex functions look depressed, then the visions and hallucinations are so vivid and realistic that some people who have experienced them seriously change their behavior.

It follows that *death is the irreversible destruction of humans as an open complex system.*

Let us estimate how much the holistic conditions must be violated for the system's destruction to become irreversible (Figure 6).

Figure 6. Conditions for the destruction of the soliton-wave system: 1. Formalization of the existence and destruction of the system. 2. Conditions under which the system is destroyed

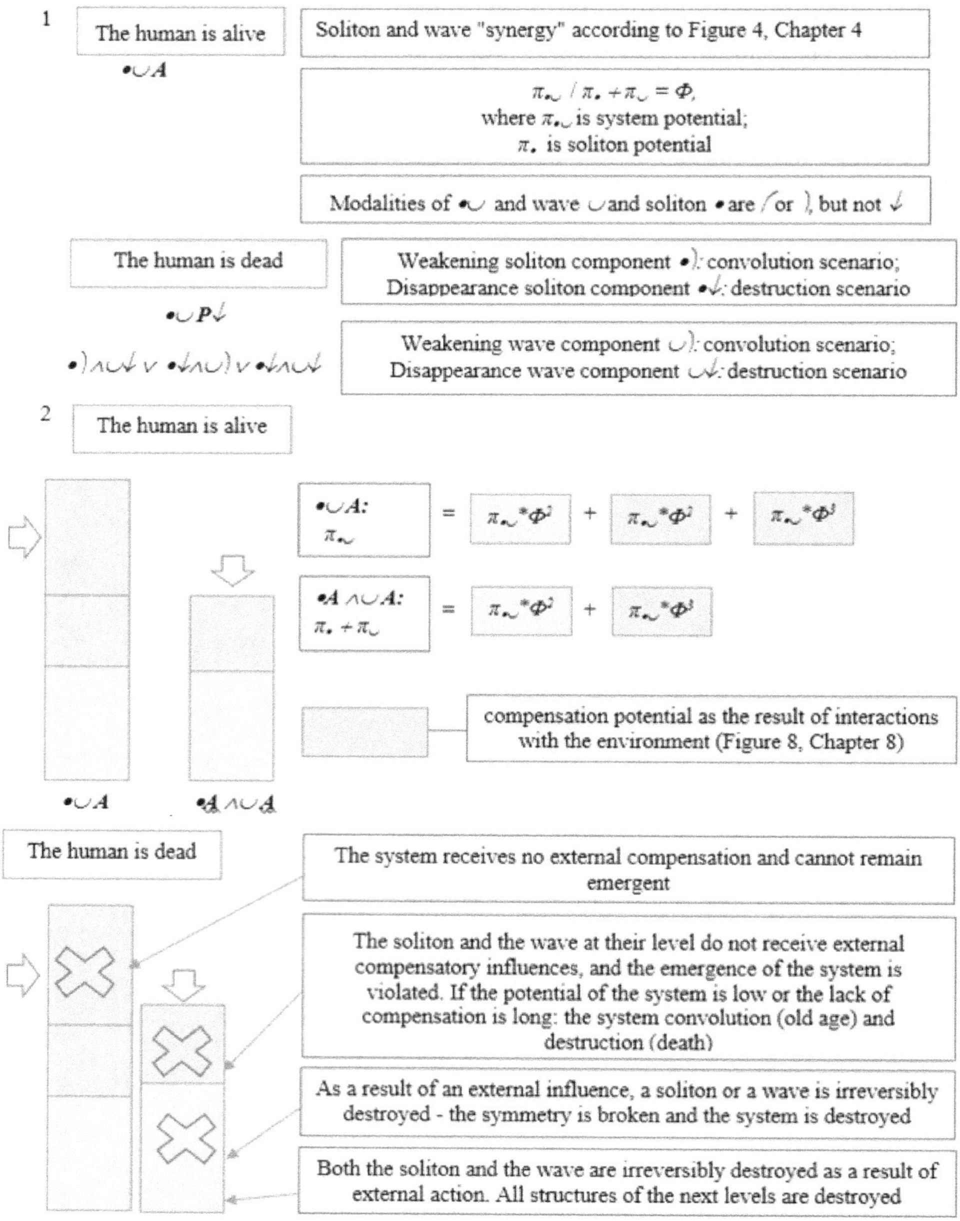

Let us consider theoretically possible variants of destruction (bottom of Figure 6-1). The development of terminal states shows a direct continuation of the aging scenarios (Chapter 9). Then the following possibilities take place (Figure 7).

Figure 7. Options for the destruction of the soliton-wave system. 1 - Destruction of the field (wave) component not compensated by internal regulation or external influences; 2 - Uncompensated destruction of the substance (soliton) component; 3 - Uniform uncompensated degradation of both components. The soliton component is indicated in black, the wave component in blue, the reversible destruction during clinical death by red wave lines, the destruction of the system components by the red cross, the blue arrows are external influences, the orange arrow is the transition to biological death.

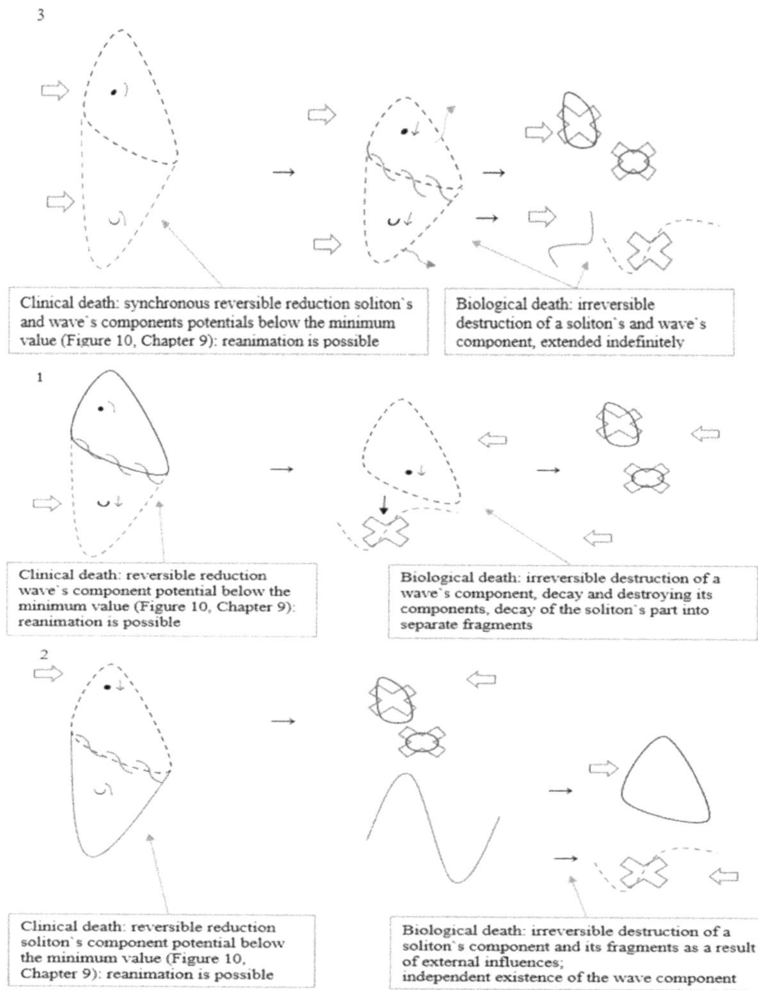

Option 1 can be interpreted as a phenomenon of "sudden death", typical, for example, for infants [a]

Option 1 can be interpreted as a phenomenon of "sudden death", typical, for example, for infants and people of other ages. There are likely more such cases than it is believed - for example, drowning, car accidents, etc. may be caused by sudden death. Measurements of vital signs, due to suddenness, are far from always carried out, resuscitation measures do not lead to success, and post-mortem pathological anatomical examination does not name the primary cause. Understanding the destruction of the wave component as the true cause has theoretical and important medical significance. If this assumption is correct, developing methods for the field component controlling and resuscitating is necessary.

Existing methods are focused on the analog amplification of various mechanical vibrations of organs (devices such as tonometer), or on the digital-to-analog conversion of bioelectric potentials and their representation in a graphical form (for example, electroencephalographs), which is far from exhausting even the list of known fields (Chapter 8).

Option 2 is best documented; medical devices and the current protocols for resuscitation, determining death, and post-mortem studies are "sharpened" for its fixation. At the same time, as discussed in detail above, the posthumous existence of "consciousness", "soul", and "psychic activity" is not fixed within the framework of these protocols and measurements. To test the feasibility of possible scenarios, it will be necessary to use other methods in addition to simulation, which is discussed in detail in Chapter 9.

Option 3 can presumably be interpreted as tukdam or other conscious or spontaneous near-death experiences. SWM allows us to understand what happens in this state (Figure 8).

So, immediately before death, a person who is well versed in the technique of meditation sharply lowers the level of his physiological activity, like how it is described in (Ebert, 1986), but to such low values that it is perceived as death by an outside observer (this moment is shown in Fig. figure as a vertical green line on the right). The Buddhist monasteries where such incidents occurred lacked the equipment to accurately determine biological death (Lott et al., 2021). At the same time, a person enters a meditative state, which is interpreted within the SWM as a convolution scenario. As a result, the threshold value of the potentials of both the soliton and the wave components, which are necessary to maintain the system's existence, is sharply reduced; the system has enough potential (shown in the figure as filled figures) to exist for some more time. Cadaverous phenomena are not observed. But its soliton-wave nature does not change, and when the theoretical limit of lifespan is reached (in fact, even earlier), it is irreversibly destroyed (this moment is shown as a vertical green line on the left). According to (Lott et al., 2021), the time from ascertaining death by monks to the onset of cadaveric phenomena is usually several days, less often weeks. When researchers with the necessary equipment reached the monastery, they also stated the biological death (Figure 5). So, there is no mysticism in this.

Figure 8. Explanation of tukdam and similar cases

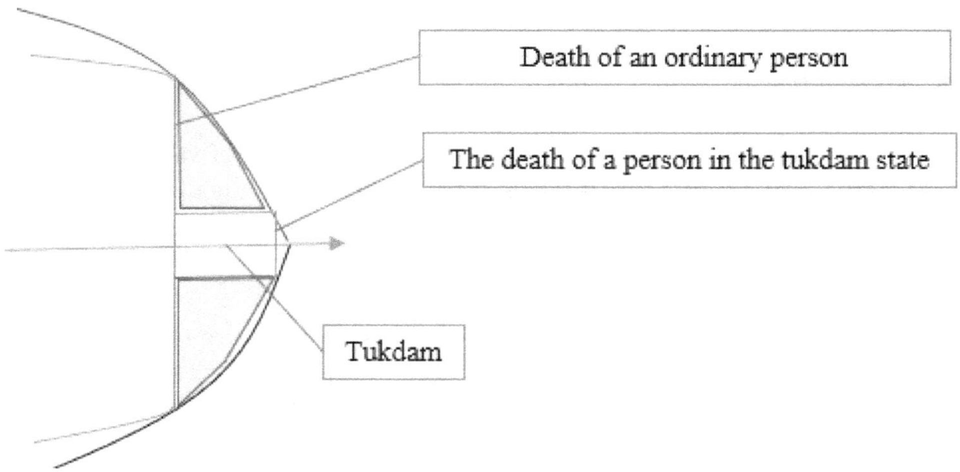

There is one interesting, documented case of the preservation of bodily remains for more than 75 years (Itigelov, 2021; Itigelov, Dashi-Dorzho, n/d), also associated with tukdam – unfortunately, the authors did not find scientific publications (Figure 9).

Figure 9. Dashi-Dorzho Itigelov (Itigelov, 2021)

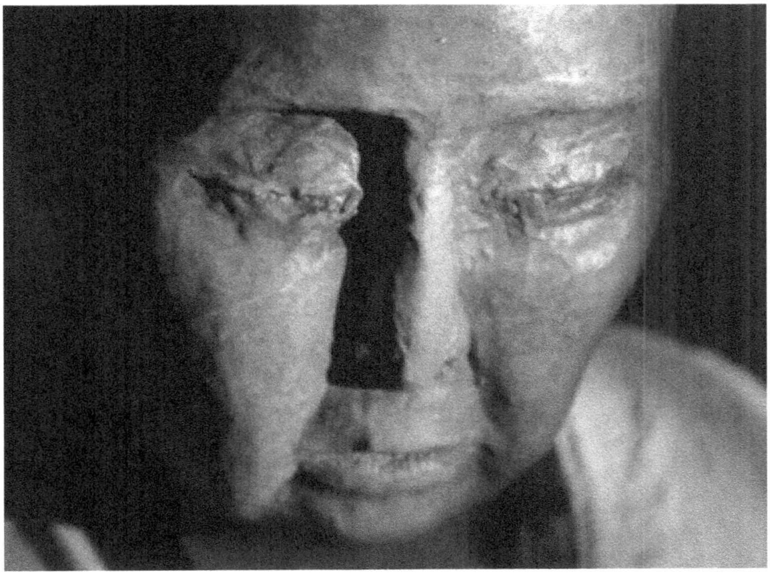

According to eyewitnesses, on June 15, 1927, Itigelov sat in the lotus position, gathered his followers, and instructed them: "You will visit and see my body in 75 years." Then he asked them to read a special prayer for the deceased. Being in a state of tukdam, Dashi-Dorzho Itigelov went into nirvana and was buried in a cedar coffin in the lotus position. The coffin was opened several times; in 2002, the "Act of the external examination of the exhumed corpse" was officially drawn up: "Any extraneous aromatic, resinous or putrid odors from the contents of the box and the corpse were not determined ... The soft tissues of the corpse are of a tightly elastic consistency, and mobility in the joints is preserved. The hairline on the head and the nail plates are preserved. The corpse's posture during the extraction from the box is maintained without any supporting and fixing devices. No traces indicated a previous opening of the body cavities for possible embalming or conservation, as well as any injuries, traces of injuries, surgical interventions, or diseases on the body of the corpse were not found. 2 mg of body samples in 2004 were studied by V. Zvyagin, Doctor of Medical Sciences, head of the Personal Identification Department of the Russian Center for Forensic Medical Examination. Infrared spectrophotometry showed that the protein fractions have in vivo characteristics, the bromine concentration exceeded the norm by 40 times; body temperature was below 20^0 C (a sign of death). Since 2005, any medical and biological studies of Itigelov's body have been suspended by order of the Khambo Lama D. Ayusheev.

It is unclear from the available data whether natural mummification occurred, or whether it is a tukdam of unusually long duration. In the second case, the diagram in Figure 7 can no longer be an explanation; probably, mechanisms of anabiosis, which are not entirely clear at present, are involved here, and they are not characteristic of higher animals, but like the process of sporulation, which allows for long-term preservation. But this is already an area of speculative assumptions.

FUTURE RESEARCH DIRECTIONS

With the use of convolution scenarios, the variants justified above will be extended to a time more distant from death. Since the SWM did not contradict the known facts of the origin and evolution of life and the events of the human life cycle, the models will likely remain correct in this situation as well. As natural science methods are exhausted, there will be a need to select other SWM verification methods, related to the humanities and justify their applicability.

CONCLUSION

A complex of physiological and psychological phenomena and factors associated with human death is considered. It has been stated that the near-death experience is lifelong and to a certain extent like meditative states.

With the help of SWM, three options for the transition from clinical death to biological and immediate postmortem events are substantiated. Having a severe illness leading to death also fits one of these scenarios.

The proposed models are naturally related to the human life cycle models (Chapter 9), as well as the human-environment models (Chapter 8) and the set of models describing the evolutionary process (Chapter 6), and they do not contradict known facts and open interesting directions of research.

When solving the problem of biological immortality, it is necessary to measure, investigate, and consider the field component during resuscitation, and the possibility of the existence of the field component after the destruction of the human body system in the course of dying.

REFERENCES

Agarwal, S., & Rimm, D. L. (2012). Making Every Cell Like HeLa: A Giant Step For Cell Culture. *American Journal of Pathology*, 180(2), 443–445. DOI: 10.1016/j.ajpath.2011.12.001 PMID: 22192626

Alcock, J. (1981). *Psychology and Near-Death Experiences*. In *Kendrick Frazier. Paranormal Borderlands of Science*. (pp. 153–169). Prometheus Books.

AnAge: The animal ageing and longevity database. (n.d.). Senescence.Info. Retrieved August 14, 2024, from https://genomics.senescence.info/species/index.html

Arrojo e Drigo, R., Lev-Ram, V., Tyagi, S., Ramachandr, R., Deerinck, T., Bushong, E., Phan, S., Orphan, V., Lechene, C., Ellisman, M. H., & Hetzer, M. W. (2019). Age Mosaicism across Multiple Scales in Adult Tissues. *Cell Metabolism*, 30(2), 343–351. DOI: 10.1016/j.cmet.2019.05.010 PMID: 31178361

Australia, H. (2024). *The process of dying*. https://www.healthdirect.gov.au/the-physical-process-of-dying

Benton, M. J. (2005). *When Life Nearly Died: The Greatest Mass Extinction of All Time*. Thames & Hudson.

Bernat, J. L. (2013). The definition and criterion of death. *Handbook of Clinical Neurology*, 118, 419–435. DOI: 10.1016/B978-0-444-53501-6.00033-0 PMID: 24182395

Blackmore, S. J. (1996). Near-death experiences. *Journal of the Royal Society of Medicine*, 89(2), 73–76. DOI: 10.1177/014107689608900204 PMID: 8683504

Bonanno, G. A., & Trauma, L. (2004). Human Resilience. Have We Underestimated the Human Capacity to Thrive After Extremely Aversive Events? *The American Psychologist*, 59(1), 20–28. DOI: 10.1037/0003-066X.59.1.20 PMID: 14736317

Bosch, H. (n.d.). *Visions of the Hereafter*. Between circa 1490 and 1516. Galleria dell'Accademia Venezia, Venice. http://boschproject.org/bosch_in_venice.html

Calhoun, J. B. (1973). Death Squared: The Explosive Growth and Demise of a Mouse Population. *Proceedings of the Royal Society of Medicine*, 66(1 Pt 2), 80–88. DOI: 10.1177/00359157730661P202 PMID: 4734760

Chawla, L. S., Akst, S., Junker, C., Jacobs, B., & Seneff, M. G. (2009). Surges of electroencephalogram activity at the time of death: A case series. *Journal of Palliative Medicine*, 12(12), 1095–1100. DOI: 10.1089/jpm.2009.0159 PMID: 19803731

Clinical Criteria for the Determination of Death. (2014). *WHO Technical Expert Consultation. WHO headquarters.*https://apps.who.int/iris/bitstream/handle/10665/254737/WHO-HIS-SDS-2017.5-eng.pdf; jsessionid=AE1305BBA7C-C4223AA944447BCFC0086?sequence=1

Dachet, F., Brown, J. B., Valyi-Nagy, T., Narayan, K. D., Serafini, A., Boley, N., Gingeras, T. R., Celniker, S. E., Mohapatra, G., & Loeb, J. A. (2021). Selective time-dependent changes in activity and cell-specific gene expression in human postmortem brain. *Scientific Reports*, 11(1), 6078. Advance online publication. DOI: 10.1038/s41598-021-85801-6 PMID: 33758256

Ebert, D. (1986). *Physiologische Aspekte des Yoga.Frontiers in Ecology and Evolution*, 9(121), 621686. Advance online publication. DOI: 10.3389/fevo.2021.621686

FREE LDCT Lung Screening Programme. (n.d.). *FREE LDCT lung screening programme*. Lung Cancer Network Malaysia._https://www.lungcancer.net.my/free-ldct-lung-screening-programme-2/

Holliday, R. (2009). The extreme arrogance of anti-aging medicine. *Biogerontology*, 10(2), 223–228. DOI: 10.1007/s10522-008-9170-6 PMID: 18726707

Kübler-Ross, E. (2008). *On Death and Dying*. Routledge. DOI: 10.4324/9780203889657

Kurtz, P. (1991). *Toward a New Enlightenment: The Philosophy of Paul Kurtz*. Transaction Publishers.

Landecker, H. (2000). *Immortality, In Vitro: A History of the HeLa Cell Line*. In Biotechnology and Culture: Bodies, Anxieties, Ethics. Bloomington (pp. 53-72) Indiana University Press. DOI: 10.1073/pnas.142393611246226

Lizza, J. P. (2018). Defining Death: Beyond Biology. *Diametros : a Journal of Philosophy*, 55(55), 1–19. DOI: 10.13153/diam.1172

Lott, D. T., Yeshi, T., Norchung, N., Dolma, S., Tsering, N., Jinpa, N., Woser, T., Dorjee, K., Desel, T., Fitch, D., Finley, A. J., Goldman, R., Ortiz Bernal, A. M., Ragazzi, R., Aroor, K., Koger, J., Francis, A., & Perlman, D. (2021). No Detectable Electroencephalographic Activity After Clinical Declaration of Death Among Tibetan Buddhist Meditators in Apparent Tukdam, a Putative Postmortem Meditation State. *Frontiers in Psychology*, 11(599190), 599190. Advance online publication. DOI: 10.3389/fpsyg.2020.599190 PMID: 33584435

Magrassi, L., Leto, K., & Rossi, F. (2013). Lifespan of neurons is uncoupled from organismal lifespan. *Proceedings of the National Academy of Sciences of the United States of America*, 110(11), 4374–4379. DOI: 10.1073/pnas.1217505110 PMID: 23440189

Martinez, D. E. (1998). Mortality patterns suggest lack of senescence in hydra. *Experimental Gerontology*, 33(3), 217–225. DOI: 10.1016/S0531-5565(97)00113-7 PMID: 9615920

Mattar, C. N., Choolani, M., Biswas, A., Waddington, S. N., & Chan, J. K. Y. (2011). Fetal gene therapy: Recent advances and current challenges. *Expert Opinion on Biological Therapy*, 11(10), 1257–1271. DOI: 10.1517/14712598.2011.585153 PMID: 21623703

Moody, R. A. (1988). *Life After Life: The Investigation of a Phenomenon Survival of Bodily Death*. Mockingbird Books.

Olshansky, S. J., Hayflick, L., & Perls, T. T. (2004). Introduction: Anti-aging medicine: The hype and the reality—part II. *The Journals of Gerontology. Series A, Biological Sciences and Medical Sciences*, 59(7), B649–B651. DOI: 10.1093/gerona/59.7.B649

Raymond, C. (2012, September 13). *What happens when you die?* Verywell Health. https://www.verywellhealth.com/what-happens-to-my-body-right-after-i-die-1132498

Ring, K. (1985). *Heading Toward Omega: In Search of the Meaning of the Near-Death Experience*. Harper Perennial.

Rolland, B., Jardri, R., Amad, A., Thomas, P., Cottencin, O., & Bordet, R. (2014). Pharmacology of Hallucinations: Several Mechanisms for One Single Symptom? *BioMed Research International*, 2014, 1–9. DOI: 10.1155/2014/307106 PMID: 24991548

Russill, P. (2006). *The Yoga of Sound: Tapping the Hidden Power of Music and Chant*. New World Library.

Sender, R., & Milo, R. (2021). The distribution of cellular turnover in the human body. *Nature Medicine*, 27(1), 45–48. DOI: 10.1038/s41591-020-01182-9 PMID: 33432173

Shermer, M. (2017). Who Are You? *Scientific American*, 317(1), 73. DOI: 10.1038/scientificamerican0717-73 PMID: 28632227

The dying process - palliative care Australia. (n.d.). Org.au. Retrieved August 14, 2024, from https://palliativecare.org.au/resource/resources-the-dying-process/

The top 10 causes of death. (2024, August 7). https://www.who.int/news-room/fact-sheets/detail/the-top-10-causes-of-death

Thienes, C., & Haley, T. J. (1972). *Clinical Toxicology* (5th ed.). Lea and Febiger.

Twycross, R., & Wilcock, A. (2016). *Introducing Palliative Care* (5th ed.). Pharmaceutical Press.

Wagner, A. M., Schoeberlein, A., & Surbek, D. (2009). Fetal gene therapy: Opportunities and risks. *Advanced Drug Delivery Reviews*, 61(10), 813–821. DOI: 10.1016/j.addr.2009.04.011 PMID: 19426772

Walker, A. E. (1978). Pathology of brain death. *Annals of the New York Academy of Sciences*, 315(1), 272–280. DOI: 10.1111/j.1749-6632.1978.tb50345.x PMID: 284739

ADDITIONAL READING

Ado, A. D., Ado, M. A., Pytsky, V. I., & Poryadin G. V. (2000). *Patologhycheskaja fyzyologhyja* [Pathological physiology]. Triada-X.

Chekman, I. S. (1980). Oslozhnenyja farmakoterapyy [Complications of pharmacotherapy]. *Health (Toronto).*

Derikot, N. (2018). *Mozhet ly chelovek statj bessmertnym? Nauchnyj analyz varyantov* [Can a person become immortal? Scientific analysis of options]. https://futurist.ru/articles/1434-mozhet-li-chelovek-staty-bessmertnim-nauchniy-razbor-variantov

Gubin, N. (2002). *Termynaljnye sostojanyja y klynycheskaja smertj* [Terminal states and clinical death]. https://n-t.ru/tp/mr/ts.htm

Lyubimova, Z. V., & Nikitina, A. A. (2019). *Vozrastnaja anatomyja y fyzyologhyja*. Age anatomy and physiology. Yurayt Publishing House.

Miklashevskaya, A. (2020). Yzrayljskye uchenye zajavyly, chto nashly sposob borotjsja so starenyem [Israeli scientists said they found a way to combat aging]. *Kommersant.*, 2020(November), 19.

Negovsky, V. A., Gurvich, A. M., & Zolotokrylina, E. S. (1987). Postreanymacyonnaja boleznj [Post-resuscitation disease]. *Medicine.*

Novitsky, V. V., & Urazova, O. I. (2020). *Patofiziologaya* [Pathophysiology] (Vols. 1–2). GEOTAR-Media.

Predvestnyky skoroj smerty. (2019). [Harbingers of imminent death]. https://pbmc.ru/vessels/pravda-chto-pered-smertyu-ochishchaetsya-kishechnik-predvestniki-skoroi-smerti/

Pristavka, E. (2021). *Zhyvye yskopaemye: kak sushhestvujut bessmertnye orghanyzmy y mozhet ly chelovek tozhe bytj bessmertnym* [Living fossils: how immortal organisms exist and whether a person can be immortal]. *Hightech 06.2021*. https://hightech.fm/2021/06/18/turritopsis-dohrnii-immortal

Priznaki pryblyzhajushhejsja smerty. (n.d.). Signs of approaching death. https://fondvera.ru/manual/priznaki-priblizhayushheysya-smerti/

Ryazantsev, S. (1994). *Tanatologia* [Thanatology]. East European Institute of Psychoanalysis.

Sorokin, Y. D. (n / d). *Termynaljnye sostojanyja* [Terminal states]. In Great Russian Encyclopedia (Vols. 1–35). Great Russian Encyclopedia.

Strukov, A. I., & Serov, V. V. (2010). *Patologhycheskaja anatomyja* [Pathological anatomy] (5th ed.). Litterra.

Zharov, A. V. (2021). *Patologhycheskaja anatomyja zhyvotnykh* [Pathological anatomy of animals]. "Doe".

ENDNOTES

1 In some forms of meditation, vivid visions are associated with a lack of oxygen.

2 Neurons do not divide, so they cannot get errors when dividing. Damage occurs due to a lack of support, toxins, oxygen, nutrition, or traumatic brain injury. Most neurons are present from birth, and only a small number are produced during later life.

Conclusion

First, let's summarize the main results.

1. Collected data on the problem under consideration, related to the cultural heritage of mankind, in particular, data from the natural sciences (here the authors did not conduct independent research), as well as modeling of open complex systems (the wave model of S-space and the theory of self-organization developed by the authors).
2. The adequacy of S-space wave model abstractions and axioms, as well as the theory of self-organization of complex systems, for the problems of modeling structures and processes at the stages of the emergence and evolution of life is shown. Also, the use of this apparatus for modeling a person's structure, life cycle, aging, death, and posthumous existence is demonstrated. Thus, the scope of its possible application has been significantly expanded.
3. It is shown that the phenomena of the origin and evolution of life, as well as the structure, life cycle, aging, death, and posthumous existence of a person, can be not only described but also understood as relationships between waves and solitons of open S-space.
4. This understanding leads to some theoretical and practical results. Obtaining them was not the main goal of this work; many of them are hypothetical and require experimental confirmation, but the authors consider it useful to mention them:
   a. A definition of life, free from binding to the specific features and limitations of the earthly version of life was given.
   b. The assumption about the place of origin and forms of intermediate links between inanimate matter and living organisms was substantiated.
   c. Define evolution as a process of limited optimization, where the conditions on the earth, time intervals, and available resources act as constraints, and optimization is reduced to scenarios of self-organization of open complex

systems of a soliton-wave nature, realized under the influence of external influences and certain restrictions.

d. This, in turn, makes it possible to define a "unit" of evolution as the whole ecological system of the earth, to explain the direction of the evolutionary process, to reduce estimates of the time required for the emergence and evolution of living organisms, to explain evolutionary phenomena, to consider the mechanisms of evolution from the molecular genetic to the ecosystem levels as manifestations of the self-organization laws, which implies their different speed and certain incoordination.

5. Understanding the structure and life cycle of a person within the framework of the relationship of waves and solitons and external interactions according to the scenario (1C, 1O) allows:
   a. To introduce the concept and propose the structure of the sensory space as a set of mental structures and channels of interaction with the environment, as well as determine their significance in the human-environment system.
   b. To determine the mechanisms and general limits of self-regulation.
   c. To suggest that external modulations and resonance are initiating factors in human reproduction.
   d. To understand the phenomena of development, stabilization, and aging, as well as calculate the main periods of the life cycle and determine their content (the test showed a good correlation with known data).
   e. To consider various theories of aging and life extension strategies and suggest options for their optimization.
   f. To model death as a process of destruction of the soliton-wave system.

Second, consider the sequence of obtaining these results chapter by chapter.

Such a test was successfully passed: the abstractions of the S-space wave model were compared with the conditions for the emergence of life on Ancient Earth, and their correspondence was shown (Chapter 1. *On the Edge of Life and the S-Space Abstractions Applicability Checking*).

Further, the axioms of the wave model were compared with modern ideas about the life processes and conditions for the existence of living organisms, and no significant inconsistencies were found (Chapter 2, *The Cell as a Complex System and the S-Space Axiomatic Wave Model Adequacy Checking*). Despite the verification nature of the first two Chapters, based on known facts (the authors did not conduct independent research in these areas), quite interesting results were obtained: 10 conditions for the emergence of life and 13 characteristics of life processes and conditions for the existence of living organisms were formulated, and a certain underestimation of systemic factors in modern research. At the same time, their influence, and especially the presence of self-organization, can help solve the important problem of lack of time for the abiogenic synthesis of complex organics. Here, the probability of obtaining high-molecular compounds and their limited stability, which requires rapid accumulation for subsequent syntheses, are important. The required time is significantly reduced if these processes do not occur spontaneously but are subject to self-organization. This possibility is an assumption that needs to be verified; its verification is beyond the scope of the present study. Also, based on a systemic understanding, a definition of life was given, free from binding to specific forms of reproduction, and general enough to extend it to ecological systems.

Chapter 3, *Self-Organization of S-Space. Description Means and Construction Methods*, introduces the ternary connective apparatus, a tool for working with texts and artifacts from different epochs.

Chapter 4, *Self-Organization of S-Space Scenarios,* considers the theory of self-organization of complex systems based on the axiomatic model of S-space. The construction of the theory was carried out together with the verification of the completeness and consistency of the axioms for a certain class of problems - following the ideas of Gödel and Tarski about the incompleteness of axiomatic systems and the discrepancy between the classes of true and provable statements.

Chapter 5, *Evolution Phenomena and Theories,* discusses ideas about macro- and micro-evolution and considers molecular-genetic and ecological factors. 15 laws of the evolutionary process are formulated. The successes and difficulties that modern evolutionary theories were depicted.

Chapter 6, *Evolution as a Self-Organization,* presents their modeling and interpretation from the standpoint of the complex system self-organization theory, which gives explanations of the mechanisms of arogenesis, allogenesis and degeneration, mass extinctions, and a few others.

Separately, we note the explanation of the existing direction of evolution according to several researchers: this is a consequence of certain scenarios of self-organization implementation, and there is no need to resort to creationist interpretations. The mechanisms of reproduction and inheritance are also modeled; all known forms can be expressed in terms of soliton-wave systems self-organization scenarios. For sexual reproduction, it has been suggested that the possible model of fertilization is superposition, and the condition for its success is the occurrence of resonance. This understanding helps explain life cycle patterns (Chapter 9) but needs further testing. The abstractness of the models used raises the question of their accuracy; the answer is pragmatic - accuracy corresponds to the work tasks.

Chapter 7, *The Human Body System Structure and Self-Regulation,* considers modern data on the structure of a person and his subsystems, self-regulation, and interactions with the outside world. These data are necessary for modeling the human structure.

In Chapter 8, *Human-Environment Model, Self-regulation and Rhythms Coordination*, the general plan of the human structure, its subjective space, mechanisms of self-regulation, and relations with the environment are modeled.

Compared with earlier publications (Mkhitaryan et al., 2004; Kovalyov et al., 2020), more attention is paid to building soliton-wave models of subjective space, human interactions with the environment, connection with the available facts about the structure and evolution of the Universe, as well as the conditions of self-regulation and the emergence of pathologies. For comparison with modern theories of the Universe's structure and evolution and the place of man in it, are given on ancient natural-philosophical systems that demonstrate a systematic (but not scientific in the modern sense) approach, and a standpoint of philosophical materialism.

Chapter 9, *Life Cycle from Conception to Senility,* discusses modern ideas about the human life cycle from birth to old age: its periodization, as well as anatomical, physiological, and psychological characteristics of different periods. Soliton-wave models are constructed for various stages of the life cycle and the cycle as a whole. The periods' life cycle durations calculations show a good correlation with known data.

Particular attention is paid to the changes that occur in old age - its empirical manifestations are described, modern theories of aging are represented, modeling is carried out and the causes of aging are considered from the viewpoint of the theory of self-organization. The results obtained have not only theoretical but also practical significance, for example, they allow optimizing interactions with various types of environments of specific focus groups of consumers (Mkhitaryan et al., 2021), assessing historical and modern strategies of life prolongation, and suggest a new plan, etc.

Chapter 10, *Near the Death*, details near-death phenomena: the clinical picture of approaching death (including physiological and psychological manifestations), as well as near-death experiences and similar trance states. A systematic analysis and modeling of these phenomena has been carried out, which allows us to assert, that they have a lifetime character confidently. Changes immediately after death are also considered, in particular the fact that cells do not die at the same time. Another aspect - the problem of biological immortality - is briefly outlined, as well as possible ways to achieve it. Based on all these data, the death phenomenon is modeled as a gradual destruction of an integral soliton-wave system (3 possible scenarios are considered), prolonging the trends that manifested themselves even in old age and directly following the life cycle model. After death, the soliton and wave components continue to exist for some time (estimated at several weeks) as separate systems.

These ideas, methods, models, and results form the basis for substantiating the main scenario of postmortem existence and its particular cases in the second part of our study – *Post-Mortem Existence Within the Complex Systems Self-Organization Theory Framework and in Traditional Cultures.*

REFERENCES

Kovalyov, Y., Mkhitaryan, N., & Nitsyn, A. (2020). *Self-organization of the Human Mind and the Transition from Paleolithic to Behavioral Modernity*. IGI Global. DOI: 10.4018/978-1-7998-1706-2

Mkhitaryan, N. M., Badeyan, G. V., & Kovalyov, Y. N. (2004). *Erghonomycheskye aspekti slozhnikh system* [Ergonomic Aspects of Complex Systems]. Naukova Dumka.

Mkhitaryan, N. M., Kovalyov, Y. M., Malik, T. V., Safronov, V. K., & Safronova, O. O. (2021). *Dyzajn seredovyshha mista: Baghatokryterialjna optymizacija ta rozumni tekhnologhiji* [Design of the City Environment: Multi-Criteria Optimization and Smart Technologies]. Naukova dumka.

Glossary

{DIKS} All Sp operations
∪,● S-elements: wave, soliton
⊥ S-elements perpendicularity
S-set
A, P Modalities of existence: actual and potential
bO Boundary-object correspondence
C Coincidence operation
Co Convolution of Sp
Cs Coordinate system
D Diffraction operation
DGM Discrete boundary model
H Observation
HEM Human-environment model
I Interference operation
NGM Continuous boundary model
ø, ö, ↓ Modalities of condition: expansion, weakening, disappearance
Pc Result of measuring
S Subject
S Superposition operation
Sb Subject-boundary correspondence
***Sp*, °** S-space
St Stratification of Sp
SWM Soliton-wave model
Ts Ternary connective
U Universe
úú S-elements parallelism
π Potential
O Object
OM Object model

Compilation of References

Adelberg, E. A., & Magee, P. T. (2015, July 8). *Amino acids*. AccessScience. https://doi.org/DOI: 10.1036/1097-8542.028100

Agarwal, S., & Rimm, D. L. (2012). Making Every Cell Like HeLa: A Giant Step For Cell Culture. *American Journal of Pathology*, 180(2), 443–445. DOI: 10.1016/j.ajpath.2011.12.001 PMID: 22192626

Ageing. (2024, August 11). Wikipedia. https://en.wikipedia.org/wiki/Ageing

Aging changes in organs, tissue, and cells. (n.d.). *Medical Encyclopedia*. https://medlineplus.gov/ency/article/004012.htmDOI: 10.1098/rstb.1991.0035

Aho, A. V., & Ullman, J. D. (1972). *The theory of parsing, translation, and compiling*. Prentice-Hall.

Alberts, B., Johnson, A., Lewis, J., Raff, M., Roberts, K., & Walter, P. (2002). *Molecular biology of the cell* (4th ed.). Garland Science., DOI: 10.1093/aob/mcg023

Albuquerque, U. P., De Medeiros, P. M., & Casas, A. (2015). *Evolutionary Ethnobiology*. Springer. DOI: 10.1007/978-3-319-19917-7

Alcock, J. (1981). *Psychology and Near-Death Experiences*. In *Kendrick Frazier. Paranormal Borderlands of Science*. (pp. 153–169). Prometheus Books.

Aleksandrov, S. Yu. (2004). Lorentz-Covariant Loop Quantum Gravity. *Theoretical and Mathematical Physics*, 139(3), 751–765. DOI: 10.1023/B:TAMP.0000029699.54716.0e

Allan, G. J., & Max, T. L. (2010) Molecular Genetic Techniques and Markers for Ecological Research. *Nature Education Knowledge, 3*(10).

Allen, D. A., & Wickramasinghe, D. T. (1981). Diffuse interstellar absorption bands between 2.9 and 4.0 μm. *Nature*, 294(5838), 239–240. DOI: 10.1038/294239a0

Amino Acids Reference Guide. (2021). https://beta-static.fishersci.com/content/dam/fishersci/en_US/images/amino-acids/AminoAcidsReference.pdfwww.stratigraphy.org/ICSchart/ChronostratChart2021-07.pdf

AnAge: The animal ageing and longevity database. (n.d.). Senescence.Info. Retrieved August 14, 2024, from https://genomics.senescence.info/species/index.html

Anatomy 3D Atlas. (n.d.). https://anatomy3datlas.com

Anatomy 3D Atlas. (n.d.). https://anatomy3datlas.com/

Anatomy of the Lymphatic and Immune Systems | Anatomy and Physiology II. (n.d.). https://courses.lumenlearning.com/suny-ap2/chapter/anatomy-of-the-lymphatic-and-immune-systems/

Aries, P. (2013). *The Hour of Our Death*. Vintage.

Arrhenius, S. (1908). *Worlds in the Making: The Evolution of the Universe*. Harper & Row.

Arrojo e Drigo, R., Lev-Ram, V., Tyagi, S., Ramachandr, R., Deerinck, T., Bushong, E., Phan, S., Orphan, V., Lechene, C., Ellisman, M. H., & Hetzer, M. W. (2019). Age Mosaicism across Multiple Scales in Adult Tissues. *Cell Metabolism*, 30(2), 343–351. DOI: 10.1016/j.cmet.2019.05.010 PMID: 31178361

Aryal, S. (Ed.). (2018). RNA- Properties, Structure, Types and Functions. In *Microbe Notes*. https://microbenotes.com/rna-properties-structure-types-and-functions/

Asarian, L., Gloy, V., & Geary, N. (2012). *Homeostasis*. In Homeostasis. In *Encyclopedia of Human Behavior* (pp. 324–333). Elsevier. DOI: 10.1016/B978-0-12-375000-6.00191-9

Australia, H. (2024). *The process of dying*. https://www.healthdirect.gov.au/the-physical-process-of-dying

Barton, N. H., Briggs, D. E. G., Eisen, J. A., Goldstein, D. B., & Patel, N. H. (2007). *Evolution*. Cold Spring Harbor Laboratory Press.

Bar-Yam, Y., & Minai, A. (Eds.). (2002). Unifying Themes in Complex Systems II. In *Proceedings of the 2nd International Conference on Complex Systems Cambridge*. Perseus Pres.

Beamish, R. (2007). *Dialectical Materialism*. In *The Blackwell Encyclopedia of Sociology*. Blackwell Pub. DOI: 10.1002/9781405165518.wbeosd062

Bednarik, R. G. (2013). *The Psychology of human behavior*. Gazelle https://www.researchgate.net/publication/291076363_The_psychology_of_human_behavior

Bell, E. A., Boehnk, P. T., Harrison, M., & Mao, W. L. (2015). Potentially biogenic carbon preserved in a 4.1-billion-year-old zircon. *Proceedings of the National Academy of Sciences of the United States of America*, 112(47), 14518–14521. DOI: 10.1073/pnas.1517557112 PMID: 26483481

Bell, P. J. L. (2001). Viral eukaryogenesis: Was the ancestor of the nucleus a complex DNA virus? *Journal of Molecular Evolution*, 53(3), 251–256. DOI: 10.1007/s002390010215 PMID: 11523012

Bell, P. J. L. (2006). Sex and the eukaryotic cell cycle is consistent with a viral ancestry for the eukaryotic nucleus. *Journal of Theoretical Biology*, 243(1), 54–63. DOI: 10.1016/j.jtbi.2006.05.015 PMID: 16846615

Benner, S. A. (2010). Defining Life. *Astrobiology*, 10(10), 1021–1030. DOI: 10.1089/ast.2010.0524 PMID: 21162682

Benton, M. J. (2005). *When Life Nearly Died: The Greatest Mass Extinction of All Time*. Thames & Hudson.

Berkemer, S. J., & McGlynn, S. E. A. (2020). New Analysis of Archaea–Bacteria Domain Separation: Variable Phylogenetic Distance and the Tempo of Early Evolution. *Molecular Biology and Evolution*, 37(8), 2332–2340. DOI: 10.1093/molbev/msaa089 PMID: 32316034

Bernat, J. L. (2013). The definition and criterion of death. *Handbook of Clinical Neurology*, 118, 419–435. DOI: 10.1016/B978-0-444-53501-6.00033-0 PMID: 24182395

Bernstein, H., Hopf, F. A., & Michod, R. E. (1987). The molecular basis of the evolution of sex. In *Advances in Genetics* (pp. 323–370). Elsevier. DOI: 10.1016/S0065-2660(08)60012-7

Best, C. H., Taylor, N. B., & West, J. B. (1985). *Best & Taylor's Physiological Basis of Medical Practice*. Williams & Wilkins.

Billman, G. E. (2020). Homeostasis: The underappreciated and far too often ignored central organizing principle of physiology. *Frontiers in Physiology*, 11, 200. Advance online publication. DOI: 10.3389/fphys.2020.00200 PMID: 32210840

Biotech Khan. (2014). DNA replication in Prokaryotes. https://biotechkhan.wordpress.com/2014/10/08/dna-replication-in-prokaryotes/

Blackmore, S. J. (1996). Near-death experiences. *Journal of the Royal Society of Medicine*, 89(2), 73–76. DOI: 10.1177/014107689608900204 PMID: 8683504

Blaustein, R. (2016). The Great Oxidation Event: Evolving understandings of how oxygenic life on Earth began. *Bioscience*, 66(3), 189–195. DOI: 10.1093/biosci/biv193

Blum, G. S. (1953). *Psychoanalytic theories of personality*. McGraw-Hill Book Company.

Bonanno, G. A., & Trauma, L. (2004). Human Resilience. Have We Underestimated the Human Capacity to Thrive After Extremely Aversive Events? *The American Psychologist*, 59(1), 20–28. DOI: 10.1037/0003-066X.59.1.20 PMID: 14736317

Bosch, H. (n.d.). *Visions of the Hereafter*. Between circa 1490 and 1516. Galleria dell'Accademia Venezia, Venice. http://boschproject.org/bosch_in_venice.html

Bose, D. (2019). Six Main Cell Functions. *Sciencing*. https://sciencing.com/six-main-cell-functions-6891800.html)

Bouvier, A., & Meenakshi, W. (2010). The age of the Solar system is redefined by the oldest Pb–Pb age of a meteoritic inclusion. *Nature Geoscience*, 3(9), 637–641. DOI: 10.1038/ngeo941

Bowler, P. J. (1979). Theodor Eimer and Orthogenesis: Evolution by 'Definitely Directed Variation'. *Journal of the History of Medicine and Allied Sciences 34*(1), 40–73._DOI: 10.1093/jhmas/XXXIV.1.40

BrainKart. (2020). Depicts the process of gametogenesis in both males and females, highlighting the importance of meiosis. *Gametogenesis*. BrainKart. Retrieved August 13, 2024, from https://www.brainkart.com/article/Gametogenesis_38015/

Bratic, A., & Larsson, N.-G. (2013). The role of mitochondria in aging. *The Journal of Clinical Investigation*, 123(3), 951–957. DOI: 10.1172/JCI64125 PMID: 23454757

Breseghello, F., & Coelho, S. A. G. (2013). Traditional and Modern Plant Breeding Methods with Examples in Rice (*Oryza sativa* L.). *Journal of Agricultural and Food Chemistry*, 61(35), 8277–8286. DOI: 10.1021/jf305531j PMID: 23551250

Brouwer, L. E. J. (1907). *Over de grondslagen der wiskunde*. Amst.-Lpz.

Brouwer, L. E. J. (1908). *De onbetrouwbaarheid der logische principes* (Vol. 2). Tijdsehz voor Wijsbegeerte.

Brown, G. (1979). *Liquid crystals and biological structures*. Elsevier Inc., DOI: 10.1016/B978-0-12-136850-0.X5001-1

Brown, J. S., & Pavlovic, N. B. (1992). Evolution in heterogeneous environments: Effects of migration on habitat specialization. *Evolutionary Ecology*, 6(5), 360–382. DOI: 10.1007/BF02270698

Brunk, C. F., & Marshall, C. R. (2021). 'Whole Organism', Systems Biology, and Top-Down Criteria for Evaluating Scenarios for the Origin of Life. *Life (Chicago, Ill.)*, 11(7), 690. DOI: 10.3390/life11070690 PMID: 34357062

Burkle, A. (2001). Physiology and pathophysiology of poly(ADP-rebosyl)ation. *BioEssays*, 23(9), 795–806. DOI: 10.1002/bies.1115 PMID: 11536292

Burns, T., & Dietzgen, J. (2002). The History of Marxism. *Science and Society*, 66(2), 202–227. DOI: 10.1521/siso.66.2.202.20997

Buzsáki, G. (2007). The structure of consciousness. *Nature*, 446(7133), 267. DOI: 10.1038/446267a PMID: 17361165

Buzsáki, G., & Watson, B. O. (2012). Brain rhythms and neural syntax: Implications for efficient coding of cognitive content and neuropsychiatric disease. *Dialogues in Clinical Neuroscience*, 14(4), 345–367. DOI: 10.31887/DCNS.2012.14.4/gbuzsaki PMID: 23393413

Calhoun, J. B. (1973). Death Squared: The Explosive Growth and Demise of a Mouse Population. *Proceedings of the Royal Society of Medicine*, 66(1 Pt 2), 80–88. DOI: 10.1177/00359157730661P202 PMID: 4734760

Çalıyurt, O. (2017). Role of Chronobiology as a Transdisciplinary Field of Research: Its Applications in Treating Mood Disorders. *Balkan Medical Journal*, 34(6), 514–521. DOI: 10.4274/balkanmedj.2017.1280 PMID: 29072179

Calvin Cycle. (n.d.). Calvin Cycle: Diagram, Definition, and Mechanism. https://researchtweet.com/calvin-cycle-diagram-definition-and-mechanism/

Cameron, D., Hartig, F., Minnuno, F. J., Oberpriller, J., Reineking, B., Van Oijen, M., & Dietze, M. (2022). Issues in calibrating models with multiple unbalanced constraints: The significance of the systematic model and data errors. *Methods in Ecology and Evolution*, 13(12), 2757–2770. DOI: 10.1111/2041-210X.14002

Carrard, G., Bulteau, A. L., Petropoulos, I., & Friguet, B. (2002). Impairment of proteasome structure and function in aging. *The International Journal of Biochemistry & Cell Biology*, 34(11), 1461–1474. DOI: 10.1016/S1357-2725(02)00085-7 PMID: 12200039

Cavalli-Sforza, L. L., & Feldman, M. W. (2003). The Application of Molecular Genetic Approaches to the Study of Human Evolution. *Nature Genetics*, 33(S3), 266–275. DOI: 10.1038/ng1113 PMID: 12610536

Cavender-Bares, J., Kothari, S. J., & Pearse, W. (2018). *Evolutionary Ecology of Communities. (2019). Evolutionary Biology*. Oxford University Press., DOI: 10.1093/obo/9780199941728-0111

Cell membrane detailed diagram. (2007). Author: LadyofHats Mariana Ruiz. Wikipedia.org. Retrieved August 12, 2024, from https://en.wikipedia.org/wiki/File:Cell_membrane_detailed_diagram_en.svg)

Chakraborty, U. (2014). Effects of different phases of the lunar month on humans. *Biological Rhythm Research*, 45(3), 383–396. DOI: 10.1080/09291016.2013.830508

Characteristics of Life. (2021). https://bio.libretexts.org/Bookshelves/Introductory_and_General_Biology/Book%3A_Introductory_Biology_(CK-12)/01%3A_Introduction_to_Biology/1.04%3A_Characteristics_of_Life

Chawla, L. S., Akst, S., Junker, C., Jacobs, B., & Seneff, M. G. (2009). Surges of electroencephalogram activity at the time of death: A case series. *Journal of Palliative Medicine*, 12(12), 1095–1100. DOI: 10.1089/jpm.2009.0159 PMID: 19803731

Chen, S.-C., Sun, G.-X., Yan, Y., Konstantinidis, K. T., Zhang, S.-Y., Deng, Y., Li, X.-M., Cui, H.-L., Musat, F., Popp, D., Rosen, B. P., & Zhu, Y.-G. (2020). The Great Oxidation Event expanded the genetic repertoire of arsenic metabolism and cycling. *Proceedings of the National Academy of Sciences of the United States of America*, 117(19), 10414–10421. DOI: 10.1073/pnas.2001063117 PMID: 32350143

Chivian, D., Brodie, E. L., Alm, E. J., Culley, D. E., Dehal, P. S., DeSantis, T. Z., Gihring, T. M., Lapidus, A., Lin, L.-H., Lowry, S. R., Moser, D. P., Richardson, P. M., Southam, G., Wanger, G., Pratt, L. M., Andersen, G. L., Hazen, T. C., Brockman, F. J., Arkin, A. P., & Onstott, T. C. (2008). Environmental Genomics Reveals a Single-Species Ecosystem Deep Within Earth. *Science*, 322(5899), 275–278. DOI: 10.1126/science.1155495 PMID: 18845759

Chomsky, N. *Knowledge of Language: Its Nature, Origin, and Use. Greenwood* Publishing Group.

Chomsky, N. (1957). *Natural language syntax*. Mouton & Co.

Chomsky, N. (1965). *Aspects of the Theory of Syntax*. MIT Press.

Chomsky, N. (1975). *The Logical Structure of Linguistic Theory*. Springer US.

Chomsky, N. (2006). *Language and Mind* (3rd ed.). Cambridge University Press. DOI: 10.1017/CBO9780511791222

Clinical Criteria for the Determination of Death. (2014). *WHO Technical Expert Consultation. WHO headquarters.*https://apps.who.int/iris/bitstream/handle/10665/254737/WHO-HIS-SDS-2017.5-eng.pdf; jsessionid=AE1305BBA7C-C4223AA944447BCFC0086?sequence=1

Cloverleaf structures of tRNA. (1998). Uploaded by Victor Serebrov. https://www.researchgate.net/figure/Primary-and-cloverleaf-structures-of-tRNA-Phe-from-Ecoli-A-and-its-gene-transcript_fig5_13692313

Coleman, W. B. & Tsongalis, G. J. (Eds.). (2018). *Molecular Pathology: The Molecular Basis of Human Disease.* Academic Press. DOI: 10.1016/C2014-0-02163-6

Conscious. (2004). 11,000-Year Sunspot Number Reconstruction. The tabulated years correspond to centers of the corresponding 10-year intervals. Negative values are artifacts and are consistent with zero within the error limits. Wikipedia contributors. https://en.wikipedia.org/wiki/File:Sunspots_11000_years.svg

Copley, S. D., Smith, E., & Morowitz, H. J. (2005). A mechanism for the association of amino acids with their codons and the origin of the genetic code. *Proceedings of the National Academy of Sciences of the United States of America*, 102(12), 4442–4447. DOI: 10.1073/pnas.0501049102 PMID: 15764708

Dachet, F., Brown, J. B., Valyi-Nagy, T., Narayan, K. D., Serafini, A., Boley, N., Gingeras, T. R., Celniker, S. E., Mohapatra, G., & Loeb, J. A. (2021). Selective time-dependent changes in activity and cell-specific gene expression in human postmortem brain. *Scientific Reports*, 11(1), 6078. Advance online publication. DOI: 10.1038/s41598-021-85801-6 PMID: 33758256

Dalrymple, G. B. (2001). The age of the Earth in the twentieth century: A problem (mostly) solved. *Special Publication - Geological Society of London*, 190(1), 205–221. DOI: 10.1144/GSL.SP.2001.190.01.14

Darwin, C. (1859). *On the Origin of Species by Means of Natural Selection, or the Preservation of Favoured Races in the Struggle for Life* (1st ed.). John Murray. DOI: 10.5962/bhl.title.82303

Darwin, C. (1871). *The Descent of Man, and Selection in Relation to Sex* (1st ed.). John Murray.

Di Giulio, M. (2008). An extension of the coevolution theory of the origin of the genetic code. *Biology Direct*, 3(1), 37–37. DOI: 10.1186/1745-6150-3-37 PMID: 18775066

Drake equation. (n.d.). Wikipedia.org. Retrieved August 12, 2024, from https://en.wikipedia.org/wiki/Drake_equation

Drenos, F., & Kirkwood, T. B. L. (2005). Modelling the disposable soma theory of aging. *Mechanisms of Ageing and Development*, 126(1), 99–103. DOI: 10.1016/j.mad.2004.09.026 PMID: 15610767

Duret, L. (2008) Neutral theory: The null hypothesis of molecular evolution. *Nature Education, 1*(1), 218. https://onlinelibrary.wiley.com/journal/20457758

Ebert, D. (1986). *Physiologische Aspekte des Yoga.Frontiers in Ecology and Evolution*, 9(121), 621686. Advance online publication. DOI: 10.3389/fevo.2021.621686

Ecoddington14. (2014). Binary_Fission. *File: Binary fission 2.Svg*https://commons.wikimedia.org/wiki/File:Binary_Fission_2.svg

Einasto, J. (2012). *Dark matter*. In *Astronomy and Astrophysics*. https://www.eolss.net/sample-chapters/c01/E6-119-21-00.pdf

Eisenstein, M. (2013). Chronobiology: Stepping out of time. *Nature*, 497(7450), S10–S12. DOI: 10.1038/497S10a PMID: 23698500

Erikson, E. H. (1950). *Childhood and Society*. W. W. Norton & Co.

Erikson, E. H., & Erikson, J. M. (1987). *The Life Cycle Completed*. W. W. Norton & Co.

Erikson, E. H., Erikson, J. M., & Kivnick, H. Q. (1986). *Vital involvement in old age: The experience of old age in our time*. W. W. Norton & Company.

Explain the process of transcription in Eukaryotes. (2021). Upload by Ankush01. https://www.sarthaks.com/1199908/explain-the-process-of-transcription-in-eukaryotes

Fasoula, D. A., Ioannides, I. M., & Omirou, M. (2020). Phenotyping and Plant Breeding: Overcoming the Barriers. *Frontiers in Plant Science*, 10, 1713. Advance online publication. DOI: 10.3389/fpls.2019.01713 PMID: 31998353

Fedo, C. M., & Whitehouse, M. J. (2002). Metasomatic Origin of Quartz-Pyroxene Rock, Akilia, Greenland, and Implications for Earth's Earliest Life. *Science*, 296(5572), 1448–1451. DOI: 10.1126/science.1070336 PMID: 12029129

Finch, C. (1990). *Senescence, Longevity, and the Genome*. The University of Chicago Press.

Fisher, R. A. (1958). *The genetic theory of natural selection* (2nd ed.). Dover Publications.

Flatt, T., & Schmidt, P. S. (2009). Integrating evolutionary and molecular genetics of aging. *Biochimica et Biophysica Acta*, 1790(10), 951–962. DOI: 10.1016/j.bbagen.2009.07.010 PMID: 19619612

Forterre, P. (2005). The two ages of the RNA world, and the transition to the DNA world: A story of viruses and cells. *Biochimie*, 87(9-10), 793–803. DOI: 10.1016/j.biochi.2005.03.015 PMID: 16164990

Forterre, P. (2006). Three RNA cells for ribosomal lineages and three DNA viruses to replicate their genomes: A hypothesis for the origin of cellular domain. *Proceedings of the National Academy of Sciences of the United States of America*, 103(10), 3669–3674. DOI: 10.1073/pnas.0510333103 PMID: 16505372

Forterre, P. (2010). Defining Life: The Virus Viewpoint. *Origins of Life and Evolution of the Biosphere*, 40(2), 151–160. DOI: 10.1007/s11084-010-9194-1 PMID: 20198436

Forterre, P., Gribaldo, S., & Brochier, C. (2005). Luca: The last universal common ancestor. *Medecine Sciences*, 21(10), 860–865. DOI: 10.1051/medsci/20052110860 PMID: 16197904

Foster, R. G., & Roenneberg, T. (2008). Human Responses to the Geophysical Daily, Annual and Lunar Cycles. *Current Biology*, 18(17), R784–R794. DOI: 10.1016/j.cub.2008.07.003 PMID: 18786384

Fox, C. W., Roff, D. A., & Fairbairn, D. J. (2001). *Evolutionary Ecology: Concepts and Case Studies*. Oxford University Press. DOI: 10.1093/oso/9780195131543.001.0001

Frank, M. G. (2008). *Brain Rhythms. Springer eBooks* (pp. 482–483). DOI: 10.1007/978-3-540-29678-2_727

Frankel, S., & Rogina, B. (2021). Evolution, Chance, and Aging. *Frontiers in Genetics*, 12, 733184. Advance online publication. DOI: 10.3389/fgene.2021.733184 PMID: 34567083

Fray, N., Bardyn, A., Cottin, H., Altwegg, K., Baklouti, D., Briois, C., Colangeli, L., Engrand, C., Fischer, H., Glasmachers, A., Grün, E., Haerendel, G., Henkel, H., Höfner, H., Hornung, K., Jessberger, E. K., Koch, A., Krüger, H., Langevin, Y., & Hilchenbach, M. (2016). High-molecular-weight organic matter in the particles of comet 67P/Churyumov–Gerasimenko. *Nature*, 538(7623), 72–74. DOI: 10.1038/nature19320 PMID: 27602514

FREE LDCT Lung Screening Programme. (n.d.). *FREE LDCT lung screening programme*. Lung Cancer Network Malaysia. https://www.lungcancer.net.my/free-ldct-lung-screening-programme-2/

Frenzel, A., Binder, H., Walter, N., Wirkner, K., Loeffler, M., & Loeffler-Wirth, H. (2020). The aging human body shape. *NPJ Aging and Mechanisms of Disease*, 6(1), 5. Advance online publication. DOI: 10.1038/s41514-020-0043-9 PMID: 32218988

Frías, L. D. (2010). Omissions in the synthetic theory of evolution. *Biological Research*, 43(3). Advance online publication. DOI: 10.4067/S0716-97602010000300006 PMID: 21249301

Furukawa, Y., Chikaraishi, Y., Ohkouchi, N., Ogawa, N. O., Glavin, D. P., Dworkin, J. P., Abe, C., & Nakamura, T. (2019). Extraterrestrial ribose and other sugars in primitive meteorites. *Proceedings of the National Academy of Sciences of the United States of America*, 116(49), 24440–24445. DOI: 10.1073/pnas.1907169116 PMID: 31740594

Gaillard, C., De Sousa, C., Amengual, J., Loriette, C., Ziane, C., Hassen, S. B. H., Di Bello, F., & Hamed, S. B. (2021). Attentional brain rhythms during prolonged cognitive activity. bioRxiv *(Cold Spring Harbor Laboratory)*. DOI: 10.1101/2021.05.26.445730

Gavrilov, L. A., & Gavrilova, N. S. (2002). Evolutionary Theories of Aging and Longevity. *TheScientificWorldJournal*, 2, 339–356. DOI: 10.1100/tsw.2002.96 PMID: 12806021

Gennis, R. B. (1989). *Biomembranes: Molecular Structure and Function*. Springer-Verlag. DOI: 10.1007/978-1-4757-2065-5

George, A. (2016). *The Epic of Gilgamesh*. Penguin.

Gibson, E. K., Westall, F., Jr., McKay, D. S., Thomas-Keprta, K., Wentworth, S., & Romanek, C. S. (1999). Evidence for ancient Martian life. *NASA Johnson Space Center.*

Gibson, E. K.Jr, Westall, F., McKay, D. S., Thomas-Keprta, K., Wentworth, S., & Romanek, C. S. (1999). *Evidence for ancient Martian life. Mail Code SN2*. NASA Johnson Space Center.

Gilbert, W. (1986). The RNA world. *Nature*, 319(6055), 618–618. DOI: 10.1038/319618a0

Glansdorff, N., Xu, Y., & Labedan, B. (2008). The Last Universal Common Ancestor: Emergence, constitution and genetic legacy of an elusive forerunner. *Biology Direct*, 3(1), 29. Advance online publication. DOI: 10.1186/1745-6150-3-29 PMID: 18613974

Gödel, K. (1995). Some basic theorems on the foundations of mathematics and their implications. In *Feferman, ed., 1995. Kurt Gödel Collected Works*, v. III, 304-323. Oxford University Press. (Original work published 1951)

Gödel, K. (1995). Some basic theorems on the foundations of mathematics and their implications. In *Kurt Gödel Collected Works*, v. III, 304-323. Oxford University Press. (Original work published 1951)

Golitsin, Y. N., & Krylov, M. C. (2010). *Cell division: theory, variants, and degradation*. Nova Science Publishers.

Gompertz, B. (1825). On the Nature of the Function Expressive of the Law of Human Mortality, and on a New Mode of Determining the Value of Life Contingencies. *Philosophical Transactions of the Royal Society of London*, 115, 513–585. DOI: 10.1098/rstl.1825.0026

Grant, V. (1985). *The Evolutionary Process. A Critical Review of Evolutionary Theory*. Columbia University Press.

Green, B. (2011). *The Elegant Universe: Superstrings, Hidden Dimensions, and the Quest for the Ultimate Theory* (Reprint edition). Norton & Company.

Greenberg, A. J., & Wu, C.-I. (2006). Molecular Genetics of Natural Populations. *Molecular Biology and Evolution*, 23(5), 883–886. DOI: 10.1093/molbev/msj077 PMID: 16361304

Grehan, J. R., & Ainsworth, R. (1985). Orthogenesis and Evolution. *Systematic Zoology*, 34(2), 174–192. DOI: 10.2307/2413325

Gukov, S. (1998). Introduction to string dualities. *Physics Uspekhi*, 41(7), 627–638. DOI: 10.1070/PU1998v041n07ABEH000416

Guyton, A. C., & Hall, J. E. (2005). *Textbook of Medical Physiology*. Saunders. (Original work published 1969)

Hachmo, Y., Hadanny, A., Abu Hamed, R., Daniel-Kotovsky, M., & Catalogna, M. (2020). Hyperbaric oxygen therapy increases telomere length and decreases immunosenescence in isolated blood cells: a prospective trial. *Research Paper 12*(22), 22445-22456. DOI: 10.18632/aging.202188

Haldane, J. B. S. (1957). The cost of natural selection. *Journal of Genetics*, 55(3), 551–524. DOI: 10.1007/BF02984069

Hames, D., & Hooper, N. (2005). *Biochemistry* (3 ed). Taylor & Francis Group.

Hamilton, W. D. (1964). The Genetical Evolution of Social Behaviour. *Journal of Theoretical Biology*, 7(1), 1–52. DOI: 10.1016/0022-5193(64)90038-4 PMID: 5875341

Hanczyc, M. M., & Monnard, P.-A. (2017). Primordial membranes: More than simple container boundaries. *Current Opinion in Chemical Biology*, 40, 78–86. DOI: 10.1016/j.cbpa.2017.07.009 PMID: 28802999

Hansen, C. D. A. (2000). *Daoist Theory of Chinese Thought: A Philosophical Interpretation.* Oxford University Press. DOI: 10.1093/oso/9780195134193.001.0001

Harman, D. (1972). The biologic clock: The mitochondria? *Journal of the American Geriatrics Society*, 20(4), 145–147. DOI: 10.1111/j.1532-5415.1972.tb00787.x PMID: 5016631

Harman, D. (1981). The aging process. *Proceedings of the National Academy of Sciences of the United States of America* 78(11), 7124-7128. DOI: 10.1073/pnas.78.11.7124

Hartung, F., & Schiemann, J. (2014). Precise plant breeding using new genome editing techniques: Opportunities, safety, and regulation in the EU. *The Plant Journal*, 78(5), 742–752. DOI: 10.1111/tpj.12413 PMID: 24330272

Harvey, J. M. (n.d.). Four major membrane protein complexes in photosynthesis: photosystem I, cytochrome b 6 f, photosystem II, and ATP synthase. Researchgate.net. Retrieved August 12, 2024, from https://www.researchgate.net/figure/Four-major-membrane-protein-complexes-in-photosynthesis-photosystem-I-cytochrome-b-6-f_fig2_237046097

Hayflick, L. (1998). How and why do we age? *Experimental Gerontology*, 33(7–8), 639–653. DOI: 10.1016/S0531-5565(98)00023-0 PMID: 9951612

Hayflick, L., & Moorhead, P. S. (1961). The serial cultivation of human diploid cell strains. *Experimental Cell Research*, 25(3), 585–621. DOI: 10.1016/0014-4827(61)90192-6 PMID: 13905658

Hein, J. E., & Blackmond, D. G. (2012). On the Origin of Single Chirality of Amino Acids and Sugars in Biogenesis. *Accounts of Chemical Research*, 45(12), 2045–2054. DOI: 10.1021/ar200316n PMID: 22353168

Hein, J. E., Tse, E., & Blackmond, D. G. (2011). A route to enantiopure RNA precursors from nearly racemic starting materials. *Nature Chemistry*, 3(9), 704–706. DOI: 10.1038/nchem.1108 PMID: 21860459

Heyting, A. (1930). *Die formalen Regeln der intuitionistischen Logik.* 3 parts, In *Sitzungsberichte der preußischen Akademie der Wissenschaften. phys.-math. Klasse.* 42-65, 57-71, 158–169.

Heyting, A. (1956). *Intuitionism. An introduction.* North-Holland Publishing Co.

Hilbert, D. (1950), *Grundlagen der Geometrie* [The Foundations of Geometry] (2nd ed.). Open Court Publishing. (Original work published 1902)

Hilbert, D. (1950). *Grundlagen der Geometrie* [The Foundations of Geometry] (2nd ed.). (Townsend, E. J., Trans.). Open Court Publishing. (Original work published 1902)

Hilbert, D., & Bernays, P. (1934). *Grundlagen der mathematic* [The Foundations of Mathematics]. Julius Springer.

Hiller-Sturmhöfel, S., & Bartke, A. (1998). The Endocrine System. *Alcohol Health and Research World*, 22(3), 153–164. PMID: 15706790

Hodgskiss, M. S. W., Crockford, P. W., Peng, Y. A., Wing, B. A., & Horner, T. J. (2019). Productivity collapse to end Earth's Great Oxidation. *Proceedings of the National Academy of Sciences of the United States of America*, 116(35), 17207–17212. DOI: 10.1073/pnas.1900325116 PMID: 31405980

Holland, H. D. (2006). The oxygenation of the atmosphere and oceans. *Philosophical Transactions of the Royal Society of London. Series B, Biological Sciences*, 361(1470), 903–915. DOI: 10.1098/rstb.2006.1838 PMID: 16754606

Holliday, R. (2009). The extreme arrogance of anti-aging medicine. *Biogerontology*, 10(2), 223–228. DOI: 10.1007/s10522-008-9170-6 PMID: 18726707

Hollis, J. (1993). *The Middle Passage: From Misery to Meaning in Midlife*. Inner City Books.

Hoover, R. B., Rozanov, A. Yu., Krasavin, E. A., Ryumin, A. K., & Kapralov, M. I. (2018). Diatoms in the Orgueil Meteorite. *Paleontological Journal*, 52(13), 1647–1650. DOI: 10.1134/S0031030118130051

Hughes, J. F., Skaletsky, H., Pyntikova, T., Minx, P. J., Graves, T., Rozen, S., Wilson, R. K., & Page, D. C. (2005). Conservation of Y-linked genes during human evolution revealed by comparative sequencing in chimpanzees. *Nature*, 437(7055), 100–103. DOI: 10.1038/nature04101 PMID: 16136134

Human Brain Project. (n.d.). https://www.humanbrainproject.eu/en/

Human Physiology/Development: birth through death. (2022, August 15). Wikibooks, Open Books for an Open World. https://en.wikibooks.org/wiki/Human_Physiology/Development:_birth_through_death

Huneman, P., & Walsh, D. M. (2017). *Challenging the Modern Synthesis: Adaptation, Development, and Inheritance.* Oxford University Press. DOI: 10.1093/oso/9780199377176.001.0001

Huxley, J. (1942). *Evolution: the Modern Synthesis.* George Allen and Unwin Ltd.

Hyde, W. T., Crowley, T. J., Baum, S. K., & Peltier, W. R. (2000). Neoproterozoic 'snowball Earth' simulations with a coupled climate/ice-sheet model. *Nature*, 405(6785), 425–429. DOI: 10.1038/35013005 PMID: 10839531

Ichihashi, N., Usui, K., Kazuta, Y., Sunami, T., Matsuura, T., & Yomo, T. (2013). Darwinian evolution in a translation-coupled RNA replication system within a cell-like compartment. *Nature Communications*, 4(1), 2494. Advance online publication. DOI: 10.1038/ncomms3494 PMID: 24088711

Jan, Y., & Jan, L. Y. (1999). Asymmetry across species. *Nature Cell Biology*, 1(2), E42–E44. DOI: 10.1038/10036 PMID: 10559895

Jiang, L., Dziedzic, P., Spacil, Z., Zhao, G.-L., Nilsson, L., Ilag, L. L., & Córdova, A. (2014). Abiotic synthesis of amino acids and self-crystallization under prebiotic conditions. *Scientific Reports*, 4(1), 6769. Advance online publication. DOI: 10.1038/srep06769 PMID: 25346284

Jmarchn (2009). Organ systems and human sense organs. Own work based on: Human body features.png. https://natworld.info/nauki-o-prirode/osnovnye-sistemy-organov-cheloveka-nazvaniya-funkczii-i-illyustracziyab

Johanson, D. C., & Wong, K. (2009). *Lucy's Legacy: The Quest for Human Origins.* Crown.

Johns Hopkins Medicine. (n.d.). Anatomy of the Urinary System. https://www.hopkinsmedicine.org/health/wellness-and-prevention/anatomy-of-the-urinary-system

Johns Hopkins Medicine. (n.d.). Renal system / Urinary system. Front view. Anatomy of the Urinary System. https://www.hopkinsmedicine.org/health/wellness-and-prevention/anatomy-of-the-urinary-system

Johnston, W., Unrau, P., Lawrence, M., Glasner, M. E., & Bartel, D. P. (2001). RNA-catalyzed RNA polymerization: Accurate and general RNA-templated primer extension. *Science*, 292(5520), 1319–1325. DOI: 10.1126/science.1060786 PMID: 11358999

Jones, R. E., & Lopez, K. (2013). *Human Reproductive Biology* (4th ed.). Academic Press. [Textbook]

Kan, S. B., Lewis, R. D., Chen, K., & Arnold, F. H. (2016). Directed evolution of cytochrome c for carbon–silicon bond formation: Bringing silicon to life. *Science*, 354(6315), 1048–1051. DOI: 10.1126/science.aah6219 PMID: 27885032

Karagyaur, M., Rostovtseva, A., Semina, E., Klimovich, P., Balabanyan, V., Makarevich, P., Popov, V., Stambolsky, D., & Tkachuk, V. (2020). A Bicistronic Plasmid Encoding Brain-Derived Neurotrophic Factor and Urokinase Plasminogen Activator Stimulates Peripheral Nerve Regeneration After Injury. *The Journal of Pharmacology and Experimental Therapeutics*, 372(3), 248–255. DOI: 10.1124/jpet.119.261594 PMID: 31888957

Kataru, R. P., Baik, J. E., Park, H. J., Wiser, I., Rehal, S., Shin, J. Y., & Mehrara, B. J. (2019). Regulation of Immune Function by the Lymphatic System in Lymphedema. *Frontiers in Immunology*, 10, 470. Advance online publication. DOI: 10.3389/fimmu.2019.00470 PMID: 30936872

Kimura, M. (1983). *The Neutral Theory of Molecular Evolution*. Cambridge University Press. DOI: 10.1017/CBO9780511623486

Kimura, M. (1991). The neutral theory of molecular evolution: A review of recent evidence. *Japanese Journal of Genetics*, 66(4), 367–386. DOI: 10.1266/jjg.66.367 PMID: 1954033

Kirkwood, T. B. L. (1977). Evolution of aging. *Nature*, 270(5635), 301–304. DOI: 10.1038/270301a0 PMID: 593350

Kirkwood, T. B. L. (2005). Understanding the Odd Science of Aging. *Cell*, 120(4), 437–447. DOI: 10.1016/j.cell.2005.01.027 PMID: 15734677

Kleaney, S. C., & Wesley, R. E. (1965). *Foundations of intuitionistic mathematics*. North-Holland Pub.

Kleene, S. C. (2002). *Mathematical Logic*. Dover Publications. (Original work published 1961)

Kleene, S. C. (2009). *Introduction to metamathematics*. Ishi Press. (Original work published 1950)

Kolmogoroff, A. (1932). Zur Deutung der intuitionistischen Logik. *Mathematische Zeitschrift*, 35(1), 58–65. DOI: 10.1007/BF01186549

Koonin, E. V. (2003). Comparative genomics, minimal gene-sets and the last universal common ancestor. *Nature Reviews. Microbiology*, 1(2), 127–136. DOI: 10.1038/nrmicro751 PMID: 15035042

Koonin, E. V. (2009). On the Origin of Cells and Viruses. Primordial Virus World Scenario. *Annals of the New York Academy of Sciences*, 1178(1), 47–64. Advance online publication. DOI: 10.1111/j.1749-6632.2009.04992.x PMID: 19845627

Koonin, E. V. (2011). *The Logic of Chance: The Nature and Origin of Biological Evolution*. FT Press.

Koonin, E. V., & Novozhilov, A. S. (2009). Origin and evolution of the genetic code: The universal enigma. *IUBMB Life*, 61(2), 99–111. DOI: 10.1002/iub.146 PMID: 19117371

Koshland, D. E.Jr. (2002). The Seven Pillars of Life. *Science*, 295(5563), 2215–2216. DOI: 10.1126/science.1068489 PMID: 11910092

Kostakis, G. E., & Tibbetts, I. (2020). Recent Bio-Advances in Metal-Organic Frameworks. *Molecules (Basel, Switzerland)*, 25(6), 1291. DOI: 10.3390/molecules25061291 PMID: 32178399

Kovalyov, Y. M., & Kalashnikova, V. V. (2023). Human life cycle modeling. *Suchasni problemy modeljuvannja, 25,* 110-122.

Kovalyov, Y., Mkhitaryan, N., & Nitsyn, A. (2020). *Self-organization of the human mind and the transition from paleolithic to behavioral modernity*. IGI Global. DOI: 10.4018/978-1-7998-1706-2

Krause, W. (1992). Measuring mental performance: An old idea and a new approach. *Zeitschrift für Experimentelle und Angewandte Psychologie*, 39(1), 114–128. https://pubmed.ncbi.nlm.nih.gov/1598763/ PMID: 1598763

Krol, L. R. (2020). Ten seconds of simulated EEG data in the five different frequency bands of neural oscillations, or brainwaves: delta, theta, alpha, beta, and gamma. *File: EEG Brainwaves.svg - Wikimedia Commons*. https://commons.wikimedia.org/wiki/File:EEG_Brainwaves.svg

Krut'ko, V. N., Dontsov, V. I., Ermakova, N. A., Makarova, V. V., Mitrokhin, O. V., Shashina, E. A., & Shcherbakov, D. V. (2021). Method and Computer System for Dialog Optimization of Aging Biomarker Panels for Biological Age Assessment. *Frontiers in Genetics*, 12, 634734. Advance online publication. DOI: 10.3389/fgene.2021.634734 PMID: 33747048

Kübler-Ross, E. (2008). *On Death and Dying*. Routledge. DOI: 10.4324/9780203889657

Kurtz, P. (1991). *Toward a New Enlightenment: The Philosophy of Paul Kurtz*. Transaction Publishers.

LadyofHats & Jmarchn. Respiratory system. (2007). *File: Respiratory system complete en.svg*. Wikipedia, The Free Encyclopedia. https://en.wikipedia.org/wiki/File:Respiratory_system_complete_en.svg

LadyofHats, & Mariana Ruiz Villarreal. (2009)., The diagram of the human Circulatory system in anterior view. Retrieved August 13, 2024, from https://upload.wikimedia.org/wikipedia/commons/thumb/2/29/Circulatory_System_en.svg/1357px-Circulatory_System_en.svg.png

Laland, K. N., Uller, T., Feldman, M. W., Sterelny, K., Müller, G. B., Moczek, A., Jablonka, E., & Odling-Smee. (2015). The extended evolutionary synthesis: its structure, assumptions and predictions. *Proceedings. Biological Sciences, 282*(1813), 20151019. DOI: 10.1098/rspb.2015.1019

Landecker, H. (2000). *Immortality, In Vitro: A History of the HeLa Cell Line*. In Biotechnology and Culture: Bodies, Anxieties, Ethics. Bloomington (pp. 53-72) Indiana University Press. DOI: 10.1073/pnas.142393611246226

Larin, N., & Warren, H. C. (1993). *Hydridic Earth: the New Geology of Our Primordially Hydrogen-rich Planet*. Polar publishing.

Le Bourg, E. (2001). A mini-review of the evolutionary theories of aging. *Demographic Research*, 4, 1–28. DOI: 10.4054/DemRes.2001.4.1

Lee, T. I., & Young, R. A. (2013). Transcriptional regulation and its misregulation in disease. *Cell*, 152(6), 1237–1251. DOI: 10.1016/j.cell.2013.02.014 PMID: 23498934

Leipe, D. D., Aravind, L., & Koonin, E. V. (1999). Did DNA replication evolve twice independently? *Nucleic Acids Research*, 27(17), 3389–3401. DOI: 10.1093/nar/27.17.3389 PMID: 10446225

Levit, G. S., & Olsson, L. (2006). 'Evolution on Rails': Mechanisms and Levels of Orthogenesis. *Annals of the History and Philosophy of Biology*, 11, 99–138.

Lewin, B. (2004). *Genes VIII*. Pearson Prentice Hall.

Liberles, D. A., Chang, B., Geiler-Samerotte, K., Goldman, A., Hey, J., Kaçar, B., Meyer, M., Murphy, W., Posada, D., & Storfer, A. (2020). Emerging Frontiers in the Study of Molecular Evolution. *Journal of Molecular Evolution*, 88(3), 211–226. DOI: 10.1007/s00239-020-09932-6 PMID: 32060574

Libretti, S., & Puckett, Y. (2022). *Physiology, Homeostasis*. StatPearls Publishing. https://pubmed.ncbi.nlm.nih.gov/32644564/

Lim, Q., Ver Steeg, G., Yu, S., & Malo, J. (2022). Functional Connectome of the Human Brain with Total Correlation. *Entropy (Basel, Switzerland)*, 24(12), 1725. DOI: 10.3390/e24121725 PMID: 36554129

Lin, Y. C., & Tung, C. (2013). Acupuncture Qi Flow and Points Measurement. In *Springer eBooks* (pp. 153–158). DOI: 10.1007/978-1-4614-5275-1_9

List of systems of the human body. (n.d.). Wikipedia, The Free Encyclopedia https://en.wikipedia.org/wiki/List_of_systems_of_the_human_body

Litwack, G. (2021). *Human Biochemistry*. Elsevier.

Litwack, G. (2022). *Hormones*. Academic Press., DOI: 10.1016/C2020-0-02802-1

Lizza, J. P. (2018). Defining Death: Beyond Biology. *Diametros : a Journal of Philosophy*, 55(55), 1–19. DOI: 10.13153/diam.1172

Lodé, T. (2001). *Les stratégies de reproduction des animaux*. Dunod.

Lott, D. T., Yeshi, T., Norchung, N., Dolma, S., Tsering, N., Jinpa, N., Woser, T., Dorjee, K., Desel, T., Fitch, D., Finley, A. J., Goldman, R., Ortiz Bernal, A. M., Ragazzi, R., Aroor, K., Koger, J., Francis, A., & Perlman, D. (2021). No Detectable Electroencephalographic Activity After Clinical Declaration of Death Among Tibetan Buddhist Meditators in Apparent Tukdam, a Putative Postmortem Meditation State. *Frontiers in Psychology*, 11(599190), 599190. Advance online publication. DOI: 10.3389/fpsyg.2020.599190 PMID: 33584435

Lovett Doust, J., & Lovett Doust, L. (1988). *Plant reproductive ecology: patterns and strategies*. Oxford University Press.

Luisi, P. L. (2006). *The Emergence of Life: From Chemical Origins to Synthetic Biology*. Cambridge University Press. DOI: 10.1017/CBO9780511817540

Lurquin, P. (2003). *The Origins of Life and the Universe*. Columbia University Press. DOI: 10.7312/lurq12654

Lüttge, U., & Souza, G. M. (2019). The Golden Section and Beauty in Nature: The Perfection of Symmetry and the Charm of Asymmetry. *Progress in Biophysics and Molecular Biology*, 146, 98–103. DOI: 10.1016/j.pbiomolbio.2018.12.008 PMID: 30557534

Mackenzie, R. J. (2020, December 18). *DNA vs. RNA – 5 key differences and comparison*. Genomics Research from Technology Networks; Technology Networks. https://www.technologynetworks.com/genomics/lists/what-are-the-key-differences-between-dna-and-rna-296719

Magee, P. T. (1976). Meiosis in protists. Some structural and physiological aspects of meiosis in algae, fungi and protozoa. *Bacteriological Reviews*, 40(1), 190–240. DOI: 10.1128/br.40.1.190-240.1976 PMID: 773364

Magrassi, L., Leto, K., & Rossi, F. (2013). Lifespan of neurons is uncoupled from organismal lifespan. *Proceedings of the National Academy of Sciences of the United States of America*, 110(11), 4374–4379. DOI: 10.1073/pnas.1217505110 PMID: 23440189

Mahmoud, S. A., Balla Elamien, M., Zaatar, A., & El Tinai, M. (2014). Low noise low power instrumentation amplifier for biomedical application. *University of Sharjah. Senior Design Project Report.* https://www.researchgate.net/publication/303487700_Low_noise_low_power_instrumentation_amplifier_for_biomedical_application

Maizels, N., & Weiner, A. M. (1987). Peptide-specific Ribosomes, Genomic Tags, and the Origin of the Genetic Code. *Cold Spring Harbor Symposia on Quantitative Biology*, 52(0), 743–749. DOI: 10.1101/SQB.1987.052.01.083 PMID: 3454285

Maloy, S., & Hughes, K. (Eds.). (2013). *Brenner's Encyclopedia of Genetics* (2nd ed.). Academic Press.

Mariana Ruiz, & Jmarchn. (2006). Diagram of the human digestive (gastrointestinal) tract. https://en.wikipedia.org/wiki/File:Digestive_system_diagram_en.svg

Marple, K. (2024, February 20). *Fetal development week by week*. BabyCenter. https://www.babycenter.com/pregnancy/your-baby/fetal-development-week-by-week_10406730

Martinez, D. E. (1998). Mortality patterns suggest lack of senescence in hydra. *Experimental Gerontology*, 33(3), 217–225. DOI: 10.1016/S0531-5565(97)00113-7 PMID: 9615920

Maruyama, S., Kurokawa, K., Ebisuzaki, T., Sawaki, Y., Suda, K., & Santosh, M. (2019). Nine requirements for the origin of Earth's life: Not at the hydrothermal vent, but in a nuclear geyser system. *Geoscience Frontiers*, 10(4), 1337–1357. DOI: 10.1016/j.gsf.2018.09.011

Masoro, E. J., & Austad, S. N. (Eds.). (2006). *Handbook of the Biology of Aging* (6th ed.). Academic Press.

Mattar, C. N., Choolani, M., Biswas, A., Waddington, S. N., & Chan, J. K. Y. (2011). Fetal gene therapy: Recent advances and current challenges. *Expert Opinion on Biological Therapy*, 11(10), 1257–1271. DOI: 10.1517/14712598.2011.585153 PMID: 21623703

Mayhew, P. J. (2006). *Discovering Evolutionary Ecology: Bringing Together Ecology and Evolution*. Oxford University Press. DOI: 10.1093/oso/9780198570608.001.0001

Maynard, S. J., & Szathmáry, E. (1995). *The Major Transitions in Evolution*. Oxford University Press.

Mayr, E. (1999). *Systematics and the Origin of Species from the Viewpoint of a Zoologist*. Harvard University Press. (Original work published 1942)

McEwen, B. S., & Wingfield, J. C. (2003). The concept of allostasis in biology and biomedicine. *Hormones and Behavior*, 43(1), 2–15. DOI: 10.1016/S0018-506X(02)00024-7 PMID: 12614627

McGeoch, M. W., Dikler, S., & McGeoch, J. E. M. (2002) Hemolithin: a Meteoritic Protein containing Iron and Lithium. https://arxiv.org/ftp/arxiv/papers/2002/2002.11688.pdf

McKay, C. P. (2004). What Is Life–and How Do We Search for It in Other Worlds? *PLoS Biology*, 2(9), E302. DOI: 10.1371/journal.pbio.0020302 PMID: 15367939

McKay, D. S., Gibson, E. K.Jr, Thomas-Keprta, K. L., Vali, H., Romanek, C. S., Clemett, S. J., Chillier, X. D., Maechling, C. R., & Zare, R. N. (1996). Search for past life on Mars: Possible relic biogenic activity in Martian meteorite ALH84001. *Science*, 273(5277), 924–930. DOI: 10.1126/science.273.5277.924 PMID: 8688069

Medawar, P. B. (1952). *An unsolved problem of Biology*. H.K Lewis.

Medium69 & Jmarchn. (2014). Nervous system diagram. Retrieved August 13, 2024, from https://commons.wikimedia.org/wiki/File:Nervous_system_diagram-en.svg

Mei, H., Liao, J.-Y., Jimenez, R. M., Wang, Y., Bala, S., McCloskey, C., Switzer, C., & Chaput, J. C. (2018). Synthesis and Evolution of a Threose Nucleic Acid Aptamer Bearing 7-Deaza-7-Substituted Guanosine Residues. *Journal of the American Chemical Society*, 140(17), 5706–5713. DOI: 10.1021/jacs.7b13031 PMID: 29667819

Mesarovic, M. D., & Takahara, Y. (1975). *General Systems Theory: Mathematical Foundations (Mathematics in Science and Engineering)*. Elsevier.

Mesarović, M. D., Takahara, Y., & Macko, D. (1972). *Theory of Hierarchical, Multilevel, Systems*. Academic Press.

Milhiet, V., Etain, B., Boudebessem, C., & Bellivier, F. (2011). Circadian biomarkers, circadian genes and bipolar disorders. *Journal of Physiology, Paris*, 105(4–6), 183–189. DOI: 10.1016/j.jphysparis.2011.07.002 PMID: 21767641

Miller, S. L., & Urey, H. C. (1959). Organic Compound Synthesis on the Primitive Earth. *Science*, 130(3370), 245–251. DOI: 10.1126/science.130.3370.245 PMID: 13668555

Miller-Urey experiment. (2008). *File: Miller-Urey experiment-en.svg - Wikipedia*. https://en.m.wikipedia.org/wiki/File:Miller-Urey_experiment-en.svg#/media/File%3AMiller-Urey_experiment-it.svg

Mitchison, J. M. (2003). Growth during the cell cycle. In *International Review of Cytology* (pp. 165–258). Elsevier. DOI: 10.1016/S0074-7696(03)01004-0

Mizuuchi, R., Ichihashi, N., Usui, K., Kazuta, Y., & Yomo, T. (2015). Adaptive Evolution of an Artificial RNA Genome to a Reduced Ribosome Environment. *ACS Synthetic Biology*, 4(3), 292–298. DOI: 10.1021/sb5000884 PMID: 24933578

Mkhitaryan, N. M., Badeyan, G. V., & Kovalyov, Y. N. (2004). *Erghonomycheskye aspekti slozhnikh system* [Ergonomic Aspects of Complex Systems]. Naukova Dumka.

Mkhitaryan, N. M., Kovalyov, Y. M., Malik, T. V., Safronov, V. K., & Safronova, O. O. (2021). *Dyzajn seredovyshha mista: Baghatokryterialjna optymizacija ta rozumni tekhnologhiji* [Design of the City Environment: Multi-Criteria Optimization and Smart Technologies]. Naukova dumka.

Molenaar, P. C. M. (2001). *Systems Modeling*. In *International Encyclopedia of the Social & Behavioral Sciences*. Pergamon National Cancer Institute. https://training.seer.cancer.gov/melanoma/anatomy/

Molinari, M. (2000). Cell cycle checkpoints and their inactivation in human cancer. *Cell Proliferation*, 33(5), 261–274. DOI: 10.1046/j.1365-2184.2000.00191.x PMID: 11063129

Moody, R. A. (1988). *Life After Life: The Investigation of a Phenomenon Survival of Bodily Death*. Mockingbird Books.

Morozov, A. (2013). Modelling biological evolution: Recent progress, current challenges and future direction. *Interface Focus*, 3(6), 20130054. DOI: 10.1098/rsfs.2013.0054

Mueller, L. D., & Rose, M. R. (1996). Evolutionary theory predicts late-life mortality plateaus. *Proceedings of the National Academy of Sciences of the United States of America*, 93(26), 15249–15253. DOI: 10.1073/pnas.93.26.15249 PMID: 8986796

Mukhi, S. (1999). The theory of strings: An introduction on JSTOR. https://www.jstor.org/stable/24104471

Mulkidjanian, A. Y. (2009). On the origin of life in the zinc world: 1. Photosynthesizing, porous edifices built of hydrothermally precipitated zinc sulfide as cradles of life on Earth. *Biology Direct*, 4(1), 26. DOI: 10.1186/1745-6150-4-26 PMID: 19703272

Mulkidjanian, A. Y., Cherepanov, D. A., & Galperin, M. Y. (2003). Survival of the fittest before the beginning of life: Selection of the first oligonucleotide-like polymers by UV light. *BMC Evolutionary Biology*, 3(1), 12. DOI: 10.1186/1471-2148-3-12 PMID: 12773209

Mulkidjanian, A. Y., & Galperin, M. Y. (2007). Physico-chemical and evolutionary constraints for the formation and selection of first biopolymers: Towards the consensus paradigm of the abiogenic origin of life. *Chemistry & Biodiversity*, 4(9), 2003–2015. DOI: 10.1002/cbdv.200790167 PMID: 17886857

Mulkidjanian, A. Y., & Galperin, M. Y. (2009). On the origin of life in the zinc world. 2. Validation of the hypothesis on the photosynthesizing zinc sulfide edifices as cradles of life on Earth. *Biology Direct*, 4(1), 27. DOI: 10.1186/1745-6150-4-27 PMID: 19703275

Mulkidjanian, A. Y., Galperin, M. Y., & Koonin, E. V. (2009). Co-evolution of primordial membranes and membrane proteins. *Trends in Biochemical Sciences*, 34(4), 206–215. DOI: 10.1016/j.tibs.2009.01.005 PMID: 19303305

Mulkidjanian, A. Y., Galperin, M. Y., Makarova, K. S., Wolf, Y. I., & Koonin, E. V. (2008). Evolutionary primacy of sodium bioenergetics. *Biology Direct*, 3(13), 13. Advance online publication. DOI: 10.1186/1745-6150-3-13 PMID: 18380897

NASA (n.d.). Mars Meteorite Home Page (JPL). *NASA*.

NASA. (2019). *First Detection of Sugars in Meteorites Gives Clues to Origin of Life*. (n / a).

NASA. (2019). *First Detection of Sugars in Meteorites Gives Clues to Origin of Life*. NASA.

National Cancer Institute. (n.d.). Introduction to the nervous system. https://training.seer.cancer.gov/anatomy/nervous/

Nei, M., Suzuki, Y., & Nozawa, M. (2010). The Neutral Theory of Molecular Evolution in the Genomic Era. *Annual Review of Genomics and Human Genetics*, 11(1), 265–289. DOI: 10.1146/annurev-genom-082908-150129 PMID: 20565254

Nelson, P., & Masel, J. (2017). Intercellular competition and the inevitability of multicellular aging. *Proceedings of the National Academy of Sciences of the United States of America*, 114(49), 12982–12987. DOI: 10.1073/pnas.1618854114 PMID: 29087299

Nikitin, M. P. (2023). Non-complementary strand commutation as a fundamental alternative for information processing by DNA and gene regulation. *Nature Chemistry*, 15(1), 70–82. DOI: 10.1038/s41557-022-01111-y PMID: 36604607

Noble, D. (2011). Neo-Darwinism, the Modern Synthesis and selfish genes: Are they of Use in Physiology? *The Journal of Physiology*, 589(5), 1007–1015. DOI: 10.1113/jphysiol.2010.201384 PMID: 21135048

O'Dwyer, J., & Chisholm, R. (2013). Neutral theory and beyond. In *Encyclopedia of Biodiversity* (pp. 510–518). Elsevier. DOI: 10.1016/B978-0-12-384719-5.00429-9

Okauchi, H., & Ichihashi, N. (2021). Continuous Cell-Free Replication and Evolution of Artificial Genomic DNA in a Compartmentalized Gene Expression System. *ACS Synthetic Biology*, 10(12), 3507–3517. DOI: 10.1021/acssynbio.1c00430 PMID: 34781676

Olejarz, J., Iwasa, Y., Knoll, A. H., & Nowak, M. A. (2021). The Great Oxygenation Event as a consequence of ecological dynamics modulated by planetary change. *Nature Communications*, 12(3985), 3985. Advance online publication. DOI: 10.1038/s41467-021-23286-7 PMID: 34183660

Olovnikov, A. M. (1973). A theory of marginotomy: The incomplete copying of template margin in enzymic synthesis of polynucleotides and biological significance of the phenomenon. *Journal of Theoretical Biology*, 41(1), 181–190. DOI: 10.1016/0022-5193(73)90198-7 PMID: 4754905

Olshansky, S. J., Hayflick, L., & Perls, T. T. (2004). Introduction: Anti-aging medicine: The hype and the reality—part II. *The Journals of Gerontology. Series A, Biological Sciences and Medical Sciences*, 59(7), B649–B651. DOI: 10.1093/gerona/59.7.B649

O'Rahilly, R., & Müller, F. (2010). Developmental stages in human embryos: revised and new measurements. *Cells Tissue Organs, 192*(2), 73-84. DOI: 10.1159/000289817

Orgel, L. E. (1968). Evolution of the genetic apparatus. *Journal of Molecular Biology*, 38(3), 381–393. DOI: 10.1016/0022-2836(68)90393-8 PMID: 5718557

Ott, E., Kawaguchi, Y., Kölbl, D., Rabbow, E., Rettberg, P., Mora, M., Moissl-Eichinger, C., Weckwerth, W., Yamagishi, A., & Milojevic, T. (2020). Molecular repertoire of Deinococcus radiodurans after 1 year of exposure outside the International Space Station within the Tanpopo mission. *Microbiome*, 8(150), 150. Advance online publication. DOI: 10.1186/s40168-020-00927-5 PMID: 33121542

Passive vs Active Membrane Transport. (2022). *File:Passive vs active membrane transport.Svg*. Wikimedia.org. Retrieved August 12, 2024, from https://commons.wikimedia.org/wiki/File:Passive_vs_Active_Membrane_Transport.svg

Pearce, B. K. D., Tupper, A. S., Pudritz, R., & Higgs, P. G. (2018). Constraining the Time Interval for the Origin of Life on Earth. *Astrobiology*, 18(3), 343–364. DOI: 10.1089/ast.2017.1674 PMID: 29570409

Pease, C. M., Lande, R., & Bull, J. J. (1989). A model of population growth, dispersal, and evolution in a changing environment. *Ecology*, 70(6), 1657–1664. DOI: 10.2307/1938100

Penrose, O. (2005). An asymmetric world. *Nature*, 438(7070), 919–919. DOI: 10.1038/438919a PMID: 16355197

Peretó, J. (2005). Controversies on the origin of life. *International Microbiology*, 8(1), 23–31. PMID: 15906258

Pianka, E. R. (2000). *Evolutionary Ecology*. Benjamin-Cummings.

Pigliucci, M., & Finkelman, L. (2014). The Extended (Evolutionary) Synthesis Debate: Where Science Meets Philosophy. *Bioscience*, 64(6), 511–516. DOI: 10.1093/biosci/biu062

Pino, S. (2015). Nucleotide. In *Encyclopedia of Astrobiology* (pp. 1736–1738). Springer Berlin Heidelberg. DOI: 10.1007/978-3-662-44185-5_1085

Planck Collaboration. (2016). Planck 2015 results: XIII. Cosmological parameters. *Astronomy and Astrophysics.*, 594, 3.

Pomeroy, E., Mirazón Lahr, M., Crivellaro, F., Farr, L., Reynolds, T., Hunt, C. O., & Barker, G. (2017). Newly discovered Neanderthal remains from Shanidar Cave, Iraqi Kurdistan, and their attribution to Shanidar 5. *Journal of Human Evolution*, 111, 102–118. DOI: 10.1016/j.jhevol.2017.07.001 PMID: 28874265

Ponnamperuma, C., Mariner, R., & Sagan, C. (1963). Formation of adenosine by ultra-violet irradiation of a solution of adenine and ribose. *Nature*, 198(4886), 1199–1200. DOI: 10.1038/1981199a0 PMID: 13972132

Powner, M. W., Gerland, B., & Sutherland, J. D. (2009). Synthesis of activated pyrimidine ribonucleotides in prebiotically plausible conditions. *Nature*, 459(7244), 239–242. DOI: 10.1038/nature08013 PMID: 19444213

Radhakrishnan, S. (2009). Indian Philosophy. Oxford University Press. (Original work published 1923)

Randall, L. (2005). *Warped Passages. Unraveling the Mysteries of the Universe's Hidden Dimensions*. Ecco.

Rauchfuss, H. (2008). *Chemical Evolution and the Origin of Life* (Mitchell, T. N., Trans.). Springer.

Ravi Kumar Patil, H. S., Makari, H. K., Gurumurthy, H., & Sowmya, S. V. (2009). *A Textbook of Human Physiology*. I K International Publishing House.

Raymond, C. (2012, September 13). *What happens when you die?* Verywell Health. https://www.verywellhealth.com/what-happens-to-my-body-right-after-i-die-1132498

Ray, S., Kassan, A., Busija, A. R., Rangamani, P., & Patel, H. H. (2016). The plasma membrane as a capacitor for energy and metabolism. *American Journal of Physiology. Cell Physiology*, 310(3), C181–C192. DOI: 10.1152/ajpcell.00087.2015 PMID: 26771520

Reif, W.-E., Junker, T., & Hossfeld, U. (2000). The synthetic theory of evolution: General problems and the German contribution to the synthesis. *Theory in Biosciences*, 119(1), 41–91. DOI: 10.1007/s12064-000-0004-6

Ring, K. (1985). *Heading Toward Omega: In Search of the Meaning of the Near-Death Experience*. Harper Perennial.

Rink, W. J., & Thompson, J. W. (Eds.). (2015). *Encyclopedia of Scientific Dating Methods*. Springer. DOI: 10.1007/978-94-007-6304-3

Rolland, B., Jardri, R., Amad, A., Thomas, P., Cottencin, O., & Bordet, R. (2014). Pharmacology of Hallucinations: Several Mechanisms for One Single Symptom? *BioMed Research International*, 2014, 1–9. DOI: 10.1155/2014/307106 PMID: 24991548

Rose, M. R., & Oakley, T. H. (2007). The new biology: Beyond the Modern Synthesis. *Biology Direct*, 2(1), 30. DOI: 10.1186/1745-6150-2-30 PMID: 18036242

Rosenzweig, M. (1995). *Species diversity in space and time*. Cambridge University Press., DOI: 10.1017/CBO9780511623387

Russill, P. (2006). *The Yoga of Sound: Tapping the Hidden Power of Music and Chant*. New World Library.

Sagaradze, G., Basalova, N., Kirpatovsky, V., Ohobotov, D., Nimiritsky, P., Grigorieva, O., Popov, V., Kamalov, A., Tkachuk, V., & Efimenko, A. (2019). A magic kick for regeneration: Role of mesenchymal stromal cell secretome in spermatogonial stem cell niche recovery. *Stem Cell Research & Therapy*, 10(1), 342. Advance online publication. DOI: 10.1186/s13287-019-1479-3 PMID: 31753023

Sakatani, Y., Yomo, T., & Ichihashi, N. (2018). Self-replication of circular DNA by a self-encoded DNA polymerase through rolling-circle replication and recombination. *Scientific Reports*, 8(1), 13089. DOI: 10.1038/s41598-018-31585-1 PMID: 30166584

Saladino, R., Costanzo, G., Crestini, C., Ciciriello, F., & Di Mauro, E. (2007). Formamide as the main building block in the origin of nucleic acids. *BMC Evolutionary Biology*, 7(4), 694–720. DOI: 10.1186/1471-2148-7-S2-S1 PMID: 17767725

Schmidt, R. F., & Thews, G. (Eds.). (1983). *Human Physiology*. Springer-Verlag.

Sciencia58 and the makers of the single images Domdomegg, Fæ, Petr94 & Manu5. (2020). The processes involved in the primary immune response. *File: Primary immune response 1.png*. Retrieved August 13, 2024, from https://commons.wikimedia.org/wiki/File:Primary_immune_response_1.png

Scientists find the 'smoking gun,' signs of life from 3.5B years ago. (n.d.). https://www.foxnews.com/science/scientists-find-smoking-gun-signs-of-life?fbclid=IwAR0CLOHoDiKNLFOaL3Qc0uP98KtB2vo-0iid1y8cUoMOIMRMVx8WQo5vy5k

Selye, H. (1976). Stress without distress. In *Psychopathology of Human Adaptation* (pp. 137–146). Springer US. DOI: 10.1007/978-1-4684-2238-2_9

Sender, R., & Milo, R. (2021). The distribution of cellular turnover in the human body. *Nature Medicine*, 27(1), 45–48. DOI: 10.1038/s41591-020-01182-9 PMID: 33432173

Serin, Y., & Acar Tek, N. (2019). Effect of Circadian Rhythm on Metabolic Processes and the Regulation of Energy Balance. *Annals of Nutrition & Metabolism*, 74(4), 322–330. DOI: 10.1159/000500071 PMID: 31013492

Shapiro, R. (1987). *Origins: A Skeptic's Guide to the Creation of Life on Earth*. Bantam New Age.

Shaw, F. H., Promislow, D. E. L., Tatar, M., Huges K. A., & Geyes, C. J. (1999). Toward reconsiling inferences concerning geetic variations in Drosophila melamogaster. *Genetics - Genetics Society of America, 152,* 553—566.

Shermer, M. (2017). Who Are You? *Scientific American*, 317(1), 73. DOI: 10.1038/scientificamerican0717-73 PMID: 28632227

Sherwood, L. (2015). *Human Physiology: From Cells to Systems* (9th ed.). Cengage Learning.

Shields, J. (2021). Here's How Plant and Animal Cells Are Different. [Illustration]. Researchgate.net. Retrieved August 12, 2024, from https://science.howstuffworks.com/life/cellular-microscopic/plant-cells-animal-cells.htm

Shmelova, T., Sikirda, Y., Rizun, N., Salem, A.-B. M., & Kovalyov, Y. (2018). *Socio-Technical Decision Support in Air Navigation Systems: Emerging Research and Opportunities*. IGI Global International Publisher of Progressive Information Science and Technology Research. DOI: 10.4018/978-1-5225-3108-1

Simpkins, C. A., & Simpkins, A. M. (2011). *Neuroscience Findings: How Meditation Can Change the Brain*. DOI: 10.1002/9781118155684.ch02

Simpson, G. G. (1953). *The major features of evolution* (3rd ed.). Columbia Univ. Press. DOI: 10.7312/simp93764

Sole, R. V., & Newman, M. (2002). Extinctions and Biodiversity in the Fossil Record. In *Encyclopedia of Global Environmental Change*. John Wiley & Sons.

Spalding K. L., Bhardwaj R. D., Buchholz B. A., Druid, H., & Frisén, J. (2005). Retrospective birth dating of cells in humans. *Cell, 15*(122), 133-43. DOI: 10.1016/j.cell.2005.04.028

Srimaharaj, W., Chaisricharoen, R., Chaising, S., & Sittiprapaporn, P. (2018). *Classification of human brain attention focused on meditation, effected by L-theanine acid in Oolong tea*. DOI: 10.1109/ICDAMT.2018.8376536

Sukel, K. (2019). Neuroanatomy: The Basics. https://www.dana.org/wp-content/uploads/2019/08/anatomy-function-brain-areas-basics-aug-2019-2024.jpg

Tamura, K., Stecher, G., & Kumar, S. (2021). MEGA11: Molecular evolutionary genetics analysis version 11. *Molecular Biology and Evolution*, 38(7), 3022–3027. DOI: 10.1093/molbev/msab120 PMID: 33892491

Taper, M. L., & Case, T. J. (1985). Quantitative genetic models for the coevolution of character displacement. *Ecology*, 66(2), 355–371. DOI: 10.2307/1940385

Tarski A. (1969). Truth and Proof. *Scientific American,220*, 63–77. (Original work published 1936)

Tarski, A. (1969). Truth and Proof. *Scientific American,* 220, 63–77. (Original work published 1936)

Tetter, S., Terasaka, N., Steinauer, A., Bingham, R. J., Clark, S., Scot, A. J. P., Patel, N., Leibundgut, M., Wroblewski, E., Ban, N., Stockley, P. G., Twarock, R., & Hilvert, D. (2021). Evolution of a virus-like architecture and packaging mechanism in a repurposed bacterial protein. *Science*, 372(6547), 1220–1224. DOI: 10.1126/science.abg2822 PMID: 34112695

The 7 Unmistakable Characteristics of Life. (n.d.). https://biologywise.com/characteristics-of-life

The Characteristics of Life. (n.d.). https://courses.lumenlearning.com/suny-wmopen-biology1/chapter/the-characteristics-of-life/

The chemical structure of adenosine triphosphate. (n.d.). *File:ATP structure revised.png*. https://commons.wikimedia.org/w/index.php?curid=5041461

The dying process - palliative care Australia. (n.d.). Org.au. Retrieved August 14, 2024, from https://palliativecare.org.au/resource/resources-the-dying-process/

The Human Body. (n.d.). https://www.healthline.com/human-body-maps

The New York School of Regional Anesthesia. (n.d.). Autonomic Nervous System. https://www.nysora.com/anesthesia/autonomic-nervous-system/

The top 10 causes of death. (2024, August 7). https://www.who.int/news-room/fact-sheets/detail/the-top-10-causes-of-death

The University of Arizona. (2013). Habitability and Biology: What are the Properties of Life? Phoenix Mars Mission. http://phoenix.lpl.arizona.edu/mars141.php

Theories of aging. (n.d.). An introduction to aging science brought to you by the American Federation for Aging Research. https://www.afar.org/imported/AFAR_INFOAGING_GUIDE_THEORIES_OF_AGING_2016.pdf

Thienes, C., & Haley, T. J. (1972). *Clinical Toxicology* (5th ed.). Lea and Febiger.

Thomas, J. A., & Rana, F. R. (2007). The influence of environmental conditions, lipid composition, and phase behavior on the origin of cell membranes. *Origins of Life and Evolution of the Biosphere*, 37(3), 267–285. DOI: 10.1007/s11084-007-9065-6 PMID: 17361322

Thurner, S., Hanel, R., & Klimek, P. (2018). *Introduction to the Theory of Complex Systems*. Oxford University Press. DOI: 10.1093/oso/9780198821939.001.0001

Tian, Y. E., Cropley, V., Maier, A. B., Lautenschlager, N. T., Breakspear, M., & Zalesky, A. (2023). Biological aging of the human body and brain systems. *Nature Medicine*, 29(5), 1221–1231. DOI: 10.1038/s41591-023-02296-6 PMID: 37024597

Tononi, G., & Edelman, G. M. (1998). Consciousness and complexity. *Science, 4*(282), 1846-51. DOI: 10.1126/science.282.5395.1846

Trigrams Meaning and Definition in Qi Men Dun Jia. (2022). https://dougleschan.com/8-trigrams/8-trigrams-meaning-and-definition-in-qi-men-dun-jia/

Tsaitgaist. (2009). Drawing of the Male Internal Sexual Anatomy. (n.d.). Wikimedia.org. Retrieved August 13, 2024, from https://commons.wikimedia.org/wiki/File:Male_anatomy_number.svg

Tsaitgaist. (2010). Genitals of a woman. Retrieved August 13, 2024, from https://commons.wikimedia.org/wiki/File:Female_Organs.png

Twycross, R., & Wilcock, A. (2016). *Introducing Palliative Care* (5th ed.). Pharmaceutical Press.

Ulett, M. A. (2014). Making the case for orthogenesis: The popularization of definitely directed evolution (1890–1926). *Studies in History and Philosophy of Science Part C Studies in History and Philosophy of Biological and Biomedical Sciences*, 45, 124–132. DOI: 10.1016/j.shpsc.2013.11.009 PMID: 24368232

Urry, L., Cain, M., Wasserman, S., Minorsky, P., & Reece, J. (2016). *Campbell biology* (11th Edn.). Pearson Higher Education.

Valero, M., Richerd, S., Perrot, V., & Destombe, C. (1992). Evolution of the alternation of haploid and diploid phases in life cycles. *Trends in Ecology & Evolution*, 7(1), 25–29. DOI: 10.1016/0169-5347(92)90195-H PMID: 21235940

Vellas, F. (1988). Mayonnaise and the Origin of Life: Thoughts of Minds and Molecules. *Biochemical Education*, 16(1), 50. DOI: 10.1016/0307-4412(88)90031-3

Villarreal, L. P. (2005). *Viruses and the Evolution of Life*. ASM Press. DOI: 10.1128/9781555817626

Vina, J., Borra's, C., & Miquel, J. (2022). Theories of Ageing: Critical Review. *IUBMB Life 59*(4 – 5), 249 – 254. https://iubmb.onlinelibrary.wiley.com/doi/pdf/10.1080/15216540601178067

Von Zglinicki, T. (2002). Cell Press. *Trends in Biochemical Sciences*, 27(7), 339–344. DOI: 10.1016/S0968-0004(02)02110-2 PMID: 12114022

Wächtershäuser, G. (2006). From volcanic origins of chemoautotrophic life to Bacteria, Archaea and Eukarya. *Philosophical Transactions of the Royal Society of London. Series B, Biological Sciences*, 361(1474), 1787–1808. DOI: 10.1098/rstb.2006.1904 PMID: 17008219

Wagner, A. M., Schoeberlein, A., & Surbek, D. (2009). Fetal gene therapy: Opportunities and risks. *Advanced Drug Delivery Reviews*, 61(10), 813–821. DOI: 10.1016/j.addr.2009.04.011 PMID: 19426772

Walker, A. E. (1978). Pathology of brain death. *Annals of the New York Academy of Sciences*, 315(1), 272–280. DOI: 10.1111/j.1749-6632.1978.tb50345.x PMID: 284739

Wang, C., Liu, C., Gao, H., & Liu, H. (2017). Order of aging of major human organs or systems and evaluation of health status based on aging. *Biomarkers*, 22(2), 145–148. DOI: 10.1080/1354750X.2016.1252951 PMID: 27800690

Wang, D., & Farhana, A. (2021). *Biochemistry, RNA Structure*. StatPearls Publishing LLC.

Weinberg, S. (1993). *The First Three Minutes: A Modern View of the Origin of the Universe*. Basic Books.

Weiss, M. C., Sousa, F. L., Mrnjavac, N., Neukirchen, S., Roettger, M., Nelson-Sathi, S., & Martin, W. F. (2016). The physiology and habitat of the last universal common ancestor. *Nature Microbiology*, 1(9), 16116. Advance online publication. DOI: 10.1038/nmicrobiol.2016.116 PMID: 27562259

Weyl, G. (1934). *Phylosofyja matematyky: sbornyk nauchnykh trudov* [Philosophy of mathematics: a collection of scientific papers]. State Technical and Theoretical Publishing House.

Whitbourne, S. K. (1985). Digestive and Excretory Systems. In *The Aging Body*. Springer. DOI: 10.1007/978-1-4612-5126-2_5

Wickramasinghe, C. (2011). Bacterial morphologies supporting cometary panspermia: A reappraisal. *International Journal of Astrobiology*, 10(1), 25–30. DOI: 10.1017/S1473550410000157

Wickramasinghe, D. T., & Allen, D. A. (1983). Three components of 3–4 μm absorption bands. *Astrophysics and Space Science*, 97(2), 369–378. DOI: 10.1007/BF00653492

Wikipedia.org. (2008). Retrieved August 12, 2024. Scheme of the S. Miller apparatus. https://en.wikipedia.org/wiki/File:Miller-Urey_experiment-en.svg

Wilde, S. A., Valley, J. W., Peck, W. H., & Graham, C. M. (2001). Evidence from detrital zircons for the existence of continental crust and oceans on the Earth 4.4 Gyr ago. *Nature*, 409(6817), 175–178. DOI: 10.1038/35051550 PMID: 11196637

Woese, C. R., Kandler, O., & Wheelis, M. L. (1990). Towards a natural system of organisms: Proposal for the domains Archaea, Bacteria, and Eucarya. *Proceedings of the National Academy of Sciences of the United States of America*, 87(12), 4576–4579. DOI: 10.1073/pnas.87.12.4576 PMID: 2112744

Wright, W. (1977). *Evolution and Genetics of Populations*. v. 3. Experimental Results and Evolutionary Deductions. University of Chicago Press.

Wright, S. (1955). Classification of the factors of evolution. *Cold Spring Harbor Symposia on Quantitative Biology*, 20(0), 16–24. DOI: 10.1101/SQB.1955.020.01.004 PMID: 13433551

Xie, Y., Tang, Q., Chen, G., Xie, M., Yu, S., Zhao, J., & Chen, L. (2019). New Insights Into the Circadian Rhythm and Its Related Diseases. *Frontiers in Physiology*, 25(10), 682. DOI: 10.3389/fphys.2019.00682 PMID: 31293431

YassineMrabet. (2007). Biological clock. Wikimedia.org. Retrieved August 13, 2024, from https://commons.wikimedia.org/wiki/File:Biological_clock_human-af.svg

Yu, H., Zhang, S., & Chaput, J. C. (2012). Darwinian evolution of an alternative genetic system provides support for TNA as an RNA progenitor. *Nature Chemistry*, 4(3), 183–187. DOI: 10.1038/nchem.1241 PMID: 22354431

Zephyris. (n.d.). The first few weeks of embryogenesis in humans. Beginning at the fertilized egg, and ending with the closing of the neural tube. https://commons.wikimedia.org/wiki/File:HumanEmbryogenesis.svg

About the Authors

Yury N. Kovalyov is Doctor of Technical Science, Professor Head of Industrial Design and Computer Technologies Department of the Mykhailo Boichuk Kyiv State Academy of Decorative Applied Arts and Design, Ukraine. Author of 268 scientific publications (8 monographs, 4 textbooks) in the fields of geometry foundations, mathematical modeling and multicriteria optimization of complex systems and human-machine systems, design, sociometry, psychology. Hobbies: poetry (author of 3 books), philosophy, yoga, sport, travel.

Nver M. Mkhitaryan graduated from Yerevan. Polytechnic. Institute (1982). In 1982-88 he worked in the trust "Armvodbud", in 1988-92 — chief engineer Yehegnut. Reinforced concrete products, 1992-96 — Art. teacher Yerevan. Institute of Architecture and Construction; in 1996-1999 — Chairman of the Board of Directors of the company, 1999 — 2002 — President of the corporation "Poznyakizhytlobud", at the same time in 2003 — 201 — Director of the Institute of Renewable Energy of the National Academy of Sciences of Ukraine (Kyiv). Member of Parliament of Ukraine (2002-06 and 2007-14). Scientific researches concern the physicochemical bases of heliothermal treatment of concrete and heliotechnology.

Andriy Y. Morozov was born in Kyiv (Ukraine), 1981, Education. Taras Shevchenko Kyiv national university, the faculty of philosophy, 1998-2003 Master degree, 2003 PH.D. (in philosophy), 2007, Doctor of philosophical sciences (2014), Associate professor of department of philosophy, sociology and political sciences (Kyiv national university of trade and economics), 2015 - til present Author of the books of more than 50 articles and 3 monographs, including: "Love and death, "Intuition in search of goodness", "Evil: spiritual and metaphysical aspects". The field of scientific research: ethics, social philosophy, metaphysics, philosophical anthropology, christian theology, existential problems of fear of death, nostalgia, intuition and transpersonal experience.

Index

www.ingramcontent.com/pod-product-compliance
Lightning Source LLC
LaVergne TN
LVHW080347011026
849161LV00009B/2222

* 9 7 8 1 6 6 8 4 8 5 0 9 5 *